Epigenetics and Human Health

Volume 15

This books series concerns the quickly expanding field of epigenetics. It evokes novel research questions and how these are linked to human health. Each of its volumes covers a particular topic that is important to clinical practice and research, and to researchers in epigenetics who study the relevant medical issues of today. The individual books of the series complement each other and underline the importance of epigenetics in health-related disciplines such as toxicology, reproduction, psychiatry, and nutritional and environmental medicine.

Epigenetics and Human Health was founded in 2011. Prof. Dr. Robert Feil, PD Dr. Mario Noyer-Weidner and Prof. Dr. Jörn Walter were the founding editors and were responsible as series editors for volumes 1-10 until 2016. In 2023, the series was relaunched together with Prof. Rasime Kalkan and Dr. Luis M. Vaschetto, who are the current series editors.

Epigenetics and Human Health is indexed in SCOPUS, EMBASE, Reaxys, EMCare, EMBiology and SCImago.

Feng Gao • Xing Zhang
Editors

Exercise, Epigenetics and Human Health

Springer

Editors
Feng Gao
Key Laboratory of Aerospace Medicine
of the Ministry of Education
School of Aerospace Medicine
The Fourth Military Medical University
Xi'an, China

Xing Zhang
Key Laboratory of Aerospace Medicine
of the Ministry of Education
School of Aerospace Medicine
The Fourth Military Medical University
Xi'an, China

ISSN 2191-2262 ISSN 2191-2270 (electronic)
Epigenetics and Human Health
ISBN 978-3-032-08044-8 ISBN 978-3-032-08045-5 (eBook)
https://doi.org/10.1007/978-3-032-08045-5

This Springer imprint is published by the registered company Springer Nature Switzerland AG
The registered company address is: Gewerbestrasse 11, 6330 Cham, Switzerland

Preface

The profound impact of exercise on health is well-documented, establishing physical activity as a fundamental strategy in the prevention of chronic diseases and the enhancement of overall well-being. Emerging evidence has revealed that the benefits of exercise extend beyond short-term effects; regular exercise can produce long-lasting, enduring impacts, some of which may even be heritable. This "exercise health memory" is thought to be partially mediated through epigenetic mechanisms. The convergence of exercise, epigenetics, and human health presents a transformative paradigm in our understanding of how physical activity reshapes our health, especially long-term health effects induced by early-life exercise.

This volume, *Exercise, Epigenetics, and Human Health*, was conceived to synthesize and present the latest development in this rapidly evolving field. It brings together a multidisciplinary body of research that explores how physical activity influences the epigenome—the collection of molecular markers that modify gene expression without altering the DNA sequence. Through epigenetic modifications such as DNA methylation, histone remodeling, and noncoding RNA regulation, exercise exerts a diverse array of long-term health benefits, including a reduced risk of chronic diseases such as diabetes, cardiovascular conditions, and certain types of cancer, enhanced cognitive function and mental well-being, healthy aging, and overall fitness. Intriguingly, some of these epigenetic effects may be transgenerational, though this remains an active area of investigation.

The book is structured to guide readers from foundational principles to specialized applications: Chapter 1 provides an overview of the epigenetic landscape of exercise in human health, and Chaps. 2 and 3 introduce the heritable and enduring benefits induced by exercise. Subsequent sections explore specialized topics, including epigenetic remodeling in cardiovascular protection (Chap. 4), mental health (Chap. 5), and skeletal muscle health (Chap. 6). The chapters on the metabolic-epigenetic axis (Chaps. 7, 8, 9, and 10) emphasize the roles of redox balance, bioenergetics, metabolites, and mitochondria as critical regulators connecting exercise with epigenetic modifications and health outcomes. Further chapters broaden the scope to interorgan communication, focusing on the roles of intestinal flora, extracellular vesicles, and exerkines (Chaps. 11, 12, and 13). Chapter 14 explores the intersection of exercise, epigenetics, and cell death, illustrating how exercise-induced epigenetic modifications can shift the balance toward cellular survival in pathological contexts. Finally, Chap. 15 discusses current challenges and future perspectives, emphasizing the importance of translating these mechanistic insights into practical lifestyle interventions and clinical applications.

We hope this volume will serve a valuable resource for researchers, educators, and graduate students in exercise science, genetics, epigenetics, and public health. General readers will also benefit from its insights, gaining the knowledge to embrace exercise as a potent tool for promoting health, longevity, and well-being.

As editors, we extend our sincere gratitude to the global community of contributors whose expertise and dedication have made this work possible. Their collective vision underscores exercise not merely as a lifestyle choice but as a dynamic epigenetic architect of health. It is our hope that this volume will inspire researchers, clinicians, and policymakers alike to leverage emerging epigenetic insights as a foundation for transforming public health strategies—advancing a new era of preventive, personalized, and precision-based care.

Xi'an, China

Feng Gao
Xing Zhang

Disclaimer

The language in the manuscript has been edited with the assistance of artificial intelligence (AI) tools DeepSeek, DeepSeek-R1, 1.0, 2025. Every section was carefully reviewed, revised, and approved by author/s to ensure that the text is appropriate concerning content and scientific correctness and maintain the integrity of the original content.

Contents

Editor and Contributors

About the Editors

Feng Gao, MD, PhD, is Professor of Physiology and Fellow of the American Physiological Society (FAPS) and the American Heart Association (FAHA). His research focuses on cardiovascular protection and exercise health benefits. His major contributions include elucidating mechanisms of exercise-induced cardioprotection via exosome-mediated cross-talk between vascular endothelial cells and the heart, as well as long-term benefits of physical activity on healthy aging. His work has been published in *Circulation*, *Cir Res*, *PNAS*, *Hypertension*, and *Nat Commun*, with over 120 peer-reviewed publications. He also serves as an Associate Editor of *FASEB J* and *Acta Physiologica Sinica*.

Xing Zhang, obtained his PhD in Physiology from the Fourth Military Medical University in 2013, and serves as a Professor in the Key Laboratory of Aerospace Medicine of the Ministry of Education at the Fourth Military Medical University, China. His research focuses on the biological effects of lifestyle (e.g., exercise and diet) on health and the underlying mechanisms, with specific interests in metabolic regulation and mitochondrial function.

Contributors

Ke Cao Center for Mitochondrial Biology and Medicine, The Key Laboratory of Biomedical Information Engineering of Ministry of Education, School of Life Science and Technology, Xi'an Jiaotong University, Xi'an, Shaanxi, China

Yangpo Cao Department of Pharmacology, Joint Laboratory of Guangdong-Hong Kong Universities for Vascular Homeostasis and Diseases, School of Medicine, Southern University of Science and Technology, Shenzhen, China

Yousen Cao School of Aerospace Medicine, Fourth Military Medical University, Xi'an, China

Xinxin Cui Cardiac Regeneration and Ageing Lab, Institute of Geriatrics (Shanghai University), Affiliated Nantong Hospital of Shanghai University (The Sixth People's Hospital of Nantong) and School of Life Science, Shanghai University, Nantong, China

Institute of Cardiovascular Sciences, Shanghai Engineering Research Center of Organ Repair, Joint International Research Laboratory of Biomaterials and Biotechnology in Organ Repair (Ministry of Education), School of Life Science, Shanghai University, Shanghai, China

Gareth W. Davison Faculty of Life and Health Sciences, Ulster University, Belfast, Northern Ireland, UK

Xiaotong Ding Cardiac Regeneration and Ageing Lab, Institute of Geriatrics (Shanghai University), Affiliated Nantong Hospital of Shanghai University (The Sixth People's Hospital of Nantong) and School of Life Science, Shanghai University, Nantong, China

Institute of Cardiovascular Sciences, Shanghai Engineering Research Center of Organ Repair, Joint International Research Laboratory of Biomaterials and Biotechnology in Organ Repair (Ministry of Education), School of Life Science, Shanghai University, Shanghai, China

Rui Duan School of Physical Education and Sports Science, South China Normal University, Guangzhou, China

Mengya Feng Key Laboratory of Aerospace Medicine of the Ministry of Education, School of Aerospace Medicine, Fourth Military Medical University, Xi'an, China

Zhihui Feng Center for Mitochondrial Biology and Medicine, The Key Laboratory of Biomedical Information Engineering of Ministry of Education, School of Life Science and Technology, Xi'an Jiaotong University, Xi'an, Shaanxi, China

Frontier Institute of Science and Technology, Xi'an Jiaotong University, Xi'an, Shaanxi, China

Feng Gao Key Laboratory of Aerospace Medicine of the Ministry of Education, School of Aerospace Medicine, Fourth Military Medical University, Xi'an, China

Shuang Hao School of Aerospace Medicine, Fourth Military Medical University, Xi'an, China

Tian Hao Department of Anesthesia, Critical Care and Pain Medicine, Massachusetts General Hospital, Harvard Medical School, Boston, MA, USA

Yihao Hou Department of Cardiology, Xijing Hospital, Air Force Medical University (Fourth Military Medical University), Xi'an, Shaanxi, China

Yang Hu Key Laboratory of Aerospace Medicine of the Ministry of Education, School of Aerospace Medicine, Fourth Military Medical University, Xi'an, China

Yumeng Hu School of Aerospace Medicine, Fourth Military Medical University, Xi'an, China

Rachelle Irwin Faculty of Life and Health Sciences, Ulster University, Belfast, Northern Ireland, UK

Yin Jia Student of the Basic Medical College, Air Force Medical University (Fourth Military Medical University), Xi'an, China

Rikard Landberg Food and Nutrition Science, Department of Life Sciences, Chalmers University of Technology, Gothenburg, Sweden

Diane Lees-Murdock Faculty of Life and Health Sciences, Ulster University, Belfast, Northern Ireland, UK

Minghong Leng Department of Pharmacology, Joint Laboratory of Guangdong-Hong Kong Universities for Vascular Homeostasis and Diseases, School of Medicine, Southern University of Science and Technology, Shenzhen, China

Xinyu Liao Center for Mitochondrial Biology and Medicine, The Key Laboratory of Biomedical Information Engineering of Ministry of Education, School of Life Science and Technology, Xi'an Jiaotong University, Xi'an, Shaanxi, China

Guoping Li Department of Anesthesia, Critical Care and Pain Medicine, Massachusetts General Hospital, Harvard Medical School, Boston, MA, USA

Han Li Key Laboratory of Aerospace Medicine of the Ministry of Education, School of Aerospace Medicine, Fourth Military Medical University, Xi'an, China

Jia Li School of Aerospace Medicine, Fourth Military Medical University, Xi'an, China

Min Li Key Laboratory of Aerospace Medicine of the Ministry of Education, School of Aerospace Medicine, Fourth Military Medical University, Xi'an, China

Tianqi Liu College of Food Engineering and Nutritional Science, Shaanxi Normal University, Xi'an, China

Xuyun Liu Center for Mitochondrial Biology and Medicine, The Key Laboratory of Biomedical Information Engineering of Ministry of Education, School of Life Science and Technology, Xi'an Jiaotong University, Xi'an, Shaanxi, China

Yongbo Li Department of Physiology and Pathophysiology, School of Basic Medical Sciences, Fudan University, Shanghai, China

Jing Lou Key Laboratory of Aerospace Medicine of the Ministry of Education, School of Aerospace Medicine, Fourth Military Medical University, Xi'an, China

Jiayao Lv College of Food Engineering and Nutritional Science, Shaanxi Normal University, Xi'an, China

Dan Meng Department of Physiology and Pathophysiology, School of Basic Medical Sciences, Fudan University, Shanghai, China

Aya Mohammed Faculty of Life and Health Sciences, Ulster University, Belfast, Northern Ireland, UK

Yizhe Qu School of Physical Education, Shaanxi Normal University, Xi'an, China

Haiwang Shi School of Physical Education and Sports Science, South China Normal University, Guangzhou, China

Lijun Shi Department of Exercise Physiology, Beijing Sport University, Beijing, China

Laboratory of Sports Stress and Adaptation of General Administration of Sport, Beijing Sport University, Beijing, China

Key Laboratory of Physical Fitness and Exercise, Ministry of Education, Beijing Sport University, Beijing, China

Lin Shi College of Food Engineering and Nutritional Science, Shaanxi Normal University, Xi'an, China

Guohua Song School of Clinical and Basic Medical Sciences, Shandong First Medical University & Shandong Academy of Medical Science, Jinan, Shandong, China

Shanhu Sun Department of Pharmacology, Joint Laboratory of Guangdong-Hong Kong Universities for Vascular Homeostasis and Diseases, School of Medicine, Southern University of Science and Technology, Shenzhen, China

Colum P. Walsh Department of Biomedical and Clinical Sciences, Linköping University, Linköping, Sweden

Di Wang School of Physical Education, Shaanxi Normal University, Xi'an, China

Lijun Wang Cardiac Regeneration and Ageing Lab, Institute of Geriatrics (Shanghai University), Affiliated Nantong Hospital of Shanghai University (The Sixth People's Hospital of Nantong) and School of Life Science, Shanghai University, Nantong, China

Institute of Cardiovascular Sciences, Shanghai Engineering Research Center of Organ Repair, Joint International Research Laboratory of Biomaterials and Biotechnology in Organ Repair (Ministry of Education), School of Life Science, Shanghai University, Shanghai, China

Shujuan Wang Department of Health Service, Air Force Medical University (Fourth Military Medical University), Xi'an, Shaanxi, China

Baile Wu Department of Exercise Physiology, Beijing Sport University, Beijing, China

Guiling Wu Key Laboratory of Aerospace Medicine of the Ministry of Education, School of Aerospace Medicine, Fourth Military Medical University, Xi'an, China

Junjie Xiao Cardiac Regeneration and Ageing Lab, Institute of Geriatrics (Shanghai University), Affiliated Nantong Hospital of Shanghai University (The Sixth People's Hospital of Nantong) and School of Life Science, Shanghai University, Nantong, China

Institute of Cardiovascular Sciences, Shanghai Engineering Research Center of Organ Repair, Joint International Research Laboratory of Biomaterials and Biotechnology in Organ Repair (Ministry of Education), School of Life Science, Shanghai University, Shanghai, China

Jie Xu Center for Mitochondrial Biology and Medicine, The Key Laboratory of Biomedical Information Engineering of Ministry of Education, School of Life Science and Technology, Xi'an Jiaotong University, Xi'an, Shaanxi, China

Xinxin Xu School of Clinical and Basic Medical Sciences, Shandong First Medical University & Shandong Academy of Medical Science, Jinan, Shandong, China

Lining Zhang Department of Rehabilitation, First Medical Center, Chinese PLA General Hospital, Beijing, China

Xing Zhang Key Laboratory of Aerospace Medicine of the Ministry of Education, School of Aerospace Medicine, Fourth Military Medical University, Xi'an, China

Yanyan Zhang Department of Exercise Physiology, Beijing Sport University, Beijing, China

Laboratory of Sports Stress and Adaptation of General Administration of Sport, Beijing Sport University, Beijing, China

Key Laboratory of Physical Fitness and Exercise, Ministry of Education, Beijing Sport University, Beijing, China

Ziang Zhang Key Laboratory of Aerospace Medicine of the Ministry of Education, School of Aerospace Medicine, Fourth Military Medical University, Xi'an, China

Hang Zhao Department of Health Service, Air Force Medical University (Fourth Military Medical University), Xi'an, Shaanxi, China

Lanqi Zhou College of Food Engineering and Nutritional Science, Shaanxi Normal University, Xi'an, China

Chapter 1
The Epigenetic Landscape of Exercise in Human Health: An Overview

Guiling Wu, Xing Zhang, and Feng Gao

Abstract Exercise has long been recognized as a cornerstone of human health, exerting wide-ranging effects on physical, mental, and metabolic well-being. Recent advances in epigenetics have revealed deeper insights into how exercise impacts gene expression without altering the DNA sequence. This review provides an overview of the epigenetic mechanisms responsible for the health benefits of exercise, involving DNA methylation, histone modifications, and RNA-based regulations. We explore how these epigenetic modifications govern key physiological processes, and discuss their implications for disease prevention and health promotion.

Keywords Exercise · Epigenetics · DNA methylation · Histone modification · Noncoding RNAs

Abbreviations

A	Adenosine
ADAR2	Adenosine deaminase acting on RNA 2
ADP	Adenosine diphosphate
A-to-I	Adenosine-to-inosine
C	Cytidine
CVD	Cardiovascular diseases
DNMTs	DNA methyltransferases
ENO1	Enolase 1
G	Guanosine
Glut4	Glucose transporter type 4
HDACs	Histone deacetylases

G. Wu · X. Zhang · F. Gao (✉)
Key Laboratory of Aerospace Medicine of the Ministry of Education, School of Aerospace Medicine, Fourth Military Medical University, Xi'an, China
e-mail: fgao@fmmu.edu.cn

F. Gao, X. Zhang (eds.), *Exercise, Epigenetics and Human Health*, Epigenetics and Human Health 15, https://doi.org/10.1007/978-3-032-08045-5_1

H3K4me3	Histone 3 lysine 4 trimethylation
ISWI	Imitation switch
Klf2	Kruppel-like factor 2
LncRNAs	Long noncoding RNAs
MEF2A	Myocyte enhancer factor 2A
METTL14	Methyltransferase like 14
miRNAs	MicroRNAs
mRNAs	Messenger RNAs
NCD	Noncommunicable diseases
ncRNA	Noncoding RNA
NRF1	Nuclear respiratory factor1
NuRD	Nucleosome remodeling and deacetylation
Phlpp2	PH domain leucine-rich repeat protein phosphatase 2
piRNAs	Piwi-interacting RNAs
PPARGC1A	Peroxisome proliferator-activated receptor gamma coactivator 1-alpha
RNAi	RNA interference
siRNAs	Small interfering RNAs
SWI/SNF	Switch/sucrose-non-fermenting
U	Uridine
WAT	White adipose tissue
WHO	World Health Organization
5mC	The fifth carbon position of cytosine
6 mA	The sixth carbon of adenine forming N6-methyladenine
7mG	The seventh carbon of guanine forming 7-methylguanine

1.1 Introduction

In modern society, physical inactive lifestyles are common due to advancements in labor-saving technology and transportation, coupled with the fast-paced nature of daily life. These inactive lifestyles pose a critical threat to population health, leading to higher rates of premature mortality and morbidity, and elevating the risks of various noncommunicable diseases (NCDs), including coronary heart disease, stroke, type 2 diabetes, cancers, and mental illness (Hou et al. 2019). Extensive research has demonstrated a negative correlation between physical activity (whether self-reported or objectively assessed) and the incidence of NCDs as well as overall mortality (Myers et al. 2002; Nocon et al. 2008; Kodama et al. 2009). Exercise is a low-cost intervention to promote human health (Carson et al. 2013). Even engaging in amounts of physical activity lower than those recommended by the World Health Organization (WHO) can yield significant health benefits. For example, in an Asian adult cohort, performing just 15 min of daily exercise lowered the risk of all-cause mortality by 14% in comparison to inactivity. Similarly, among U.S. adults, running for 5–10 min per day at a moderate speed (<6 miles per hour) substantially decreased the likelihood of mortality from cardiovascular diseases (CVDs) (Wen et al. 2011). In addition to short-term protective effects, recent studies show that exercise also exerts long-lasting benefits. It is reported that regular exercise during early life delivers lasting positive effects on inflammatory immune responses (Zhang et al. 2024). Research also indicates that these benefits may even extend beyond the individual, influencing future generations through what is referred to as transgenerational effects. Such findings highlight the possibility that exercise may induce lasting changes at the molecular level, potentially through epigenetic mechanisms.

While the biological pathways driving benefits of exercise have been extensively studied, the molecular processes responsible for the protective effects of exercise remain incompletely elucidated. Recent research has highlighted the role of epigenetic modifications in mediating the health-promoting effects of exercise (Chen et al. 2021; Yan et al. 2022; Galkin et al. 2023; Moulton et al. 2024). Epigenetics refers to almost all biochemical changes that influence gene expression independent of genetic code changes. Lifestyle and environmental factors have been shown to influence health via epigenetic modifications, and exercise is one of the critical factors responsible for playing a positive role in human health and preventing disease (Denham 2018). Moreover, recent research has highlighted that maternal or paternal exercise counteracts metabolic dysfunction in offspring caused by aging or high-fat diets. Since metabolic dysfunction is a known risk factor for NCDs, this indicates that the advantages conferred by exercise can be inherited (Kusuyama et al. 2022). These observations further emphasize the critical role of epigenetics in driving the beneficial impacts of exercise. Accordingly, this chapter explores the epigenetic landscape of exercise in relation to human health and diseases.

1.2 Exercise Benefits Health

1.2.1 Exercise Has Been Shown to Contribute to Disease Prevention

Consistent exposure to repeated exercise leads to long-term adaptations across multiple tissues and triggers numerous well-documented effects of physical activity, including decreased blood pressure, enhanced vascularization, mitochondrial biogenesis, improved cardiac and immune cell performance, and optimized substrate metabolism (McGee and Hargreaves 2020). Over time, these modifications establish a preconditioned state, equipping trained individuals with resilience against future physiological or pathological challenges, such as intense physical exertion or chronic diseases (Ruegsegger and Booth 2018). Consistent subjection to repeated exercise leads to sustained modifications across multiple tissues and triggers a wide range of well-documented impacts, including reduced blood pressure, enhanced vascularization and mitochondrial biogenesis, improved cardiac performance and immune cell activity, as well as better substrate utilization. Collectively, these adaptations establish a preconditioned physiological state that equips trained individuals to better withstand future (patho)physiological challenges, such as physical exertion or chronic diseases. Exercise is widely recognized for its extensive health benefits, playing a pivotal role in the primary prevention of over 35 chronic diseases, such as type 2 diabetes mellitus, nonalcoholic fatty liver disease, cardiovascular disease, and various types of cancer (Booth et al. 2012). Consistent engagement in exercise induces a range of physiological adaptations that collectively enhance overall health (Egan and Zierath 2013). While a large proportion of these beneficial adaptations occur within skeletal muscle, exercise also exerts positive influences on numerous other tissues throughout the body (Febbraio 2017). The underlying biological mechanisms of preventive effects of exercise include increasing insulin sensitivity, controlling blood glucose, reducing lipid profile improving endothelial function, etc.

1.2.2 Exercise Has Been Shown to Play a Therapeutic Role in Treating the Diseases

Exercise is increasingly acknowledged as an effective therapeutic strategy in the management and treatment of a wide range of diseases. Its role extends beyond prevention, offering significant benefits for patients already diagnosed with chronic conditions. For individuals with cardiovascular diseases, exercise-based cardiac rehabilitation is proven to improve heart function, reduce symptoms, and lower the risk of recurrent events (Dibben et al. 2023). Similarly, in metabolic disorders such as type 2 diabetes, regular physical activity enhances glycemic control, improves insulin sensitivity, and reduces the progression of complications (Hamasaki 2016). Exercise also demonstrates therapeutic effects in managing musculoskeletal disorders, such as osteoarthritis and chronic back pain, by strengthening muscles, improving joint mobility, and alleviating pain (De la Corte-Rodriguez et al. 2024). In neurological conditions like Parkinson's disease and multiple sclerosis, exercise has shown promise in slowing disease progression, improving motor function, and enhancing overall quality of life (Sanaeifar et al. 2024). Additionally, exercise is widely used in oncology, where it mitigates treatment-related side effects, like fatigue and depression, while improving immune function and promoting posttreatment recovery (Baumann et al. 2024). From a psychological standpoint, exercise has been extensively studied for its positive effects on mental health. It is a recommended adjunct therapy for depression, anxiety, and post-traumatic stress disorder, as it boosts endorphin levels, enhances mood, and improves cognitive function (Yan et al. 2024). Furthermore, emerging research explores the anti-inflammatory, neuroprotective, and epigenetic effects of exercise, shedding light on its ability to support healing processes at the cellular and systemic levels.

1.2.3 Exercise Has Long-Term Protective Effects

Besides short-term protective effects, long-term protective effects of exercise even though the exercise was ceased have been shown. Accumulative research indicates that sustaining consistent physical activity or adequate aerobic capacity during the early stages of life can markedly influence long-term health outcomes. Findings from the Amsterdam Growth and Health Longitudinal Study highlight that promoting exercise in adolescence leads to various favorable health advantages in adulthood, such as enhanced cardiorespiratory fitness, improved bone mineral density, lower body mass index, and diminished fat mass (Kemper and Monyeki 2019). In parallel, a retrospective investigation showed that regular involvement in physical activity throughout youth is associated with reduced incidences of arterial hypertension and type 2 diabetes mellitus later in life (Fernandes and Zanesco 2010). Recently, our study has shown that regular exercise during early life significantly improves metabolism, cardiovascular function, muscle strength in aged mice, as well as exerts long-term beneficial effects on inflammatory immunity via epigenetic immunometabolic modulation (Feng et al. 2025; Zhang et al. 2024).

1.2.4 Exercise Has Transgenerational Protective Effects

Research also indicates that these benefits may even extend beyond the individual, influencing future generations in what is referred to as transgenerational effects. Recent studies in both humans and animals have revealed that parental exercise can produce transgenerational health benefits. Research conducted by Stanford et al. demonstrated that maternal physical activity both prior to and during pregnancy mitigated the adverse metabolic effects of a maternal high-fat diet on offspring. This intervention significantly lowered the risk of developing obesity and insulin resistance, which are key factors contributing to numerous noncommunicable diseases (Stanford et al. 2015). The underlying mechanism involves preventing hypermethylation and promoting the expression of peroxisome proliferator-activated receptor gamma coactivator 1-alpha (PPARGC1A) through maternal exercise (Laker et al. 2014). Furthermore, findings by Stanford et al. indicate that paternal exercise positively influenced glucose metabolism in the offspring (Stanford et al. 2018), underscoring the heritability of exercise-induced advantages. Notably, it has also been shown that maternal voluntary physical activity, whether early or later in life, can mitigate maternal age-associated risks of congenital heart defects in their newborns (Schulkey et al. 2015). In addition, a separate study in humans revealed that maternal exercise during obesity improved adult male offspring's cardiovascular health and increased levels of calcium-handling proteins (Beeson et al. 2018). Such findings highlight the possibility that consistent exercise may induce lasting changes at the molecular level, potentially through epigenetic mechanisms.

1.3 Epigenetics Mechanisms Underlying Exercise-Induced Benefits

Historically, the concept of "epigenetics" was introduced in 1942 by Conrad Waddington, who used the term to explain heritable changes in gene activity that occur without altering the underlying DNA sequence, ultimately influencing biological phenotypes. Over the course of nearly a century of scientific exploration, researchers have uncovered a wide variety of enzymes involved in epigenetic modifications and have gained significant insights into the molecular mechanisms that drive these processes. Today, epigenetics is recognized as an intricate system regulating chromatin states, which involves five primary mechanisms: DNA modifications, histone modifications, RNA modifications, chromatin remodeling, and regulation mediated by noncoding RNA (ncRNA). These processes function independently of the DNA sequence itself, allowing specific regions of the genome to be activated or silenced in response to various physiological or pathological signals. Furthermore, epigenetics is dynamic and influenced by environmental factors like diet and exercise, with some marks even being heritable across generations. This expanded view highlights epigenetics as a comprehensive framework for understanding gene regulation and its adaptability.

1.3.1 Chromatin Accessibility

Chromatin accessibility is one of the important epigenetic mechanisms. Regions with limited accessibility are primarily found in heterochromatin, which is tightly compacted and largely inaccessible. In contrast, accessible regions are generally localized within euchromatin, characterized by lower nucleosome density and heightened regulatory activity. Chromatin opening is a critical prerequisite for essential processes such as DNA transcription, replication, and damage repair. These processes are influenced by various factors, including genetic, epigenetic, and environmental cues, and play significant roles in numerous biological events. Nucleosome positioning, as a fundamental component of chromatin structure, is intricately linked to chromatin accessibility. ATP-dependent chromatin remodeling complexes serve as key regulators of nucleosome dynamics. By utilizing energy from ATP hydrolysis, these complexes modulate chromatin between "closed" and "open" states. Based on their distinct core domains, these complexes are classified into four families: switch/sucrose-non-fermenting (SWI/SNF), nucleosome remodeling and deacetylation (NuRD), imitation switch (ISWI), and INO80 (Chen et al. 2024b).

1.3.2 DNA Methylation

DNA methylation is the only established epigenetic mechanism that directly modifies DNA, involving the addition of a methyl group to specific sites on nucleotides. This process includes methylation at the fifth carbon position of cytosine (5mC), the sixth carbon of adenine forming N6-methyladenine (6 mA), and the seventh carbon of guanine forming 7-methylguanine (7mG). Among these, 5mC has been the most extensively investigated. The impact of DNA methylation on gene expression depends heavily on its specific distribution. Methylation occurring at promoter regions or enhancer sites typically suppresses gene transcription, whereas methylation in intergenic regions is often associated with increased gene expression. DNA methyltransferases (DNMTs) are the key enzymes responsible for establishing and maintaining DNA methylation patterns (Jurkowska and Jeltsch 2022). The DNMT family includes DNMT1, DNMT2, DNMT3A, DNMT3B, and DNMT3L. Among them, DNMT1, DNMT3A, and DNMT3B are classified as canonical DNMTs, responsible for catalyzing the addition of methyl groups to genomic DNA. In contrast, DNMT2 and DNMT3L are considered noncanonical members due to their lack of catalytic DNMT activity. DNMT3A and DNMT3B are primarily involved in de novo methylation, establishing new methylation marks on previously unmethylated DNA. Meanwhile, DNMT1 is essential for maintaining preexisting DNA methylation patterns during the replication process (Okano et al. 1999; Schermelleh et al. 2007). In adults, DNA methylation patterns are generally stable. However, two critical phases of extensive demethylation are observed during gametogenesis and embryogenesis. The precise processes and regulatory mechanisms driving DNA methylation reprogramming in these stages are not yet fully clarified.

1.3.3 RNA Modification

RNA modification refers to the chemical alterations made to RNA molecules, were first discovered around 1960 (Amos and Korn 1958; Cohn 1960), and more than 150 distinct RNA modifications have been described since (Boccaletto et al. 2022). These modifications play crucial roles in fine-tuning RNA stability, structure, localization, and translation efficiency. Biochemical modifications occur in all four nucleosides: adenosine (A), guanosine (G), cytidine (C), and uridine (U). Among the diverse modifications observed in both coding and noncoding RNAs, the most prevalent types include methylation, pseudouridylation, and adenosine-to-inosine (A-to-I) editing. These alterations are dynamically regulated by enzymes often referred to as "writers" (which add the modifications), "erasers" (which remove them), and "readers" (which recognize and bind to the modified residues). RNA modifications are essential for cellular processes such as gene expression regulation, stress responses, and maintenance of genome integrity. Dysregulation of RNA modification pathways has been implicated in various human diseases, including cancer, neurological disorders, and metabolic syndromes, highlighting their critical importance in biology and potential as therapeutic targets.

1.3.4 Histone Modification

Histone modification refers to a type of covalent posttranslational alteration of histone proteins, encompassing processes such as methylation, phosphorylation, acetylation, ubiquitylation, sumoylation, biotinylation, and adenosine diphosphate (ADP)-ribosylation (Strahl and Allis 2000; Vaquero et al. 2003). These modifications can influence gene expression by modulating chromatin architecture or by facilitating the binding and activity of histone-modifying enzymes. Despite considerable advancements in studying the roles of histone modifications, the precise mechanisms through which these modifications regulate gene expression remain complex and not yet fully elucidated. Both the pattern of distribution (location on specific histones or chromatin regions) and the abundance of these modifications are critical for the regulation of genetic expression. For example, the same type of modification in different contexts can lead to either activation or repression of transcription. This underscores the intricate interplay between various modifications, a phenomenon often referred to as the "histone code". Among these modifications, histone acetylation and methylation are particularly well-studied. Histone acetylation is facilitated by histone acetyltransferases, while histone deacetylation is mediated by histone deacetylases (HDACs), which are categorized into four groups: Class I (HDACs 1, 3, and 8), Class II (HDACs 4, 7, 9, and 10), Class III (sirtuin or SIRT family), and Class IV (HDAC 11) (Dokmanovic et al. 2007; Bannister and Kouzarides 2011). Histone methylation is carried out by histone methyltransferases, whereas histone demethylation is executed by histone demethylases, which include enzymes from the Jumonji family and lysine-specific demethylase-1. Histone modifications represent a dynamic and complex layer of epigenetic regulation with widespread implications in cellular function, development, and disease. The interplay between various modifications and their corresponding enzymes underscores the precise molecular control required for proper gene regulation. Aberrations in histone modification pathways are linked to numerous diseases such as cancer, neurodegenerative diseases, and developmental disorders, highlighting their importance as potential therapeutic targets.

1.3.5 NcRNAs

NcRNAs are RNA molecules that are transcribed from the genome but do not code for proteins. Despite lacking protein-coding potential, ncRNAs play crucial regulatory roles in a wide range of biological processes by influencing gene expression at the transcriptional, posttranscriptional, and epigenetic levels. They are broadly classified into two categories based on their size: small ncRNAs, which are typically less than 200 nucleotides long (e.g., microRNAs (miRNAs), small interfering RNAs (siRNAs), and piwi-interacting RNAs (piRNAs)), and long noncoding RNAs (lncRNAs), which are longer than 200 nucleotides. miRNAs primarily function by binding to complementary sequences in target messenger RNAs (mRNAs), leading to their degradation or translational inhibition, thereby fine-tuning gene expression. siRNAs also mediate posttranscriptional silencing through the RNA interference (RNAi) pathway, while piRNAs are mostly involved in protecting genomic integrity by suppressing transposon activity in germline cells. lncRNAs, on the other hand, are involved in a variety of regulatory mechanisms, including chromatin remodeling, transcriptional activation or repression, RNA splicing, and mRNA stabilization or degradation. They can act as molecular scaffolds, guides, sponges for miRNAs, or decoys to sequester transcription factors, influencing diverse cellular processes such as development, differentiation, and immune responses. ncRNAs have emerged as key players in both normal physiological processes and pathological conditions, such as cancer, neurodegenerative diseases, and cardiovascular disorders. Their highly specific expression patterns and regulatory functions make them promising biomarkers and therapeutic targets for various diseases. As research into ncRNAs continues to expand, their importance in understanding gene regulatory networks and cellular function becomes increasingly evident.

1.4 Exercise Benefits Health Through Epigenetics Mechanism

Exercise exerts profound health benefits through various epigenetic mechanisms, including DNA methylation, histone modification, RNA modification, and the regulation of ncRNAs. These mechanisms contribute to metabolic health, cardiovascular function, neuroprotection, and overall physiological well-being, underscoring the critical role of epigenetics in exercise-induced health benefits (Fig. 1.1).

One major pathway through which exercise promotes health is DNA methylation. Exercise has been shown to induce genome-wide alterations in DNA methylation across multiple tissues, including skeletal muscle, adipose tissue, and blood. These modifications are particularly important in metabolic tissues such as skeletal muscle and adipose tissue, as they help mitigate metabolic syndrome. For example, in a population of diabetic patients, 16-week exercise was found to induce hypomethylation of the nuclear respiratory factor1 (NRF1) gene, a change associated with increased mitochondrial density (Rowlands et al. 2014). This highlights the role of DNA methylation in mediating exercise-induced improvements in metabolic health. Another critical aspect of exercise-regulated epigenetics lies in histone modifications. Recent research has unveiled the potential of exercise in mediating histone modifications. For instance, we recently found that early-life exercise enhanced anti-inflammatory immunity in middle-aged male mice by increasing histone 3

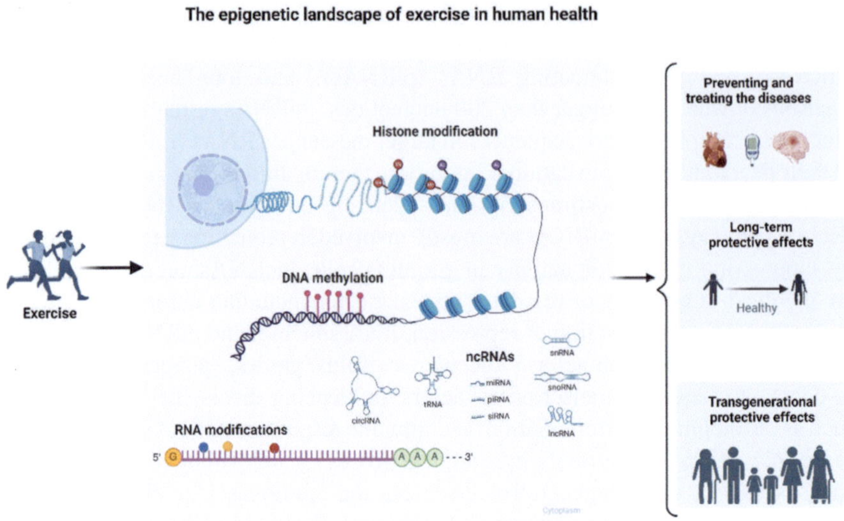

Fig. 1.1 The epigenetic landscape of exercise in human health. Exercise drives changes in histone modifications, DNA and RNA methylation patterns, as well as noncoding RNA regulation across various tissues. These molecular adaptations are fundamental to the health-enhancing and disease-preventive effects of exercise

lysine 4 trimethylation (H3K4me3) at the promoter of the *Crym* in the liver, which regulates pipecolic acid production. Exercise-induced histone modifications in skeletal muscle have also been linked to reduced cardiac disease risk through metabolic remodeling (Zhang et al. 2024). In a rat swimming study, researchers observed an increase in H3K9/14 acetylation at the promoter region of the *glucose transporter type 4* (*Glut4*) gene in skeletal muscle. This acetylation facilitates binding of the transcription factor myocyte enhancer factor 2A (MEF2A), leading to enhanced *Glut4* expression, which is vital for glucose metabolism and cardiac health (Smith et al. 2008). Although studies on cardiac-specific regions remain limited, these findings emphasize histone modification as a key mediator of exercise-associated metabolic and cardiovascular benefits. RNA modifications serve as another layer of epigenetic regulation through which exercise improves health. These modifications, such as m6A and A-to-I editing, are dynamic and reversible processes mediated by specific enzymes. A single bout of acute exercise has been shown to elevate m6A modifications, which promote glycolysis in white adipose tissue (WAT), contributing to enhanced metabolic activity (Chen et al. 2024a). Similarly, in the heart, exercise upregulates RNA editing enzyme adenosine deaminase acting on RNA 2 (ADAR2), which promotes adenosine-to-inosine editing (Wu et al. 2022). This modification suppresses miR-34a expression, thereby enhancing cardiomyocyte survival and ameliorating myocardial damage. Furthermore, swimming-induced cardiac physiological hypertrophy was associated with reduced m6A levels and the involvement of RNA m6A methyltransferase like 14 (METTL14), which regulates *PH domain leucine-rich repeat protein phosphatase 2* (*Phlpp2*) mRNA and activates the AKT signaling pathway, promoting cardiomyocyte growth and survival

(Wang et al. 2022). These findings highlight the role of RNA modifications in regulating critical exercise-induced physiological adaptations. Finally, ncRNAs, including miRNAs, lncRNAs, and circRNAs, are key regulatory molecules involved in exercise-induced health benefits. ncRNAs regulate gene expression and influence numerous physiological and pathological processes. Exercise has been shown to alter the expression of various ncRNAs, thereby activating signaling pathways that improve cellular function and health outcomes. For instance, swimming restores the aberrant expression levels of miR-214, miR-1, miR-29a, and miR-29c in hearts affected by myocardial infarction, alleviating cardiac dysfunction by targeting downstream pathways (Melo et al. 2014, 2015). In addition, the lncRNA *CPhar* is essential for swim-induced cardiac hypertrophy, and circRNA *CircFndc3b* mediates the anti-pyroptotic effects of exercise through the enolase 1 (ENO1)/Kruppel-like factor 2 (Klf2) axis, offering neuroprotection and supporting cardiac function (Gao et al. 2021). These findings underscore the importance of ncRNAs in mediating the beneficial effects of exercise across various systems. Therefore, exercise benefits health through a multitude of interconnected epigenetic mechanisms. More detailed evidence for the role of epigenetics in exercise-induced benefits will be discussed in the following chapters.

1.5 Clinical Application and Perspectives

Although many studies have explored how exercise promotes health through epigenetic mechanisms, there are still numerous questions that remain to be clarified. For instance, many studies have shown that exercise induces epigenetic changes, but how exercise induces epigenetic changes remains to be explored. Furthermore, the interactions among various epigenetic mechanisms as well as the identification of the primary initiating factor in the mechanisms underlying disease progression and exercise-induced protective effects remain poorly understood. Further exploration of the epigenetic changes triggered by various types of exercise and their effects on different tissues is essential to deepen our understanding of how exercise promotes health. Moreover, uncovering the epigenetic mechanisms that contribute to the positive impacts of exercise can potentially identify new therapeutic targets for disease treatment and health enhancement. Additionally, the role of exercise in intergenerational cardioprotection and its associated molecular mechanisms warrants further investigation. Particular attention should be given to whether exercise-induced epigenetic alterations in germ cells significantly influence cardiac function in offspring.

Compliance with Ethical Standards

Conflict of Interest The authors declare that there is no conflict of interest.

Ethical Approval This chapter does not contain any studies with human participants or animals performed by any of the authors.

References

Amos H, Korn M (1958) 5-methyl cytosine in the RNA of Escherichia coli. Biochim Biophys Acta 29(2):444–445. https://doi.org/10.1016/0006-3002(58)90214-2

Bannister AJ, Kouzarides T (2011) Regulation of chromatin by histone modifications. Cell Res 21(3):381–395. https://doi.org/10.1038/cr.2011.22

Baumann FT, Jensen W, Berling-Ernst A, Theurich S, Leitzmann M, Götte M (2024) Exercise therapy in oncology – the impact on quality of life and side effects. Dtsch Arztebl Int. https://doi.org/10.3238/arztebl.m2024.0038

Beeson JH, Blackmore HL, Carr SK, Dearden L, Duque-Guimarães DE, Kusinski LC, Pantaleão LC, Pinnock AG, Aiken CE, Giussani DA, Fernandez-Twinn DS, Ozanne SE (2018) Maternal exercise intervention in obese pregnancy improves the cardiovascular health of the adult male offspring. Mol Metab 16:35–44. https://doi.org/10.1016/j.molmet.2018.06.009

Boccaletto P, Stefaniak F, Ray A, Cappannini A, Mukherjee S, Purta E, Kurkowska M, Shirvanizadeh N, Destefanis E, Groza P, Avşar G, Romitelli A, Pir P, Dassi E, Conticello SG, Aguilo F, Bujnicki JM (2022) MODOMICS: a database of RNA modification pathways. Nucleic Acids Res 50(D1):D231–d235. https://doi.org/10.1093/nar/gkab1083

Booth FW, Roberts CK, Laye MJ (2012) Lack of exercise is a major cause of chronic diseases. Compr Physiol 2(2):1143–1211. https://doi.org/10.1002/cphy.c110025

Carson V, Ridgers ND, Howard BJ, Winkler EA, Healy GN, Owen N, Dunstan DW, Salmon J (2013) Light-intensity physical activity and cardiometabolic biomarkers in US adolescents. PLoS One 8(8):e71417. https://doi.org/10.1371/journal.pone.0071417

Chen X, Zhu X, Wei A, Chen F, Gao Q, Lu K, Jiang Q, Cao W (2021) Nrf2 epigenetic derepression induced by running exercise protects against osteoporosis. Bone Res 9(1):15. https://doi.org/10.1038/s41413-020-00128-8

Chen W, Liu Y, Liu J, Chen Y, Wang X (2024a) Acute exercise promotes WAT browning by remodeling mRNA m6A methylation. Life Sci 123269. https://doi.org/10.1016/j.lfs.2024.123269

Chen Y, Liang R, Li Y, Jiang L, Ma D, Luo Q, Song G (2024b) Chromatin accessibility: biological functions, molecular mechanisms and therapeutic application. Signal Transduct Target Ther 9(1):340. https://doi.org/10.1038/s41392-024-02030-9

Cohn WE (1960) Pseudouridine, a carbon-carbon linked ribonucleoside in ribonucleic acids: isolation, structure, and chemical characteristics. J Biol Chem 235:1488–1498

De la Corte-Rodriguez H, Roman-Belmonte JM, Resino-Luis C, Madrid-Gonzalez J, Rodriguez-Merchan EC (2024) The role of physical exercise in chronic musculoskeletal pain: best medicine-a narrative review. Healthcare (Basel) 12(2). https://doi.org/10.3390/healthcare12020242

Denham J (2018) Exercise and epigenetic inheritance of disease risk. Acta Physiol (Oxf) 222(1). https://doi.org/10.1111/apha.12881

Dibben GO, Faulkner J, Oldridge N, Rees K, Thompson DR, Zwisler AD, Taylor RS (2023) Exercise-based cardiac rehabilitation for coronary heart disease: a meta-analysis. Eur Heart J 44(6):452–469. https://doi.org/10.1093/eurheartj/ehac747

Dokmanovic M, Clarke C, Marks PA (2007) Histone deacetylase inhibitors: overview and perspectives. Mol Cancer Res 5(10):981–989. https://doi.org/10.1158/1541-7786.Mcr-07-0324

Egan B, Zierath JR (2013) Exercise metabolism and the molecular regulation of skeletal muscle adaptation. Cell Metab 17(2):162–184. https://doi.org/10.1016/j.cmet.2012.12.012

Febbraio MA (2017) Exercise metabolism in 2016: health benefits of exercise – more than meets the eye! Nat Rev Endocrinol 13(2):72–74. https://doi.org/10.1038/nrendo.2016.218

Fernandes RA, Zanesco A (2010) Early physical activity promotes lower prevalence of chronic diseases in adulthood. Hypertens Res 33(9):926–931. https://doi.org/10.1038/hr.2010.106

Galkin F, Kovalchuk O, Koldasbayeva D, Zhavoronkov A, Bischof E (2023) Stress, diet, exercise: common environmental factors and their impact on epigenetic age. Ageing Res Rev 88:101956. https://doi.org/10.1016/j.arr.2023.101956

Gao R, Wang L, Bei Y, Wu X, Wang J, Zhou Q, Tao L, Das S, Li X, Xiao J (2021) Long noncoding RNA cardiac physiological hypertrophy-associated regulator induces cardiac physiological hypertrophy and promotes functional recovery after myocardial ischemia-reperfusion injury. Circulation 144(4):303–317. https://doi.org/10.1161/circulationaha.120.050446

Hamasaki H (2016) Daily physical activity and type 2 diabetes: a review. World J Diabetes 7(12):243–251. https://doi.org/10.4239/wjd.v7.i12.243

Hou Z, Qin X, Hu Y, Zhang X, Li G, Wu J, Li J, Sha J, Chen J, Xia J, Wang L, Gao F (2019) Longterm exercise-derived Exosomal miR-342-5p: a novel Exerkine for Cardioprotection. Circ Res 124(9):1386–1400. https://doi.org/10.1161/circresaha.118.314635

Jurkowska RZ, Jeltsch A (2022) Mechanisms and biological roles of DNA Methyltransferases and DNA methylation: from past achievements to future challenges. Adv Exp Med Biol 1389:1–19. https://doi.org/10.1007/978-3-031-11454-0_1

Kemper HC, Monyeki KD (2019) The Amsterdam growth and health longitudinal study: how important is physical activity in youth for later health? (ELS 33). Cardiovasc J Afr 30(3):138–141. https://doi.org/10.5830/cvja-2018-057

Kodama S, Saito K, Tanaka S, Maki M, Yachi Y, Asumi M, Sugawara A, Totsuka K, Shimano H, Ohashi Y, Yamada N, Sone H (2009) Cardiorespiratory fitness as a quantitative predictor of all-cause mortality and cardiovascular events in healthy men and women: a meta-analysis. JAMA 301(19):2024–2035. https://doi.org/10.1001/jama.2009.681

Kusuyama J, Makarewicz NS, Albertson BG, Alves-Wagner AB, Conlin RH, Prince NB, Alves CRR, Ramachandran K, Kozuka C, Xiudong Y, Xia Y, Hirshman MF, Hatta T, Nagatomi R, Nozik ES, Goodyear LJ (2022) Maternal exercise-induced SOD3 reverses the deleterious effects of maternal high-fat diet on offspring metabolism through stabilization of H3K4me3 and protection against WDR82 Carbonylation. Diabetes 71(6):1170–1181. https://doi.org/10.2337/db21-0706

Laker RC, Lillard TS, Okutsu M, Zhang M, Hoehn KL, Connelly JJ, Yan Z (2014) Exercise prevents maternal high-fat diet-induced hypermethylation of the Pgc-1α gene and age-dependent metabolic dysfunction in the offspring. Diabetes 63(5):1605–1611. https://doi.org/10.2337/db13-1614

McGee SL, Hargreaves M (2020) Exercise adaptations: molecular mechanisms and potential targets for therapeutic benefit. Nat Rev Endocrinol 16(9):495–505. https://doi.org/10.1038/s41574-020-0377-1

Melo SF, Fernandes T, Baraúna VG, Matos KC, Santos AA, Tucci PJ, Oliveira EM (2014) Expression of MicroRNA-29 and collagen in cardiac muscle after swimming training in myocardial-infarcted rats. Cell Physiol Biochem 33(3):657–669. https://doi.org/10.1159/000358642

Melo SF, Barauna VG, Neves VJ, Fernandes T, Lara Lda S, Mazzotti DR, Oliveira EM (2015) Exercise training restores the cardiac microRNA-1 and -214 levels regulating Ca2+ handling after myocardial infarction. BMC Cardiovasc Disord 15:166. https://doi.org/10.1186/s12872-015-0156-4

Mengya, Feng Min, Li Jing, Lou Guiling, Wu Tian, Gao Fangqin, Wu Yanzhen, Tan Nini, Zhang Yong, Zhao Lin, Zhao Jia, Li Changhong, Shi Xing, Zhang Jiankang, Liu Feng, Gao (2025) Early-life exercise extends healthspan but not lifespan in mice Nature Communications 16(1). https://doi.org/10.1038/s41467-025-61443-4

Moulton C, Murri A, Benotti G, Fantini C, Duranti G, Ceci R, Grazioli E, Cerulli C, Sgrò P, Rossi C, Magno S, Di Luigi L, Caporossi D, Parisi A, Dimauro I (2024) The impact of physical activity on promoter-specific methylation of genes involved in the redox-status and disease progression: a longitudinal study on post-surgery female breast cancer patients undergoing medical treatment. Redox Biol 70:103033. https://doi.org/10.1016/j.redox.2024.103033

Myers J, Prakash M, Froelicher V, Do D, Partington S, Atwood JE (2002) Exercise capacity and mortality among men referred for exercise testing. N Engl J Med 346(11):793–801. https://doi.org/10.1056/NEJMoa011858

Nocon M, Hiemann T, Müller-Riemenschneider F, Thalau F, Roll S, Willich SN (2008) Association of physical activity with all-cause and cardiovascular mortality: a systematic review and meta-analysis. Eur J Cardiovasc Prev Rehabil 15(3):239–246. https://doi.org/10.1097/HJR.0b013e3282f55e09

Okano M, Bell DW, Haber DA, Li E (1999) DNA methyltransferases Dnmt3a and Dnmt3b are essential for de novo methylation and mammalian development. Cell 99(3):247–257. https://doi.org/10.1016/s0092-8674(00)81656-6

Rowlands DS, Page RA, Sukala WR, Giri M, Ghimbovschi SD, Hayat I, Cheema BS, Lys I, Leikis M, Sheard PW, Wakefield SJ, Breier B, Hathout Y, Brown K, Marathi R, Orkunoglu-Suer FE, Devaney JM, Leiken B, Many G, Krebs J, Hopkins WG, Hoffman EP (2014) Multi-omic integrated networks connect DNA methylation and miRNA with skeletal muscle plasticity to chronic exercise in type 2 diabetic obesity. Physiol Genomics 46(20):747–765. https://doi.org/10.1152/physiolgenomics.00024.2014

Ruegsegger GN, Booth FW (2018) Health benefits of exercise. Cold Spring Harb Perspect Med 8(7). https://doi.org/10.1101/cshperspect.a029694

Sanaeifar F, Pourranjbar S, Pourranjbar M, Ramezani S, Mehr SR, Wadan AS, Khazeifard F (2024) Beneficial effects of physical exercise on cognitive-behavioral impairments and brain-derived neurotrophic factor alteration in the limbic system induced by neurodegeneration. Exp Gerontol 112539. https://doi.org/10.1016/j.exger.2024.112539

Schermelleh L, Haemmer A, Spada F, Rösing N, Meilinger D, Rothbauer U, Cardoso MC, Leonhardt H (2007) Dynamics of Dnmt1 interaction with the replication machinery and its role in postreplicative maintenance of DNA methylation. Nucleic Acids Res 35(13):4301–4312. https://doi.org/10.1093/nar/gkm432

Schulkey CE, Regmi SD, Magnan RA, Danzo MT, Luther H, Hutchinson AK, Panzer AA, Grady MM, Wilson DB, Jay PY (2015) The maternal-age-associated risk of congenital heart disease is modifiable. Nature 520(7546):230–233. https://doi.org/10.1038/nature14361

Smith JA, Kohn TA, Chetty AK, Ojuka EO (2008) CaMK activation during exercise is required for histone hyperacetylation and MEF2A binding at the MEF2 site on the Glut4 gene. Am J Physiol Endocrinol Metab 295(3):E698–E704. https://doi.org/10.1152/ajpendo.00747.2007

Stanford KI, Lee MY, Getchell KM, So K, Hirshman MF, Goodyear LJ (2015) Exercise before and during pregnancy prevents the deleterious effects of maternal high-fat feeding on metabolic health of male offspring. Diabetes 64(2):427–433. https://doi.org/10.2337/db13-1848

Stanford KI, Rasmussen M, Baer LA, Lehnig AC, Rowland LA, White JD, So K, De Sousa-Coelho AL, Hirshman MF, Patti ME, Rando OJ, Goodyear LJ (2018) Paternal exercise improves glucose metabolism in adult offspring. Diabetes 67(12):2530–2540. https://doi.org/10.2337/db18-0667

Strahl BD, Allis CD (2000) The language of covalent histone modifications. Nature 403(6765):41–45. https://doi.org/10.1038/47412

Vaquero A, Loyola A, Reinberg D (2003) The constantly changing face of chromatin. Sci Aging Knowl Environ 2003(14):Re4. https://doi.org/10.1126/sageke.2003.14.re4

Wang L, Wang J, Yu P, Feng J, Xu GE, Zhao X, Wang T, Lehmann HI, Li G, Sluijter JPG, Xiao J (2022) METTL14 is required for exercise-induced cardiac hypertrophy and protects against myocardial ischemia-reperfusion injury. Nat Commun 13(1):6762. https://doi.org/10.1038/s41467-022-34434-y

Wen CP, Wai JP, Tsai MK, Yang YC, Cheng TY, Lee MC, Chan HT, Tsao CK, Tsai SP, Wu X (2011) Minimum amount of physical activity for reduced mortality and extended life expectancy: a prospective cohort study. Lancet 378(9798):1244–1253. https://doi.org/10.1016/s0140-6736(11)60749-6

Wu X, Wang L, Wang K, Li J, Chen R, Wu X, Ni G, Liu C, Das S, Sluijter JPG, Li X, Xiao J (2022) ADAR2 increases in exercised heart and protects against myocardial infarction and doxorubicin-induced cardiotoxicity. Mol Ther 30(1):400–414. https://doi.org/10.1016/j.ymthe.2021.07.004

Yan L, Wei JA, Yang F, Wang M, Wang S, Cheng T, Liu X, Jia Y, So KF, Zhang L (2022) Physical exercise prevented stress-induced anxiety via improving brain RNA methylation. Adv Sci (Weinh) 9(24):e2105731. https://doi.org/10.1002/advs.202105731

Yan L, Wang Y, Hu H, Yang D, Wang W, Luo Z, Wang Y, Yang F, So KF, Zhang L (2024) Physical exercise mediates cortical synaptic protein lactylation to improve stress resilience. Cell Metab. https://doi.org/10.1016/j.cmet.2024.07.018

Zhang N, Wang X, Feng M, Li M, Wang J, Yang H, He S, Xia Z, Shang L, Jiang X, Sun M, Wu Y, Ren C, Zhang X, Li J, Gao F (2024) Early-life exercise induces immunometabolic epigenetic modification enhancing anti-inflammatory immunity in middle-aged male mice. Nat Commun 15(1):3103. https://doi.org/10.1038/s41467-024-47458-3

Chapter 2
Long-Term Health Benefits Induced by Exercise: A Role of Epigenetic Memory

Mengya Feng, Min Li, Jing Lou, Han Li, and Xing Zhang

Abstract Emerging evidence suggests a long-term health benefit induced by exercise, persisting beyond exercise cessation. For example, in childhood and adolescence, regular physical activity establishes foundational metabolic and epigenetic programming, reducing risks of obesity, hypertension, type 2 diabetes, and cardiovascular morbidity in adulthood. Exercise in adults also enhances cardiovascular health, metabolic homeostasis, and neurocognitive resilience, delaying age-related cognitive decline and neurodegenerative disorders in aging. These advances indicate an enduring potential of exercise to positively impact health over time. At the molecular level, exercise induces extensive epigenetic modifications (DNA methylation, histone modification, and noncoding RNAs), metabolic reprogramming, and skeletal muscle adaptations ("muscle memory"), which collectively underpin its long-term biological effects. This chapter discusses the evidence on the long-term health benefits of exercise across the lifespan, and advocates for public health policies promoting physical activity across all life stages to combat diseases and support healthy aging.

Keywords Exercise · Long-term health benefits · Muscle memory · Epigenetic regulation

Abbreviations

EPOC	Excess postexercise oxygen consumption
PFC	Prefrontal cortex
BDNF	Brain-derived neurotrophic factor
IGF-1	Insulin-like growth factor-1

M. Feng · M. Li · J. Lou · H. Li · X. Zhang (✉)
Key Laboratory of Aerospace Medicine of the Ministry of Education, School of Aerospace Medicine, Fourth Military Medical University, Xi'an, China
e-mail: zhangxing@fmmu.edu.cn

F. Gao, X. Zhang (eds.), *Exercise, Epigenetics and Human Health*, Epigenetics and Human Health 15, https://doi.org/10.1007/978-3-032-08045-5_2

FEP	First-episode psychosis
PD	Parkinson's disease
AAA	Abdominal aortic aneurysm
HIT	High-intensity training
MIT	Moderate-intensity training
LMI	Low-to-moderate intensity
PM	Polymyositis
DM	Dermatomyositis
CNSLBP	Chronic nonspecific low back pain
IIM	Idiopathic inflammatory myopathies
$V(O_2)$ peak	Peak aerobic capacity
NCDs	Noncommunicable diseases

2.1 Introduction

Globally, the prevalence of noncommunicable diseases (NCDs), including cardiovascular diseases, diabetes, obesity, and cancer, is escalating, and these diseases constitute over 70% of the causes of death worldwide (NCD Countdown 2030 2018). Physical inactivity exacerbates this burden (Santos et al. 2023). It is well established that exercise is beneficial in a variety of physiological and pathophysiological states, and it plays a crucial role in the prevention and management of multiple NCDs (Heine et al. 2019). Additionally, regular exercise can reduce symptoms of depression and anxiety, enhancing cognitive function, and potentially combating the progression of neurodegenerative diseases (Zhao 2024). As a low-cost, non-pharmacological intervention, exercise holds irreplaceable value in public health (Ueshima et al. 2025).

Emerging evidence highlights that regular physical exercise benefits short-term and long-term health. Many studies on exercise interventions predominantly report short-term outcomes rather than long-term effects (Schootemeijer et al. 2020). Over the last decade evidence has emerged supporting the positive impact that physical activity exerts long-term health outcomes (Dimitri et al. 2020). A series of epidemiological studies highlight that higher level of physical activity in childhood is associated with benefits on aerobic fitness, osteoporosis and metabolic diseases, including obesity, hypertension, type 2 diabetes mellitus, and cardiac disease, later in life (Yang et al. 2006; Ford et al. 2008; da Silva et al. 2023). A recent study demonstrated that early-life exercise enhances anti-inflammatory immunity in middle-aged mice (Zhang et al. 2024). These findings accentuate the specificity of the long-term effects of exercise and the exigency for research. As the global population ages, it is of significance to study the long-term effects of exercise for delaying aging and prolonging the healthy lifespan of the elderly (Pahor et al. 2014). This chapter will discuss the long-term advantages of exercise, encompassing its influence on cardiovascular health, metabolic function, and cognitive function.

2.2 Long-Term Benefits of Exercise

Regular exercise promotes metabolic homeostasis and delays systemic aging through comprehensive physiological benefits. While acute health-promoting effects of exercise (spanning hours to days) are well recognized, sustained weekly regimens are generally required for optimal outcomes. Notably, emerging evidence suggests persistent protective effects persisting beyond exercise cessation. Animal studies demonstrate that prediabetic rats initiating voluntary wheel running at 5 weeks for 14 weeks maintained reduced body weight and elevated serum glycolytic enzymes along with skeletal muscle uncoupling protein 3 (UCP3) expression at 45 weeks post-interruption compared to sedentary controls (Shindo et al. 2014). Similarly, rats exposed to postnatal high-fat diets exhibited persistent weight reduction after 6 weeks of juvenile wheel training followed by 7-week detraining, outperforming non-exercised counterparts (Patterson et al. 2008). Remarkably, transient exercise training during early life (4 weeks) induces lasting improvements in adult insulin sensitivity via epigenetic reprogramming of glucose-regulatory pathways (Falcão-Tebas et al. 2019). Human cohort analyses corroborate these findings: individuals maintaining ≥180 min weekly physical activity for ≥9 months between ages 7 and 17 show significantly lower incidence of hypertension and type 2 diabetes in later life (Fernandes and Zanesco 2010). These findings collectively underscore the enduring legacy of exercise interventions. Although acute exercise mechanisms are well-characterized, systematic investigation into long-term biological embedding effects through cross-disciplinary approaches will advance our understanding of exercise-induced health preservation across the lifespan.

2.2.1 Childhood and Adolescence: Foundational Programming

Childhood and adolescence constitute a critical developmental window characterized by rapid physical and cognitive maturation, establishing foundational patterns for lifelong health behaviors. Compelling evidence now identifies early-life physiological experiences as modulators of long-term health trajectories through epigenetic and molecular programming mechanisms. Contemporary epidemiological analyses reveal consistent positive associations between pediatric physical activity engagement (ages 5–18 years) and enhanced aerobic capacity, improved bone mineral density (BMD), along with attenuated risks of metabolic disorders spanning obesity, hypertension, type 2 diabetes, and subsequent cardiovascular morbidity (Yang et al. 2006; Ford et al. 2008; da Silva et al. 2023). Our recent experimental work demonstrates that early-life exercise (1–4 months of age) improved health span in both male and female mice, as evidenced by enhanced systemic metabolism, cardiovascular function, and muscle strength, as well as reduced systemic inflammation and frailty in aged mice, which does not affect lifespan in mice (Feng et al. 2025). Additionally, early-life exercise potentiates anti-inflammatory immunocompetence in middle-aged male mice, suggesting conserved biological memory of early exercise stimuli (Zhang et al. 2024). Mechanistic insights from epigenetic studies further indicate that exercise-induced DNA methylation modifications in key metabolic genes may mediate these persistent effects. The converging evidence positions pediatric physical activity as a potent epigenetic primer that exerts durable cardiometabolic benefits extending decades beyond initial exposure.

Decades of epidemiological research underscore the prophylactic role of structured physical activity in pediatric populations. A review from 1986 demonstrated that active life and physical fitness may represent the most effective strategies to prevent chronic diseases and to improve growth and development for children (Giannini et al. 2006), and one of the primary reasons is long-term health benefits of exercise. The Amsterdam Growth and Health Longitudinal Study (AGAHLS) offers compelling evidence for the long-term advantages of exercise, particularly during adolescence. Over a 25-year period, the study documented longitudinal changes in health and lifestyle among three birth cohorts in the Amsterdam region. Physical activity during youth, especially weight-bearing exercises, was associated with a variety of beneficial health outcomes in adulthood, encompassing enhanced aerobic fitness, increased bone mineral density, lower body mass index, and decreased body fat mass (Kemper and Monyeki 2019). Similar results were found in Young Finns Study. The 21-year follow-up study involved 1319 subjects and revealed that maintaining a high level of physical activity was inversely associated with the risk of obesity in later life. Notably, individuals who were persistently active had lower waist circumferences compared to those who were decreasingly active or persistently inactive, which means youth physical activity has an indirect effect on adult obesity through the maintenance of physical activity into adulthood and by reducing body weight in youth (Yang et al. 2006, 2007). A retrospective study has emphasized the long-term advantages of early physical activity on the incidence of chronic diseases in adulthood. Analyzing data from 1436 adults, the study found that physical activity during childhood and adolescence was associated with lower rates of arterial hypertension and type 2 diabetes mellitus in adulthood. Specifically, individuals who were physically active in their youth had an odds ratio of 0.42 for arterial hypertension and 0.29 for type 2 diabetes mellitus, indicating a considerable reduction in disease risk (Fernandes and Zanesco 2010). Another retrospective population-based survey carried out in eight Brazilian cities involved 2720 adults yielded congruent findings: Early sport practice during childhood and adolescence was associated with lower occurrence of arterial hypertension (OR, 0.49; 95% CI, 0.33–0.73) and type 2 diabetes (OR, 0.46; 95% CI, 0.24–0.88) in adulthood (Fernandes and Zanesco 2015). These studies highlight that early engagement in physical activity can exert a lasting influence on health, reinforcing the significance of public health policies that promote physical activity from a young age.

2.2.2 Adulthood: Sustaining Health and Resilience

Physical exercise also exerts long-term health effects in adults, conferring sustained improvements in metabolic regulation. A follow-up study of a 6-year randomized controlled trial was devised to evaluate the impacts of a prenatal exercise program on long-term health and lifestyle habits. The study contrasted the disparities in weight, weight maintenance, and body mass index (BMI) between the intervention and control groups and appraised the effects on levels of physical activity and leisure exercise. The outcomes demonstrated that for women adhering to the prenatal exercise program (attending at least 24 prenatal exercise classes), their body weight and BMI were conspicuously lower than those of the control group during the 6-year follow-up, and none gained more than 5 kg from their prepregnancy weight. This implies that women following prenatal exercise interventions might have better control overweight and BMI in the long run (Haakstad et al. 2021).

Complementary evidence from geriatric populations demonstrates exercise-induced neuroplasticity and musculoskeletal resilience. The study involving 25 healthy older adults (with a mean age of 70–82 years) compared to the changes in physical (cardiovascular health, lower limb strength, and agility) and cognitive (executive function) status during training and withdrawal. It was found that both physical and cognitive status improved significantly within 2 years. Even after the cessation of exercise, physical and cognitive status continued to improve from the baseline (Blasco-Lafarga et al. 2020). This highlights the significance of exercise for sustainably maintaining physical and cognitive health in adults.

2.3 Exercise Shows Long-Term Benefits for Managing Diseases

Physical activity confers long-term protection to lower the risk of numerous diseases, with residual benefits persisting beyond intervention periods (Fig. 2.1). Exercise attenuates musculoskeletal degeneration through distinct biomechanical pathways. In knee osteoarthritis, clinical trials demonstrate that patients suffering from knee osteoarthritis and undergoing a specialized exercise intervention three times a week for 12 weeks exhibited enhanced treatment outcomes with sustained quality-of-life improvements at 6 and 12 months after the intervention (Torstensen et al. 2023). Another comprehensive study involving 12,796 Danish patients with knee osteoarthritis has underscored the long-term beneficial effects of exercise on pain relief. The research revealed that regardless of initial physical activity levels, patients experienced similar and persistent long-term pain relief (Skou et al. 2018). For idiopathic inflammatory myopathies (IIM), a group of diseases characterized by muscle weakness and inflammation. Historically, exercise was discouraged for IIM patients due to concerns it could exacerbate inflammation. Similarly, recent studies suggest that exercise training can improve or stabilize muscle strength and

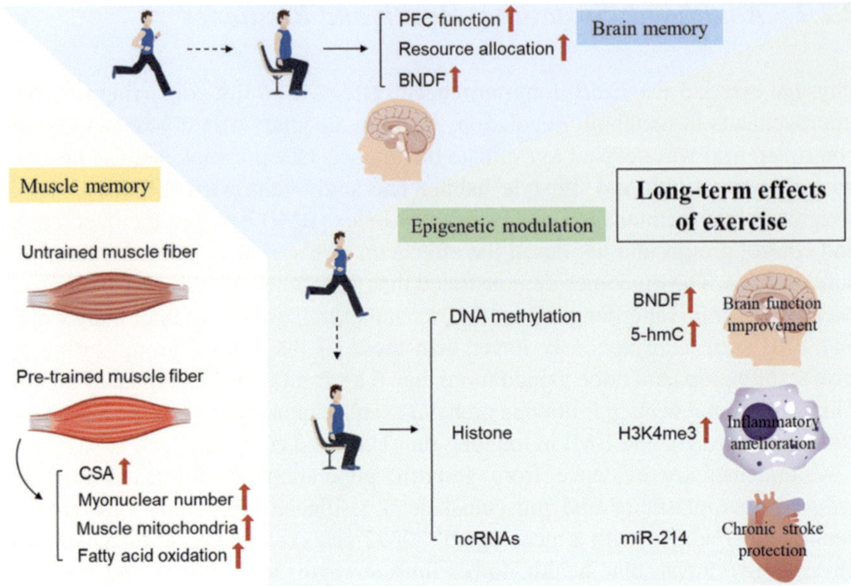

Fig. 2.1 Mechanisms underlie exercise-induced long-term benefits, including muscle memory, brain memory, and epigenetic modulation**.** Physical exercise enhances PFC function and increases BNDF level in the long-term. For muscle memory, pre-trained muscle fibers acquire strength and volume much faster, with increased CSA and nuclear number. Exercise also benefits mitochondria function in the long-term. Physical exercise also exerts long-term benefits through epigenetic mechanisms, including DNA, histone methylation and ncRNAs. *CSA* cross-sectional area, *PFC* prefrontal cortex, *BNDF* brain-derived neurotrophic factors, *5-hmC* 5-hydroxymethylcytosine

functional ability without triggering disease flares on polymyositis (PM) and dermatomyositis (DM). Notably, these benefits were observed both during the chronic phase and in the active course of the disease (de Salles Painelli et al. 2009). High intensity training (HIT) is proved to be an effective exercise to improve disability and physical performance after cessation of a 12-week supervised exercise, in patients with chronic nonspecific low back pain (CNSLBP) (Verbrugghe et al. 2021).

Exercise confers long-term neuroprotective effects, mitigating disease progression and enhancing functional outcomes across neurodegenerative and psychiatric disorders. Long-term healthy physical activity leads to physical benefits in Pompe disease (Ismailova et al. 2023). Significant improvements in first-episode psychosis (FEP) psychiatric symptoms have been found after exercise with a 6-month cessation (p value = 0.001) (Firth et al. 2018). Parkinson's disease (PD) is a progressive neurodegenerative movement disorder characterized by symptoms such as postural instability, gait disturbance, immobility, and falls. Beyond pharmacological and surgical management, exercise and physical therapy interventions are actively investigated for PD. Research indicates that a minimum of 4 weeks of gait training or 8 weeks of balance training can confer benefits persisting for 3–12 months post-intervention. Furthermore, sustained interventions (including strength training, aerobic training, tai chi, or dance therapy) lasting at least 12 weeks can lead to long-lasting improvements in motor symptoms (Mak et al. 2017). These advances indicate that exercise induces sustained neurotrophic and anti-inflammatory cascades, decelerating neurodegenerative progression.

The influence of exercise on NCDs has been globally acknowledged. NCDs typically have a prolonged duration, and exercise can offer protection via long-term health effects. A 4-year follow-up study of a randomized controlled trial evaluated the long-term benefits of progressive resistance training in individuals with chronic stroke. Eighteen participants (mean age 66 ± 4 years; 11 in the training group, seven in the control group) underwent assessment. The training group had completed 10 weeks of supervised progressive resistance training targeting the knee extensors and flexors (at 80% of maximum load), performed twice weekly. The control group maintained their usual daily activities. At the 4-year follow-up, the training group retained the improvements in muscle strength achieved during the intervention. While the control group did not exhibit significant strength declines compared to their post-intervention baseline, significant between-group differences persisted for both isotonic and isokinetic strength measures, relative to the study's initial baseline (Flansbjer et al. 2012). These findings demonstrate that exercise training confers sustained improvements in functional capacity among individuals with noncommunicable diseases (NCDs), highlighting its potential as a long-term strategy for mitigating age-related functional decline. A large-scale study evaluated the comparative effects of high-intensity interval training (HIT) versus low-to-moderate intensity (LMI) exercise on health-related quality of life (HRQoL) in patients with breast, colorectal, or prostate cancer within 18 months post-tumor treatment. This randomized controlled trial allocated 577 patients initiating adjuvant therapy to a 6-month combined resistance and endurance exercise program, comparing HIT and LMI against usual care. The findings robustly demonstrate that structured exercise, regardless of intensity (HIT or LMI), significantly enhances HRQoL not only during active tumor treatment but also for a sustained period following the completion of the intervention, supporting its role in preserving well-being and functional status during survivorship (Ax et al. 2022). A longitudinal study involving 209 women treated for breast cancer has demonstrated the significant long-term benefits of a comprehensive exercise and education program on quality of life and symptom management. Over a 24-month follow-up period, participants in the experimental group, who underwent a 3-month rehabilitation program incorporating physical training and psychoeducational sessions, exhibited substantial improvements in overall quality of life, functional role, emotional state, physical functions, fatigue, pain, insomnia, diarrhea, and anxiety and depression levels. These improvements were made throughout the follow-up period and were significantly greater compared to the control group (Leclerc et al. 2018). A preoperative supervised exercise program (SEP) improves cardiorespiratory fitness and perioperative outcomes for patients undergoing elective abdominal aortic aneurysm (AAA) repair. To assess the effect of a preoperative SEP on long-term survival of patients, 124 patients were followed up to 5 years post-surgery. The per protocol analysis demonstrated a significant survival benefit associated with SEP attendance (4 vs. 16 deaths, p

value = 0.01), which suggests a long-term mortality benefit for patients attending a SEP prior to elective AAA repair (Sethi et al. 2024).

These results demonstrate that exercise improves a variety of diseases in the long term, encompassing relieving pain in multiple areas, improving neurological function, and effectively controlling the occurrence and development of multiple NCDs, including cardiovascular disease and cancer.

2.4 Mechanisms of Long-Term Effects of Exercise

Exercise has been demonstrated to have profound and lasting impacts that persist long after the cessation of training. These long-term effects of exercise are not merely a result of sustained physical activity but are underpinned by complex physiological, biochemical, and molecular mechanisms that reshape the body and mind. Further exploration of these mechanisms will assist in understanding how exercise creates a "memory" protective effect in the body that persists even after exercise cessation.

2.4.1 Neurocognitive Adaptations: Exercise-Induced "Memory" in the Brain

Physical exercise induces persistent neural adaptations through molecular reorganization and structural remodeling, particularly within the prefrontal cortex (PFC) and neuroendocrine signaling pathways. The PFC, which undergoes continued structural and functional maturation throughout adolescence, is critical for maintaining goal-relevant information and exerting top-down control over bottom-up sensory processing. Reviewed findings indicate that mixed and coordinative exercise elicit both quantitative (e.g., enhanced resource allocation) and qualitative (e.g., adoption of efficient cognitive control strategies) changes in PFC function. Beyond neurocognitive mechanisms, exercise modulates concentrations of neuroendocrine markers implicated in cognitive development, such as brain-derived neurotrophic factor (BDNF) and insulin-like growth factor-1 (IGF-1). Notably, the ability to manipulate information within working memory exhibits delayed maturation compared to the maintenance of information. This developmental trajectory arises from the former component's greater dependence on frontal lobe functions, which undergo protracted specialization and connectivity refinement during adolescence (Ludyga et al. 2022). This indicates a greater sensitivity of working memory components and stimulus modalities that undergo a delayed maturation.

2.4.2 Skeletal Muscle Memory: Molecular Basis of Long-Term Benefits

Emerging investigations into resistance training's persistent effects demonstrate its capacity to induce lasting musculoskeletal remodeling. A study examining the long-term effects of strength training of female cohorts has revealed that exercise has enduring benefits, particularly in terms of muscle adaptation and recovery (Staron et al. 1991). Previously trained muscles required some parameters (i.e., fiber area, maximal dynamic strength) at an accelerated rate when retraining even after a prolonged detraining period (Staron et al. 1991; Taaffe and Marcus 1997). This "muscle memory" phenomenon was originally perceived to be the result of a distinct motor learning capacity (Rutherford and Jones 1986).

Another study has revealed the long-term beneficial impacts of exercise, with a particular emphasis on the notion of muscle memory established during juvenile development. Researchers discovered that male juvenile rats undergoing a 5-week climbing exercise regime exhibited an augmented cross-sectional area of muscle fibers, an enhancement in nuclear accretion, as well as a reduction in intraperitoneal fat content and body weight. These physiological alterations persisted even after a 10-week detraining period, suggesting a persistent "muscle memory" effect. Remarkably, when re-exposed to training in adulthood, these preconditioned rats demonstrated amplified hypertrophic responses compared to exercise-naïve counterparts, suggesting developmental exercise enhances lifelong muscle plasticity (Eftestøl et al. 2022). For muscle memory, previously trained muscles acquire strength and volume much faster than naive muscles. The number of myonuclear and the protein level of muscle energy-sensing network both increased and were retained during a long-term detraining period, which means that exposing young muscles to resistance training would help to restore age-related muscle loss coupled with mitochondrial dysfunction in later life. The acquired myonuclear were related to hypertrophic muscle and following mitochondrial biogenesis processes, where previously trained muscles confer greater mitochondrial adaptations when retraining (Bruusgaard et al. 2010; Bae et al. 2018; Lee et al. 2018). The neutrophil enzyme myeloperoxidase (MPO) in the muscle of male albino rats, which was determined as a measure of neutrophil migration into muscles, exhibited a marked increase following high-intensity acute exercise. A single bout of exercise to exhaustion produced a greater increase in MPO content in untrained compared to trained rats. Smaller neutrophil infiltration in muscles of trained animals after exhaustive exercise suggests a protective effect of previous training to muscle injury (Morozov et al. 2006). But other studies found that increased myonuclear number and differentially expressed genes related to muscle performance and development in the previously trained muscle did not translate into clearly superior responses during retraining (Dungan et al. 2019; Cumming et al. 2024). Because of the unclear effect on the subsequent hypertrophy and muscle strength gain with retraining, the physiological benefit remains to be determined. Our lab recently investigated long-term effects of exercise, especially how early-life exercise impacts longevity and health of aging in mice. C57BL/6J mice were subjected to swimming exercise from 1 to 4 months of age, followed by detraining for the remainder of their lives. We found that early-life exercise did not extend the overall lifespan, it promoted healthy aging in both male and female mice with enhanced fatty acid metabolism in aged skeletal muscles. These findings posit myonuclear-epigenetic crosstalk as the cornerstone of muscle memory, yet unresolved discordance in retraining efficacy necessitates deeper interrogation of satellite cell quiescence dynamics and NLRP3 inflammasome modulation.

2.4.3 Metabolic Reprogramming: Benefits to Multiple Tissues

Exercise also exerts long-term effects sustained metabolic reprogramming. The long-term beneficial effects of exercise are reflected in muscle inflammation and immune response. Experimental models reveal that trained rats exhibited a smaller neutrophil infiltration in muscles following exhaustive exercise, suggesting training-induced protection against exercise-associated myocyte damage (Morozov et al. 2006). Regular exercise can result in significant adaptations that safeguard against muscle damage and enhance recovery. A study conducted in 2006 probed into the long-term advantages of higher intensity training in ameliorating insulin sensitivity among older women. Twenty-five healthy elderly women were randomly allocated to one of three training groups: (1) high-intensity [80% peak aerobic capacity ($V(O_2)$ peak)] aerobic training; (2) moderate-intensity (65% $V(O_2)$ peak) aerobic training; (3) low-intensity (stretching) placebo control group. The findings indicated that the high-intensity training group demonstrated significant improvements in insulin sensitivity after training in contrast to the moderate and low-intensity training groups, and these enhancements remained significant 72 h after the conclusion of training, suggesting that high-intensity training has a more sustained benefit in improving insulin sensitivity (DiPietro et al. 2006). These results provide insight into the long-term benefits of exercise by affecting metabolism.

Chronic physical activity induces systemic adaptations across multiple organ systems. Adipose tissue exhibits divergent response patterns compared to skeletal muscle. The acute and repeated reduction in transcripts of lipid storage and lipogenesis, interconnected with a modulated circadian rhythm, can counteract metabolic syndrome progression toward diabetes (Zucker and Musch 2018; Dreher et al. 2023). Endurance training has a long-term impact on the cardiovascular system. Juvenile exercise leads to physiological cardiac hypertrophy that is sustained into adulthood long after exercise training has ceased. Furthermore, this cardiac reprogramming is largely due to a 36% increase in cardiomyocyte number (Asif et al. 2018). Mechanistically, these cross-tissue adaptations arise from exercise-mediated epigenetic reprogramming and metabolic pathway modulation. The coordinated regulation of gene expression networks across diverse tissues—spanning muscular, adipose, and cardiac systems—establishes a biological foundation for exercise's prophylactic effects against chronic diseases. Such multisystemic interactions highlight the necessity for holistic approaches in understanding exercise physiology.

2.4.4 *Epigenetic Priming: Lifelong Health Affected by Exercise*

Exercise exerts long-term health effects probably through the epigenetic process. For brain aging, regular exercise demonstrates neuroprotective properties by counteracting age-associated epigenetic alterations. BDNF is a circulating factor and plays critical roles in maintaining cerebral homeostasis. Study shows that aging elevates methylation levels in *Bdnf* exons/promoters I, II, and IV in the orbital frontal cortex and exercise could influence BDNF expression level to maintain brain health (McKinney et al. 2015). Furthermore, exercise mitigates neurodegenerative progression by regulating various enzymes that control DNA methylation level. A previous study showed that a 2-week, voluntary wheel training in sedentary rats increases the level of hippocampal Tet1. Additionally, 4-week physical exercise counteracts the decreased Tet1 and Tet2 expression induced by aging, through enhancing the hippocampal 5-hmC content, thus promoting cognitive functions of aged mice (Xu et al. 2021). The alterations in ncRNA induced by exercise benefits heart diseases during aging. For instance, swimming restores the aberrant expression of miR-214, miR-1, miR-29a, and miR-29c in myocardial infarction (MI) hearts. All these restorations modulate downstream targets that are implicated in alleviating cardiac dysfunction (Wu et al. 2021). Exercise attenuates skeletal muscle aging as evidence by epigenetic aging clock analysis through methylation examination. Study showed that the epigenetic age of muscle from old mice that resistance exercise trained for 8 weeks was approximately 8 weeks younger than 24-month-old sedentary counterparts, which represents ~8% of the expected murine lifespan (Murach et al. 2022). We recently found that early-life regular exercise exerts long-term beneficial effects via epigenetic immunometabolism modulation. Pipecolic acid was identified as a pivotal metabolite responding to early-life regular exercise. Moreover, early-life exercise increases histone 3 lysine 4 trimethylation at the promoter of *Crym* in the liver, an enzyme responsible for catalyzing pipecolic acid production (Zhang et al. 2024). This indicates that epigenetic modification might constitute a highly significant mechanism for the long-term impacts of exercise, and further studies are requisite to elucidate the long-term effects of exercise.

2.5 Conclusion and Perspective

Accumulating evidence delineates exercise's long-term impacts on both health and diseases. Specifically, animal experiments show that early-life exercise exerts lifelong benefits. Studies demonstrate that exercise sustainably reduces the risks of numerous chronic diseases, including cardiovascular disease, diabetes, obesity, and certain types of cancer in a long-term. Exercise also plays a crucial role in maintaining musculoskeletal health, effectively preventing osteoporosis. Emerging evidence further underscores the sustainably neuroprotective effects of aerobic exercise and strength training, which delay the onset of neurodegenerative diseases such as Parkinson's while enhancing cognitive performance.

Despite the extensive research on the health benefits of exercise, few studies have focused on the mechanisms underlying their long-term effects. Several pathways have been identified: (1) Neurocognitive adaptations: Exercise induces lasting improvements in working memory via prefrontal cortex neuroplasticity and hypothalamic-pituitary-adrenal axis modulation. (2) Muscle memory: Previously trained muscles gain strength and volume more rapidly when retrained than untrained muscles. This implies that exposing young muscles to resistance training can assist in recovering age-related muscle loss and mitochondrial dysfunction in later life. (3) Epigenetic regulation: Physical activity triggers DNA methylation changes and histone modifications across diverse tissues, persistently altering metabolic gene expression. These findings collectively elucidate how exercise confers long-term benefits at multiple biological scales.

Future research should focus on several aspects. Firstly, larger and longer longitudinal studies are necessary to better evaluate the impacts of different types of exercise on long-term health, particularly for individuals such as the elderly and those with chronic diseases. Secondly, combinatorial exercise protocols must be systematically evaluated to identify synergistic effects and enable precision prescription based on individual biomarkers. Thirdly, mechanistic studies should dissect exercise-induced alterations in gut microbiota composition, cellular senescence pathways, and tissue-specific transcriptomic landscapes. Finally, given exercise's expanding role in mental health management, rigorous trials are needed to validate its long-term efficacy in treating mental disorders and cognitive decline.

In the exploration of the long-term health benefits of exercise, epigenetic regulation emerges as a pivotal mechanism, revealing how exercise shapes physiological and pathological states. By identification of exercise-responsive epigenetic signatures, we can formulate new prevention and treatment strategies to enhance the long-term health benefits of exercise. Such advancements promise to revolutionize chronic disease prevention by optimizing exercise prescriptions according to individual genetic predispositions. As molecular profiling technologies evolve, decoding the intricate crosstalk between exercise and epigenetic machinery will likely yield transformative insights for both clinical practice and population health initiatives.

Compliance with Ethical Standards

Conflict of Interest All authors declare no conflict of interest.

Ethical Approval This chapter is a review of previously published accounts; as such, no animal or human studies were performed.

References

Asif Y, Wlodek ME, Black MJ, Russell AP, Soeding PF, Wadley GD (2018) Sustained cardiac programming by short-term juvenile exercise training in male rats. J Physiol 596(2):163–180. https://doi.org/10.1113/jp275339

Ax AK, Johansson B, Lyth J, Nordin K, Börjeson S (2022) Short- and long-term effect of high versus low-to-moderate intensity exercise to optimise health-related quality of life after oncological treatment-results from the Phys-can project. Support Care Cancer 30(7):5949–5963. https://doi.org/10.1007/s00520-022-07016-3

Bae JY, Woo J, Kang S, Shin KO (2018) Effects of detraining and retraining on muscle energy-sensing network and meteorin-like levels in obese mice. Lipids Health Dis 17(1):97. https://doi.org/10.1186/s12944-018-0751-3

Blasco-Lafarga C, Cordellat A, Forte A, Roldán A, Monteagudo P (2020) Short and long-term trainability in older adults: training and detraining following two years of multicomponent cognitive-physical exercise training. Int J Environ Res Public Health 17(16). https://doi.org/10.3390/ijerph17165984

Bruusgaard JC, Johansen IB, Egner IM, Rana ZA, Gundersen K (2010) Myonuclei acquired by overload exercise precede hypertrophy and are not lost on detraining. Proc Natl Acad Sci USA 107(34):15111–15116. https://doi.org/10.1073/pnas.0913935107

Cumming KT, Reitzner SM, Hanslien M, Skilnand K, Seynnes OR, Horwath O, Psilander N, Sundberg CJ, Raastad T (2024) Muscle memory in humans: evidence for myonuclear permanence and long-term transcriptional regulation after strength training. J Physiol 602(17):4171–4193. https://doi.org/10.1113/jp285675

da Silva GCR, Tebar WR, Saraiva BTC, Farah BQ, Vanderlei LCM, Ferrari G, Christofaro DGD (2023) Association of Early Sports Practice with cardiovascular risk factors in community-dwelling adults: a retrospective epidemiological study. Sports Med Open 9(1):15. https://doi.org/10.1186/s40798-023-00562-y

de Salles Painelli V, Gualano B, Artioli GG, de Sá Pinto AL, Bonfá E, Lancha Junior AH, Lima FR (2009) The possible role of physical exercise on the treatment of idiopathic inflammatory myopathies. Autoimmun Rev 8(5):355–359. https://doi.org/10.1016/j.autrev.2008.11.008

Dimitri P, Joshi K, Jones N (2020) Moving more: physical activity and its positive effects on long term conditions in children and young people. Arch Dis Child 105(11):1035–1040. https://doi.org/10.1136/archdischild-2019-318017

DiPietro L, Dziura J, Yeckel CW, Neufer PD (2006) Exercise and improved insulin sensitivity in older women: evidence of the enduring benefits of higher intensity training. J Appl Physiol 100(1):142–149. https://doi.org/10.1152/japplphysiol.00474.2005

Dreher SI, Irmler M, Pivovarova-Ramich O, Kessler K, Jürchott K, Sticht C, Fritsche L, Schneeweiss P, Machann J, Pfeiffer AFH, Hrabě de Angelis M, Beckers J, Birkenfeld AL, Peter A, Niess AM, Weigert C, Moller A (2023) Acute and long-term exercise adaptation of adipose tissue and skeletal muscle in humans: a matched transcriptomics approach after 8-week training-intervention. Int J Obes 47(4):313–324. https://doi.org/10.1038/s41366-023-01271-y

Dungan CM, Murach KA, Frick KK, Jones SR, Crow SE, Englund DA, Vechetti IJ Jr, Figueiredo VC, Levitan BM, Satin J, McCarthy JJ, Peterson CA (2019) Elevated myonuclear density during skeletal muscle hypertrophy in response to training is reversed during detraining. Am J Physiol Cell Physiol 316(5):C649–c654. https://doi.org/10.1152/ajpcell.00050.2019

Eftestøl E, Ochi E, Juvkam IS, Hansson KA, Gundersen K (2022) A juvenile climbing exercise establishes a muscle memory boosting the effects of exercise in adult rats. Acta Physiol (Oxf) 236(3):e13879. https://doi.org/10.1111/apha.13879

Falcão-Tebas F, Kuang J, Arceri C, Kerris JP, Andrikopoulos S, Marin EC, McConell GK (2019) Four weeks of exercise early in life reprograms adult skeletal muscle insulin resistance caused by a paternal high-fat diet. J Physiol 597(1):121–136. https://doi.org/10.1113/jp276386

Feng M, Li M, Lou J, Wu G, Gao T, Wu F, Tan Y, Zhang N, Zhao Y, Zhao L, Li J, Shi C, Zhang X, Liu J, Gao F (2025) Early-life exercise extends healthspan but not lifespan in mice. Nat Commun. https://doi.org/10.1038/s41467-025-61443-4

Fernandes RA, Zanesco A (2010) Early physical activity promotes lower prevalence of chronic diseases in adulthood. Hypertens Res 33(9):926–931. https://doi.org/10.1038/hr.2010.106

Fernandes RA, Zanesco A (2015) Early sport practice is related to lower prevalence of cardiovascular and metabolic outcomes in adults independently of overweight and current physical activity. Medicina (Kaunas) 51(6):336–342. https://doi.org/10.1016/j.medici.2015.10.003

Firth J, Carney R, French P, Elliott R, Cotter J, Yung AR (2018) Long-term maintenance and effects of exercise in early psychosis. Early Interv Psychiatry 12(4):578–585. https://doi.org/10.1111/eip.12365

Flansbjer UB, Lexell J, Brogårdh C (2012) Long-term benefits of progressive resistance training in chronic stroke: a 4-year follow-up. J Rehabil Med 44(3):218–221. https://doi.org/10.2340/16501977-0936

Ford CA, Nonnemaker JM, Wirth KE (2008) The influence of adolescent body mass index, physical activity, and tobacco use on blood pressure and cholesterol in young adulthood. J Adolesc Health 43(6):576–583. https://doi.org/10.1016/j.jadohealth.2008.06.010

Giannini C, Mohn A, Chiarelli F (2006) Physical exercise and diabetes during childhood. Acta Biomed 77(Suppl 1):18–25

Haakstad LAH, Kissel I, Bø K (2021) Long-term effects of participation in a prenatal exercise intervention on body weight, body mass index, and physical activity level: a 6-year follow-up study of a randomized controlled trial. J Matern Fetal Neonatal Med 34(9):1347–1355. https://doi.org/10.1080/14767058.2019.1636028

Heine M, Lupton-Smith A, Pakosh M, Grace SL, Derman W, Hanekom SD (2019) Exercise-based rehabilitation for major non-communicable diseases in low-resource settings: a scoping review. BMJ Glob Health 4(6):e001833. https://doi.org/10.1136/bmjgh-2019-001833

Ismailova G, Wagenmakers M, Brusse E, van der Ploeg AT, Favejee MM, van der Beek N, van den Berg LEM (2023) Long-term benefits of physical activity in adult patients with late onset Pompe disease: a retrospective cohort study with 10 years of follow-up. Orphanet J Rare Dis 18(1):319. https://doi.org/10.1186/s13023-023-02924-x

Kemper HC, Monyeki KD (2019) The Amsterdam growth and health longitudinal study: how important is physical activity in youth for later health? (ELS 33). Cardiovasc J Afr 30(3):138–141. https://doi.org/10.5830/cvja-2018-057

Leclerc AF, Slomian J, Jerusalem G, Coucke P, Bury T, Deflandre D, Devos M, Bruyère O, Foidart-Dessalle M, Kaux JF, Crielaard JM, Maquet D (2018) Exercise and education program after breast cancer: benefits on quality of life and symptoms at 3, 6, 12, and 24 months' follow-up. Clin Breast Cancer 18(5):e1189–e1204. https://doi.org/10.1016/j.clbc.2018.05.004

Lee H, Kim K, Kim B, Shin J, Rajan S, Wu J, Chen X, Brown MD, Lee S, Park JY (2018) A cellular mechanism of muscle memory facilitates mitochondrial remodelling following resistance training. J Physiol 596(18):4413–4426. https://doi.org/10.1113/jp275308

Ludyga S, Gerber M, Kamijo K (2022) Exercise types and working memory components during development. Trends Cogn Sci 26(3):191–203. https://doi.org/10.1016/j.tics.2021.12.004

Mak MK, Wong-Yu IS, Shen X, Chung CL (2017) Long-term effects of exercise and physical therapy in people with Parkinson disease. Nat Rev Neurol 13(11):689–703. https://doi.org/10.1038/nrneurol.2017.128

McKinney BC, Lin CW, Oh H, Tseng GC, Lewis DA, Sibille E (2015) Hypermethylation of BDNF and SST genes in the orbital frontal cortex of older individuals: a putative mechanism for declining gene expression with age. Neuropsychopharmacology 40(11):2604–2613. https://doi.org/10.1038/npp.2015.107

Morozov VI, Tsyplenkov PV, Golberg ND, Kalinski MI (2006) The effects of high-intensity exercise on skeletal muscle neutrophil myeloperoxidase in untrained and trained rats. Eur J Appl Physiol 97(6):716–722. https://doi.org/10.1007/s00421-006-0193-x

Murach KA, Dimet-Wiley AL, Wen Y, Brightwell CR, Latham CM, Dungan CM, Fry CS, Watowich SJ (2022) Late-life exercise mitigates skeletal muscle epigenetic aging. Aging Cell 21(1):e13527. https://doi.org/10.1111/acel.13527

NCD Countdown 2030 (2018) NCD countdown 2030: worldwide trends in non-communicable disease mortality and progress towards sustainable development goal target 3.4. Lancet 392(10152):1072–1088. https://doi.org/10.1016/s0140-6736(18)31992-5

Pahor M, Guralnik JM, Ambrosius WT, Blair S, Bonds DE, Church TS, Espeland MA, Fielding RA, Gill TM, Groessl EJ, King AC, Kritchevsky SB, Manini TM, McDermott MM, Miller ME, Newman AB, Rejeski WJ, Sink KM, Williamson JD (2014) Effect of structured physical activity on prevention of major mobility disability in older adults: the LIFE study randomized clinical trial. JAMA 311(23):2387–2396. https://doi.org/10.1001/jama.2014.5616

Patterson CM, Dunn-Meynell AA, Levin BE (2008) Three weeks of early-onset exercise prolongs obesity resistance in DIO rats after exercise cessation. Am J Physiol Regul Integr Comp Physiol 294(2):R290–R301. https://doi.org/10.1152/ajpregu.00661.2007

Rutherford OM, Jones DA (1986) The role of learning and coordination in strength training. Eur J Appl Physiol Occup Physiol 55(1):100–105. https://doi.org/10.1007/bf00422902

Santos AC, Willumsen J, Meheus F, Ilbawi A, Bull FC (2023) The cost of inaction on physical inactivity to public health-care systems: a population-attributable fraction analysis. Lancet Glob Health 11(1):e32–e39. https://doi.org/10.1016/s2214-109x(22)00464-8

Schootemeijer S, van der Kolk NM, Bloem BR, de Vries NM (2020) Current perspectives on aerobic exercise in people with Parkinson's disease. Neurotherapeutics 17(4):1418–1433. https://doi.org/10.1007/s13311-020-00904-8

Sethi S, Ravindhran B, Long J, Gurung R, Huang C, Smith GE, Carradice D, Wallace T, Ibeggazene S, Chetter IC, Pymer S (2024) A preoperative supervised exercise program potentially improves long-term survival after elective abdominal aortic aneurysm repair. J Vasc Surg 79(1):15–23.e3. https://doi.org/10.1016/j.jvs.2023.09.004

Shindo D, Matsuura T, Suzuki M (2014) Effects of prepubertal-onset exercise on body weight changes up to middle age in rats. J Appl Physiol 116(6):674–682. https://doi.org/10.1152/japplphysiol.00405.2013

Skou ST, Bricca A, Roos EM (2018) The impact of physical activity level on the short- and long-term pain relief from supervised exercise therapy and education: a study of 12,796 Danish patients with knee osteoarthritis. Osteoarthr Cartil 26(11):1474–1478. https://doi.org/10.1016/j.joca.2018.07.010

Staron RS, Leonardi MJ, Karapondo DL, Malicky ES, Falkel JE, Hagerman FC, Hikida RS (1991) Strength and skeletal muscle adaptations in heavy-resistance-trained women after detraining and retraining. J Appl Physiol 70(2):631–640. https://doi.org/10.1152/jappl.1991.70.2.631

Taaffe DR, Marcus R (1997) Dynamic muscle strength alterations to detraining and retraining in elderly men. Clin Physiol 17(3):311–324. https://doi.org/10.1111/j.1365-2281.1997.tb00010.x

Torstensen TA, Østerås H, LoMartire R, Rugelbak GM, Grooten WJA, Äng BO (2023) High- versus low-dose exercise therapy for knee osteoarthritis : a randomized controlled Multicenter trial. Ann Intern Med 176(2):154–165. https://doi.org/10.7326/m22-2348

Ueshima J, Nagano F, Wakabayashi H, Maeda K, Arai H (2025) Effectiveness of non-pharmacological therapies for preventing frailty in older people: an umbrella review. Arch Gerontol Geriatr 128:105628. https://doi.org/10.1016/j.archger.2024.105628

Verbrugghe J, Hansen D, Demoulin C, Verbunt J, Roussel NA, Timmermans A (2021) High intensity training is an effective modality to improve long-term disability and exercise capacity in chronic nonspecific low Back pain: a randomized controlled trial. Int J Environ Res Public Health 18(20). https://doi.org/10.3390/ijerph182010779

Wu G, Zhang X, Gao F (2021) The epigenetic landscape of exercise in cardiac health and disease. J Sport Health Sci 10(6):648–659. https://doi.org/10.1016/j.jshs.2020.12.003

Xu M, Zhu J, Liu XD, Luo MY, Xu NJ (2021) Roles of physical exercise in neurodegeneration: reversal of epigenetic clock. Transl Neurodegener 10(1):30. https://doi.org/10.1186/s40035-021-00254-1

Yang X, Telama R, Viikari J, Raitakari OT (2006) Risk of obesity in relation to physical activity tracking from youth to adulthood. Med Sci Sports Exerc 38(5):919–925. https://doi.org/10.1249/01.mss.0000218121.19703.f7

Yang X, Telama R, Leskinen E, Mansikkaniemi K, Viikari J, Raitakari OT (2007) Testing a model of physical activity and obesity tracking from youth to adulthood: the cardiovascular risk in young Finns study. Int J Obes 31(3):521–527. https://doi.org/10.1038/sj.ijo.0803459

Zhang N, Wang X, Feng M, Li M, Wang J, Yang H, He S, Xia Z, Shang L, Jiang X, Sun M, Wu Y, Ren C, Zhang X, Li J, Gao F (2024) Early-life exercise induces immunometabolic epigenetic modification enhancing anti-inflammatory immunity in middle-aged male mice. Nat Commun 15(1):3103. https://doi.org/10.1038/s41467-024-47458-3

Zhao R (2024) Can exercise benefits be harnessed with drugs? A new way to combat neurodegenerative diseases by boosting neurogenesis. Transl Neurodegener 13(1):36. https://doi.org/10.1186/s40035-024-00428-7

Zucker IH, Musch TI (2018) Benefits of exercise training on cardiovascular dysfunction: molecular and integrative. Am J Physiol Heart Circ Physiol 315(4):H1027–h1031. https://doi.org/10.1152/ajpheart.00516.2018

Chapter 3
Exercise Benefits for Mother and Offspring: Molecular Mechanisms and Advances

Baile Wu, Yanyan Zhang, and Lijun Shi

Abstract Maternal exercise during pregnancy is a powerful non-pharmacological intervention to enhance maternal and fetal health. It improves maternal metabolism, cardiovascular function, and neuroendocrine regulation, reducing the risks of conditions like gestational diabetes, obesity, and hypertension. Exercise also enhances placental structure and function, optimizing nutrient transport, angiogenesis, and mitochondrial activity, which support fetal development and reduce risks of growth restriction. For offspring, maternal exercise provides short-term benefits, such as improved birth outcomes, and long-term advantages, including enhanced metabolic, cardiovascular, and cognitive health into adulthood. These effects are mediated by exerkines (e.g., irisin, SOD3, apelin) and epigenetic modifications. Despite these promising findings, further research is needed to standardize exercise protocols and fully understand the molecular mechanisms involved.

Keywords Maternal exercise · Offspring development · Placenta · Exerkine · Epigenetic

B. Wu
Department of Exercise Physiology, Beijing Sport University, Beijing, China
e-mail: Chinawubaile@bsu.edu.cn

Y. Zhang · L. Shi (✉)
Department of Exercise Physiology, Beijing Sport University, Beijing, China

Laboratory of Sports Stress and Adaptation of General Administration of Sport, Beijing Sport University, Beijing, China

Key Laboratory of Physical Fitness and Exercise, Ministry of Education, Beijing Sport University, Beijing, China
e-mail: yanyanzhang@bsu.edu.cn; shilj@bsu.edu.cn

F. Gao, X. Zhang (eds.), *Exercise, Epigenetics and Human Health*, Epigenetics and Human Health 15, https://doi.org/10.1007/978-3-032-08045-5_3

Abbreviations

AMPK	AMP-activated protein kinase
BMI	Body mass index
BDNF	Brain-derived neurotrophic factor
CRP	C-reactive protein
DNA	Deoxyribonucleic acid
FATP4	Fatty acid transport protein 4
FGR	Fetal growth restriction
GDM	Gestational diabetes mellitus
GLUT3	Glucose transporter 3
GLUT4	Glucose transporter 4
HPA	Hypothalamic-pituitary-adrenal axis
HDL-C	High-density lipoprotein cholesterol
HDP	Hypertensive disorders of pregnancy
IGF1	Insulin-like growth factor 1
IL-6	Interleukin 6
IL-10	Interleukin 10
IL-15	Interleukin 15
LGA	Large-for-gestational-age
LDL-C	Low-density lipoprotein cholesterol
mTOR	Mechanistic target of rapamycin
NO	Nitric oxide
PGC-1α	Peroxisome proliferator-activated receptor gamma coactivator 1-alpha
PIH	Pregnancy-induced hypertension
ROS	Reactive oxygen species
SOD3	Superoxide dismutase 3
SGA	Small-for-gestational-age
SNAT2	Sodium-coupled neutral amino acid transporter 2
TG	Triglycerides
UCP1	Uncoupling protein 1
VEGF	Vascular endothelial growth factor
VEGFR-1	Vascular endothelial growth factor receptor 1

3.1 Introduction

The intrauterine environment plays a pivotal role in determining offspring health trajectories, with mounting evidence demonstrating that maternal lifestyle factors during pregnancy significantly influence fetal development and long-term health outcomes (Moholdt and Hawley 2020). Among these factors, maternal exercise has emerged as a potent non-pharmacological intervention to enhance maternal-fetal health through multiple physiological systems (Harris et al. 2018; Kusuyama et al. 2020). Recent epidemiological studies highlight concerning trends in maternal health, with rising prevalence of obesity, gestational diabetes mellitus (GDM), and pregnancy-induced hypertension (PIH) among women of reproductive age. These metabolic and cardiovascular complications during pregnancy not only jeopardize maternal well-being but also substantially increase the risk of adverse pregnancy outcomes and predispose offspring to chronic diseases later in life.

Historically, medical perspectives regarding maternal physical activity during pregnancy were predominantly restrictive, with pregnant women advised to avoid exertion until the early twentieth century (Downs et al. 2012). These recommendations, lacking scientific foundation, perpetuated misconceptions about female vulnerability during pregnancy (Lotgering 2014). However, systematic reviews and meta-analyses have demonstrated that physical activity neither increases miscarriage risk nor leads to preterm birth or FGR (Davenport et al. 2018). Instead, exercise represents a potent intervention for optimizing maternal-fetal health. Current evidence indicates that maternal exercise provides multiple protective effects against pregnancy-related complications and adverse birth outcomes (Morita et al. 2015) (Fig. 3.1). Regular physical activity during pregnancy significantly reduces the risk of cesarean section, excessive gestational weight gain (GWG), depression and anxiety, PIH, GDM, preeclampsia, large-for-gestational-age (LGA), and small-for-gestational-age (SGA) births (Morita et al. 2015). Furthermore, maternal exercise is associated with reduced susceptibility to various postpartum conditions, including type 2 diabetes (T2D), cardiovascular diseases, and postpartum depression (Pritchett et al. 2017).

The biological mechanisms underlying these beneficial effects involve complex adaptations across multiple systems. Maternal exercise induces significant modifications in maternal metabolism, cardiovascular function, and neuroendocrine regulation. The placenta, as a critical maternal-fetal interface, exhibits remarkable plasticity in response to maternal exercise, undergoing structural and functional modifications that optimize fetal development (Chae et al. 2022). These adaptations are mediated through various pathways, including the regulation of exercise-induced factors (exerkines), epigenetic modifications, and optimization of mitochondrial-vascular coupling (Kusuyama et al. 2020). Collectively, these adaptations contribute to enhance placental function and create favorable fetal development conditions, potentially influencing metabolic, cardiovascular, and cognitive outcomes in adulthood.

This review synthesizes current evidence on the mechanisms through which maternal exercise influences maternal-fetal health outcomes. It examines adaptive responses across three principal domains: (1) maternal physiological adaptations, including metabolic, cardiovascular, and neuroendocrine changes; (2) placental adaptations, focusing on structural remodeling, functional optimization, and molecular signaling pathways; and (3) offspring health outcomes, both immediate and long-term. Furthermore, it explores the molecular mediators and epigenetic mechanisms underlying these adaptations, offering insights into potential therapeutic targets and preventive strategies for improving maternal-fetal health outcomes.

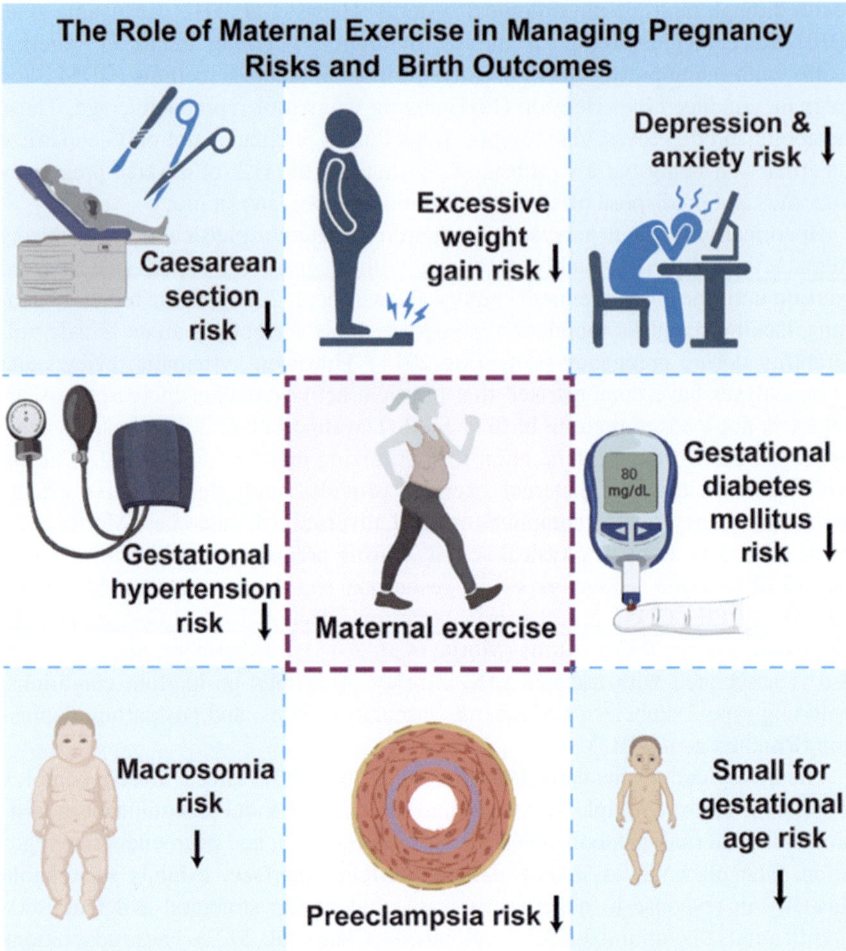

Fig. 3.1 Associations between maternal exercise and pregnancy-related outcomes. Regular physical activity during pregnancy is associated with reduced risk of multiple pregnancy complications and adverse birth outcomes

3.2 Maternal Adaptive Responses to Exercise

3.2.1 Metabolic Adaptations to Exercise in Pregnancy

Significant adaptive metabolic changes occur in the maternal metabolism during pregnancy to meet the growing developmental needs of the fetus (Zeng et al. 2017). These changes are characterized by decreased insulin sensitivity, increased lipolysis, and a preferential shift in glucose utilization toward the placenta (Zeng et al. 2017). Dysregulation of maternal metabolic adaptations, such as excessive obesity, insulin resistance, uncontrolled hyperglycemia, and lipid metabolism disorders, can lead to GDM and PIH (Plows et al. 2018). These metabolic disturbances are closely associated with unhealthy lifestyles, prepregnancy obesity, and excessive weight gain during pregnancy, potentially resulting in placental dysfunction, which negatively impacts fetal development and long-term health (Catalano and Shankar 2017). For example, a study involving 15,065 pregnant women demonstrated that overweight and obese women prior to pregnancy have a two- to threefold increased risk of developing GDM compared to women of normal weight (Song et al. 2022). Furthermore, mothers with GDM face a significantly elevated risk of adverse pregnancy outcomes, including hypertensive disorders of pregnancy (HDP; 2.5-fold increase), preterm birth (1.8-fold increase), LGA infants (2.3-fold increase), SGA infants (1.6-fold increase), and a higher cesarean section rate (2.0-fold increase) (Song et al. 2022). A systematic review by Damm et al. (2016) demonstrated that offspring of mothers with GDM are 2–8 times more likely to develop obesity, metabolic syndrome, and T2D compared to offspring of non-GDM mothers. Additionally, women who have had GDM are more than 7 times more likely to develop diabetes later in life, with approximately 50% of women who experienced GDM developing diabetes within 10 years (Damm et al. 2016). Alarmingly, the prevalence of obesity among women of reproductive age is increasing, which, if left uncontrolled, may severely impact the health of the next generation (Poston et al. 2016). Therefore, there is an urgent need for effective strategies to prevent the intergenerational transmission of metabolic dysfunction.

Exercise is an effective strategy for the prevention and improvement of metabolic diseases. It has been shown to reduce the risk of metabolic disorders during pregnancy and adverse pregnancy outcomes associated with maternal metabolic dysregulation (Cordero et al. 2015; Davenport et al. 2018). Maternal metabolic disturbances are often closely linked to placental dysfunction (Delhaes et al. 2018; Eastman et al. 2021; Sferruzzi-Perri 2021). High-fat and high-sugar environments are considered detrimental factors that disrupt placental structure and function (Delhaes et al. 2018; Eastman et al. 2021; Sferruzzi-Perri 2021). Furthermore, placental dysfunction induced by maternal metabolic disturbances is associated with adverse pregnancy outcomes (Ilekis et al. 2016). Therefore, exercise may improve fetal development by modulating maternal glucose and lipid metabolism, optimizing the intrauterine environment. A meta-analysis encompassing 106 studies reported that, for women with normal weight, exercise during pregnancy alone can reduce the risk of GDM by 25% and reduce excessive weight gain during pregnancy (Davenport et al. 2018). For women with GDM, exercise interventions effectively reduce both fasting and postprandial blood glucose levels, while improving insulin sensitivity (Xu and Liu 2024). Although there is insufficient high-quality evidence to definitively determine the impact of exercise on pregnancy outcomes in women with GDM, the studies above indicate that exercise helps maintain maternal glucose homeostasis and, to some extent, reduces the risk of fetal exposure to hyperglycemic conditions.

Adaptive changes in maternal lipid metabolism during pregnancy are critical for fetal growth and development (Zeng et al. 2017). However, lipid metabolism dysregulation is closely associated with adverse pregnancy outcomes (Eastman et al. 2021). Exercise has been shown to significantly improve maternal lipid metabolism markers, including reducing triglyceride (TG) and low-density lipoprotein cholesterol (LDL-C) levels (Ramírez-Vélez et al. 2017), while increasing high-density lipoprotein cholesterol (HDL-C) levels (Lovelady et al. 1995). Elevated TG and LDL-C levels are significantly associated with an increased risk of LGA infants (Wang et al. 2018). Therefore, exercise may reduce the risk of LGA by mitigating lipid excess. Disruptions in fatty acid oxidation, leading to fatty acid accumulation and excessive oxidation, are major contributors to chronic inflammation (Fritsche 2015). Chronic inflammation is particularly prevalent in obese and GDM-affected mothers and is considered a detrimental factor for placental and fetal development (Brombach et al. 2022). Although the precise mechanisms through which exercise improves fatty acid oxidation to directly reduce inflammation in the maternal system remain unclear, studies suggest that exercise significantly decreases levels of pro-inflammatory cytokines (e.g., CRP, TNF-α, and IL-6) in the circulation of obese mothers (Tinius et al. 2016). This further suggests that exercise may indirectly alleviate chronic inflammation by optimizing fatty acid metabolism, thereby providing critical support for improved placental function and fetal health development.

3.2.2 Cardiovascular Adaptations to Exercise in Pregnancy

Adaptive cardiovascular regulation in the maternal system during pregnancy is crucial for establishing an optimal fetal developmental environment. During pregnancy, maternal blood volume increases by 30–50%, and cardiac output significantly rises to meet the growing demands for oxygen and nutrients by the fetus (Osol et al. 2019). However, when hypertension, increased vascular resistance, or inadequate hemodynamic regulation occur, placental perfusion may be compromised, leading to IUGR, preeclampsia, and other adverse pregnancy outcomes (Qu and Khalil 2020). Insufficient placental perfusion reduces the fetus's ability to obtain oxygen and critical nutrients (such as glucose and amino acids), disrupting the homeostasis of fetal organ development and increasing the risk to fetal health (Nüsken et al. 2016). Therefore, optimizing maternal cardiovascular function is essential for the stability of placental function and fetal development. A randomized controlled trial involving 765 pregnant women found that regular physical activity reduced the risk of gestational hypertension by approximately threefold (Barakat et al. 2016). Several meta-analyses have consistently concluded that exercise improves hemodynamics and reduces the risk of developing PIH (Davenport et al. 2018; Xie et al. 2022). Although there is currently no definitive guideline regarding exercise intensity for women with gestational hypertension, existing studies suggest that moderate-intensity exercise (50–70% of maximal heart rate) is sufficient to improve hemodynamics, reduce peripheral resistance, and increase cardiac output (Srilatha and Sujatha 2019), without reporting adverse effects in either the mother or fetus. Furthermore, nitric oxide (NO) plays a pivotal role in maternal cardiovascular adaptations. Studies indicate that exercise-induced increases in NO production enhance endothelial vasodilation, improving uterine artery blood flow perfusion (Bigonnesse et al. 2018). This mechanism may help alleviate placental hypoxia and inflammation resulting from low perfusion (Renaud et al. 2011). These findings suggest that maternal exercise may improve placental blood flow through cardiovascular modulation, which is closely related to the stability of the fetal developmental environment, though the precise mechanisms require further investigation.

3.2.3 *Maternal Exercise and Neuroendocrine Function*

During pregnancy, the hypothalamic-pituitary-adrenal (HPA) axis is gradually activated under the positive feedback loop of placental hormone secretion, resulting in a significant increase in maternal cortisol levels (Zijlmans et al. 2015). This adaptive mechanism helps regulate placental function and fetal development; however, excessive activation of the HPA axis can lead to chronic elevated cortisol secretion, which is closely associated with the risks of IUGR and SGA infants (Iwata et al. 2019). Prolonged fetal exposure to high cortisol levels may also lead to negative programming effects on HPA axis development, increasing the risk of mood disorders and metabolic diseases in offspring (De Blasio et al. 2007; Isaksson et al. 2015). Regarding how maternal exercise modulates HPA axis function and its impact on fetal health outcomes, current research findings are controversial due to variations in exercise intensity. Low-intensity exercise (e.g., yoga, walking) is generally considered to reduce cortisol levels (Kumorojati et al. 2021), potentially by alleviating overactivation of the HPA axis and optimizing its negative feedback regulation, while effectively reducing anxiety, depression, and other emotional disturbances in pregnant women (Kumorojati et al. 2021).

However, the effects of high-intensity exercise are more complex. On one hand, high-intensity exercise increases maternal cortisol levels (Wilczyńska et al. 2024), which may be viewed as a negative factor, as elevated cortisol is typically associated with risks to fetal development (Jang et al. 2018). On the other hand, several systematic reviews indicate that high-intensity exercise does not significantly affect fetal birth outcomes (e.g., weight, gestational age) (Beetham et al. 2019). This suggests that the cortisol increase induced by high-intensity exercise may represent a normal adaptive response of the HPA axis to acute physiological stress, in contrast to pathological high cortisol levels resulting from chronic stress. Wilczyńska et al. (2024) found that cortisol levels in hair were elevated in pregnant women who engaged in high-intensity interval training. Despite this, the training did not significantly affect depressive symptoms or childbirth fear but led to notable improvements in overall mental health. These results suggest that cortisol increases induced by high-intensity exercise do not constitute a risk factor for emotional harm. Clapp (1996) further disproved the notion that high-intensity exercise is harmful to pregnant women. Their study on women who regularly engaged in high-intensity exercise during pregnancy revealed that high-intensity exercise did not affect fetal birth parameters (e.g., weight, head circumference, gestational age, body length); instead, offspring of mothers who participated in high-intensity exercise performed better on the Wechsler intelligence scale and language ability tests at age five (Clapp 1996). Although these studies suggest the potential benefits of high-intensity exercise during pregnancy, its applicability should vary depending on individual circumstances. For high-risk populations (e.g., those with pregnancy complications or placental dysfunction), high-intensity exercise may increase the risk of adverse pregnancy outcomes. Therefore, these findings may not apply to all pregnant women, and personalized exercise plans should be developed under professional guidance. Future research should involve larger sample sizes and more rigorous experimental designs to systematically evaluate the effects of different exercise intensities and types on HPA axis function and maternal-fetal health.

3.3 Placental Adaptations to Maternal Exercise

The placenta, as a central organ at the maternal-fetal interface, is responsible not only for the exchange of substances between the mother and fetus but also exhibits a high sensitivity to changes in maternal metabolic status and external stimuli, such as exercise. The adaptive changes in the placenta induced by exercise contribute to improving the fetal developmental environment and have a profound impact on the long-term health of offspring through placental signaling. In recent years, research has revealed the multidimensional effects of maternal exercise on placental morphology, function, and gene expression. These changes offer new perspectives on the relationship between maternal exercise and offspring health.

3.3.1 Effects of Exercise on Placental Structure and Nutrient Transport Function

The morphology and function of the placenta determine the supply of nutrients and metabolic environment to the fetus. Maternal exercise improves placental structure and function, thereby optimizing conditions for fetal development. Human studies have shown that pregnant women who are physically active exhibit a significantly higher placental growth rate during the second trimester, along with improved placental efficiency (the ratio of fetal weight to placental weight) (Jackson et al. 1995). Regular exercise during early pregnancy increases placental villous volume or density, thereby enhancing the placental surface area for oxygen exchange, which benefits fetal development (Bergmann et al. 2004). However, prolonged exercise does not further improve placental adaptations, suggesting that the fetus may have already received adequate oxygen supply (Bergmann et al. 2004). This indicates that maternal exercise may achieve a dynamic balance between promoting placental function and meeting fetal demands, thus reducing the risk of low birth weight (Bergmann et al. 2004).

Animal models further support this view. Under high-fat diet conditions, placental efficiency declines, and the thickness of the labyrinthine layer is reduced (Song et al. 2020). Maternal exercise reverses these adverse changes while maintaining normal fetal growth (Song et al. 2020), indicating that exercise enables the placenta to adapt to metabolic stress without compromising its core function of supporting fetal development. Exercise also optimizes placental gene expression, enhancing its metabolic capacity and growth regulation to provide a stable environment for fetal development. For instance, exercise regulates the expression of fatty acid transport protein 4 (FATP4) and sodium-coupled neutral amino acid transporter 2 (SNAT2) to meet fetal nutritional demands (Hutchinson et al. 2020; Song et al. 2020). In high-fat diet conditions, exercise reverses the elevated expression of placental lipoprotein lipase (Lpl), CD36, glucose transporter 3 (Glut3), and leptin receptor B (Leprb) (Mangwiro et al. 2019; Son et al. 2019a, 2020). Additionally, exercise upregulates the gene expression of growth-related factors, including insulin-like growth factor 1 (IGF-1) and its receptor (Mangwiro et al. 2018), fibroblast growth factor 2 (FGF2) and its receptor (Son et al. 2019b), and neurotrophin 4 (NT-4) (Fragoso et al. 2020), which are critical for placental development. In conclusion, the placenta is highly sensitive to maternal exercise, and its dynamic functional regulation not only optimizes the fetal developmental environment but also significantly reduces the risk of fetal developmental disorders.

3.3.2 Exercise-Induced Placental Mitochondrial Adaptations and Angiogenesis

The efficient functioning of the placenta relies on mitochondrial biogenesis and angiogenesis, which form a synergistic regulatory system referred to as the "mitochondria-vascular axis." Mitochondria are central to placental energy metabolism, as ATP produced through oxidative phosphorylation provides energy for substrate exchange, nutrient transport, and endocrine functions (Fisher 2020). Additionally, reactive oxygen species (ROS), generated as byproducts of mitochondrial metabolism, serve as critical signaling molecules that drive angiogenesis and maintain placental metabolic homeostasis (Yang et al. 2022). Angiogenesis is primarily governed by vascular endothelial growth factor (VEGF) and its signaling pathways (Huang et al. 2021), where ROS act as key regulatory factors by amplifying VEGF signaling and promoting the formation of capillary networks (Yoo et al. 2009; Yang et al. 2022). The expansion of the vascular network further supports mitochondrial metabolic function by optimizing oxygen and nutrient transport, creating a bidirectional dynamic adaptation. Mitochondrial dysfunction and impaired angiogenesis often co-occur and are associated with the development of pregnancy-related disorders (Correia et al. 2021), such as preeclampsia (Suksai et al. 2024; Toledano et al. 2024), GDM (Sferruzzi-Perri 2021), and FGR (Xu et al. 2021). Therefore, the "mitochondria-vascular axis" is considered a critical mechanism supporting placental and fetal development.

Current research on how maternal exercise improves placental function through the "mitochondria-vascular axis" remains limited, and the underlying mechanisms have yet to be thoroughly explored. However, existing evidence suggests that exercise may indirectly optimize placental function by enhancing mitochondrial activity and promoting angiogenesis. Studies have shown that exercise during pregnancy increases placental mitochondrial DNA copy number and alleviates oxidative stress-related defects, thereby improving mitochondrial function (Ekelund et al. 2015). Although direct evidence is lacking for maternal exercise upregulating key molecules involved in mitochondrial biogenesis in the placenta, such as PGC-1α and TFAM, research in other tissues, such as skeletal muscle, suggests that exercise regulates mitochondrial adaptation through PGC-1α (Zechner et al. 2010), a mechanism that warrants further investigation in the placenta. Regarding angiogenesis, maternal exercise has been shown to promote the development of vascular networks by upregulating the expression of ANG1, VEGF, and its receptor VEGFR-1 in the placenta (Son et al. 2019a). At the same time, exercise reduces the levels of low molecular weight fibroblast growth factor 2 (LMW-FGF2) (Goudreau et al. 2023), thereby mitigating inflammation and improving the placental microenvironment. These effects not only enhance vascular permeability and nutrient transport efficiency but may also support placental function by alleviating inflammatory responses.

Within the mitochondria-vascular axis, nitric oxide (NO) emerges as a critical signaling molecule coordinating the functions of these two systems. In terms of mitochondria, NO regulates ATP production by modulating the activity of mitochondrial respiratory chain complex IV while preventing excessive ROS accumulation, thereby maintaining mitochondrial metabolic balance (Davidson and Duchen 2006). Additionally, NO may promote the expression of key regulators of mitochondrial biogenesis, such as PGC-1α (Davidson and Duchen 2006). In terms of angiogenesis, NO directly participates in vascular formation by promoting endothelial cell proliferation and migration and synergizes with VEGF signaling to optimize the development of the placental vascular network (Zachary 2003). Notably, studies have found that pregnant women engaged in aerobic and resistance exercise exhibit a twofold increase in the expression of endothelial nitric oxide synthase (eNOS) in placental cytoplasm, with a fourfold increase in NO production (Ramírez-Vélez et al. 2013). However, whether these changes improve placental function through NO-mediated regulation of mitochondrial metabolism and angiogenesis requires further experimental validation.

3.3.3 *Exercise and Placental Inflammation and Immune Regulation*

During pregnancy, maternal inflammatory status plays a crucial role in placental function and fetal development. Maternal exercise, as a non-pharmacological intervention, optimizes the placental immune microenvironment by regulating the balance between pro-inflammatory and anti-inflammatory factors, thereby providing healthier developmental conditions for the fetus. Studies have shown that regular exercise significantly reduces levels of pro-inflammatory cytokines, such as IL-1β and TNF-α, in umbilical cord blood (Acosta-Manzano et al. 2020). Interestingly, acute exercise (e.g., treadmill walking) significantly increases IL-15 levels in pregnant individuals, an effect not observed in nonpregnant individuals (Huertas et al. 2017). The changes in anti-inflammatory factors exhibit complexity across studies. Some research has indicated that moderate-to-high intensity exercise may be associated with increased IL-10 levels during early pregnancy, although this effect varies significantly among individuals (Hutchinson et al. 2020; Steckle et al. 2021). Similarly, the pleiotropic cytokine IL-6 shows highly variable responses depending on exercise intensity, frequency, and individual characteristics, with studies reporting divergent effects (Acosta-Manzano et al. 2020). The specific role of IL-6 in the placental immune microenvironment remains unclear, highlighting the need to optimize exercise interventions based on individual characteristics to achieve effective immune modulation. The regulatory effects of maternal exercise on the placental immune microenvironment may not only depend on the direct action of cytokines but may also involve long-term influences on placental gene expression through epigenetic mechanisms. DNA methylation and miRNA regulation may mediate changes in the expression of pro-inflammatory and anti-inflammatory factors, further stabilizing placental function (Tomlinson et al. 2017). However, direct evidence elucidating the molecular mechanisms underlying this regulatory pathway remains limited.

3.4 Short- and Long-Term Effects of Maternal Exercise on Offspring Health

3.4.1 Short-Term Impacts of Maternal Exercise on Offspring

Maternal exercise during pregnancy has profound implications for offspring birth outcomes and early developmental trajectories, although methodological limitations warrant cautious interpretation. Meta-analyses have highlighted the nuanced relationship between physical activity during pregnancy and birth weight parameters—while average birth weight remains largely unaffected (Xu and Liu 2024), exercise is associated with a reduced risk of both LGA and SGA outcomes (Chen et al. 2021). Exercise during pregnancy among mothers with normal prepregnancy weight has been linked to a 53% reduction in the risk of childhood obesity in offspring (Chen et al. 2021). A longitudinal cohort study involving 5125 children demonstrated that offspring of exercising mothers exhibited lower BMI at 8 years of age (Mourtakos et al. 2015). However, mothers who maintain an exercise regimen may also engage in other systematically different health behaviors, underscoring the need for more controlled prospective studies to validate these findings.

Maternal exercise exhibits potential cardioprotective effects for offspring, particularly in pregnancies affected by metabolic dysfunction. In diabetic rat models, voluntary maternal exercise during pregnancy significantly reduced the incidence of congenital heart defects in offspring from 59.5% to 25%, while also mitigating cardiac oxidative stress in the offspring (Saiyin et al. 2019). Similar cardioprotective effects have been observed in mouse models (Schulkey et al. 2015). Given the rising prevalence of maternal metabolic disorders and their established association with adverse cardiovascular outcomes in offspring, these findings are particularly significant.

Maternal exercise may positively influence offspring neurodevelopment. Although research directly investigating the effects of maternal exercise on fetal neurodevelopment is highly limited, exercise-induced placental adaptations, such as increased placental volume and enhanced angiogenesis, provide a physiological foundation that supports fetal neurodevelopment (Chae et al. 2022). These adaptations, particularly in the context of maternal metabolic disorders, may help mitigate the risks of placental insufficiency (PI) and FGR, both of which are associated with adverse neurodevelopmental outcomes (Miller et al. 2016; Burger et al. 2023).

Brain-derived neurotrophic factor (BDNF), a critical regulator of neuronal growth, survival, and plasticity, is influenced by maternal exercise (Bathina and Das 2015). In humans, maternal exercise during pregnancy has been shown to increase serum BDNF levels (Rojas Vega et al. 2011), and physically active women typically exhibit higher plasma BDNF concentrations compared to sedentary women (Kodomari et al. 2009). Additionally, exercise has been shown to improve BDNF levels under adverse conditions such as obesity (Ceylan et al. 2024), suggesting a potential role for BDNF in exercise-mediated neurodevelopmental benefits. Animal studies have demonstrated that maternal exercise increases BDNF levels in the fetal brain (Gomes et al. 2016; Wu et al. 2018; Gorgij et al. 2021). Some studies report that maternal exercise enhances hippocampal BDNF expression in offspring, improving cognitive performance during childhood (Gomes et al. 2016; Gorgij et al. 2021). However, other studies have failed to observe similar effects (Segabinazi et al. 2019), indicating that this outcome may be influenced by various factors, such as animal models, exercise type and intensity, and sample processing methods. It is noteworthy that the ability of BDNF to cross the placenta remains controversial (Sonmez et al. 2019; Dingsdale et al. 2021). Thus, whether maternal BDNF levels directly influence fetal BDNF levels remains uncertain. Gomez et al. (Gomez-Pinilla et al. 2011) demonstrated that exercise induces reduced DNA methylation and increased histone acetylation at exon IV of the BDNF promoter in the hippocampus, suggesting that maternal exercise may regulate offspring BDNF expression through epigenetic mechanisms. However, the precise mechanisms underlying these effects require further investigation.

3.4.2 Long-Term Impacts of Maternal Exercise on Offspring

Maternal exercise during pregnancy has the potential to influence offspring's health into adulthood, particularly by reducing the risk of metabolic and cardiovascular diseases. Studies have shown that offspring of mothers who engage in regular exercise during pregnancy exhibit a lower risk of developing diabetes, obesity, and other metabolic and cardiovascular disorders later in life (American College of Sports Medicine 2006; Donovan 2015). These findings suggest that the health benefits of maternal exercise may persist long-term, contributing to the prevention of chronic diseases in adulthood. However, research on the impact of maternal exercise on offspring health in adulthood faces certain limitations. Ethical constraints, individual variability, and the challenges of long-term follow-up make human trials difficult to conduct. Consequently, much of the evidence supporting the effects of maternal exercise comes from animal studies.

The most significant impact of maternal exercise on offspring health appears to lie in the domain of metabolic health. Extensive evidence from animal studies suggests that maternal physical activity during pregnancy exerts positive effects on metabolic outcomes in offspring during adulthood (Carter et al. 2013; Stanford et al. 2017; Kusuyama et al. 2021; Jevtovic et al. 2023).

Offspring of mothers who exercised during pregnancy showed improved high-fat diet tolerance in adulthood, with better glucose tolerance, insulin sensitivity, and lower body fat (Carter et al. 2013; Stanford et al. 2017; Kusuyama et al. 2021; Jevtovic et al. 2023). Notably, these effects exhibit sex-specific patterns, with male offspring showing more pronounced improvements compared to females (Carter et al. 2013; Stanford et al. 2017; Kusuyama et al. 2021; Jevtovic et al. 2023; Zhang et al. 2023); however, the mechanisms underlying these sex differences remain unclear. Furthermore, maternal exercise can counteract the adverse metabolic consequences of maternal high-fat diets and nutritional deficiencies (Seneviratne et al. 2014; Beeson et al. 2018; Kusuyama et al. 2020). For example, maternal physical activity during pregnancy has been shown to reverse the development of obesity, glucose intolerance, and hypercholesterolemia in the offspring of obese mothers, thereby mitigating the risks associated with maternal metabolic dysfunction (Seneviratne et al. 2014; Beeson et al. 2018; Kusuyama et al. 2020). The beneficial effects of maternal exercise on offspring metabolism may be attributed to improved mitochondrial function in the metabolic organs of the offspring. Studies have reported that male offspring of exercising mothers exhibit higher mitochondrial density and greater oxidative metabolic capacity, particularly in skeletal muscle and liver tissues (Siti et al. 2019; Stevanović-Silva et al. 2021). Maternal exercise appears to enhance physical activity levels in offspring, a benefit that extends into adulthood. Offspring of exercising mothers demonstrate increased physical activity levels in adulthood, with this effect emerging earlier in females at sexual maturity (Eclarinal et al. 2016). Female offspring also display more significant reductions in fat mass following a 3-week voluntary exercise intervention compared to male offspring (Eclarinal et al. 2016). In summary, current findings support the role of maternal exercise in improving metabolic outcomes in offspring during adulthood. However, inconsistencies across studies hinder a comprehensive understanding of the effects of maternal exercise on offspring metabolic health. These inconsistencies may stem from differences in study design, exercise type and intensity, and evaluation methods. Future research should adopt more standardized designs and methodologies to elucidate better the mechanisms and effects of maternal exercise on offspring health.

Growing evidence suggests maternal exercise may positively influence offspring cognitive function in adulthood (Gomes et al. 2016; Meireles et al. 2021; Yang et al. 2021). A meta-analysis of 4786 rodents reported that maternal exercise enhanced offspring brain development and improved cognitive performance in adulthood (Yang et al. 2021). However, significant heterogeneity in the results, potentially driven by factors such as parental and offspring sex, age, exercise modality, and timing, challenges the generalizability and reliability of the findings. Maternal exercise has also been shown to prevent Alzheimer's-like brain pathology in adult offspring, indicating its potential neuroprotective effects (Herring et al. 2012).

Our team focuses on elucidating the mechanisms by which maternal exercise reprograms vascular function in hypertensive offspring. Our findings demonstrate that maternal exercise improves vascular function in hypertensive offspring during adulthood and prevents age-dependent blood pressure elevation (Li et al. 2021; Shan et al. 2023, 2025; Zhang et al. 2023). Specifically, maternal exercise significantly reduces blood pressure in the early developmental stages of hypertensive offspring by regulating key pathways involved in vascular tone, such as the CaV1.2 channel and AT1R signaling pathways, delaying the progression of blood pressure elevation in spontaneously hypertensive rat (SHR) offspring (Li et al. 2021; Shan et al. 2023). However, the regulatory effects of maternal exercise on blood pressure diminish after 6 months of age in SHR offspring, highlighting the need for further investigation into follow-up interventions (Li et al. 2021). Additionally, maternal exercise reduces oxidative stress in offspring vasculature, reverses vascular smooth muscle cell phenotypic switching and vascular remodeling, and thereby enhances vascular function in hypertensive offspring (Zhang et al. 2023; Shan et al. 2025).

Overall, evidence suggests that the benefits of maternal exercise on offspring health persist into adulthood. Despite these promising findings, significant limitations remain in the current body of research. Most evidence is derived from animal models, which may not fully translate to human populations due to species-specific differences in physiology, exercise responses, and genetic factors.

3.5 Potential Mediators of Offspring Health via Maternal Exercise: The Role of Exerkines

In recent years, the concept of exerkines has garnered significant attention. These factors, including proteins, peptides, metabolites, and nucleic acids secreted in response to exercise, are widely present in maternal circulation, placental tissue, and other metabolism-related organs (Chae et al. 2022). As potential mediators linking maternal exercise to fetal health, exerkines may play a pivotal role in improving maternal-fetal health, reducing the risk of pregnancy complications, and promoting long-term health outcomes in offspring (Chae et al. 2022). Here, we review several key exerkines, including irisin, superoxide dismutase 3 (SOD3), apelin, and adiponectin, which are considered critical mediators of the effects of maternal exercise on placental function and maternal-fetal metabolic regulation.

Irisin, a prototypical exercise-induced factor driven by PGC-1α activation, is primarily secreted by skeletal muscle but is also expressed in placental tissue (Boström et al. 2012). Studies suggest that irisin promotes the differentiation and development of placental trophoblast cells, potentially enhancing the efficiency of nutrient exchange between the mother and fetus (Drewlo et al. 2020). Clinical research has identified a negative correlation between late pregnancy serum irisin levels and LDL cholesterol and total cholesterol levels (Bousmpoula et al. 2019). Additionally, serum irisin levels are significantly reduced in patients with GDM and preeclampsia (Ebert et al. 2014; Garcés et al. 2014). These observations provide valuable insights into the potential role of irisin in mediating the beneficial effects of exercise on pregnancy outcomes.

SOD3, a secreted factor upregulated by maternal exercise via activation of the vitamin D receptor (VDR), may play a critical role in metabolic regulation in the mother, placenta, and offspring (Kusuyama et al. 2021). Research has shown that SOD3 activates the AMPK/TET signaling pathway in the offspring liver, promoting DNA demethylation of glucose metabolism-related genes and thereby improving metabolic function (Kusuyama et al. 2021). Additionally, SOD3 regulates the expression of key placental nutrient transporters, such as FATP4, SNAT2, and GLUT3, optimizing fetal nutrient supply (Kusuyama et al. 2021). Under high-fat diet conditions, SOD3 restores H3K4 trimethylation on histones, alleviating aberrant gene expression and reducing oxidative stress-associated metabolic abnormalities (Kusuyama et al. 2022). Moreover, pregnant women with higher physical activity levels exhibit significantly elevated SOD3 levels in both serum and placental tissue (Kusuyama et al. 2021), suggesting a potential role for SOD3 in mediating exercise-induced improvements in placental function.

The apelin signaling system is also thought to play a significant role in exercise-mediated placental adaptations. Studies have demonstrated that exercise increases apelin levels in maternal circulation, which promotes placental angiogenesis and nutrient transport through the AMPK signaling pathway (Son et al. 2020b). Additionally, apelin may activate TET enzymes to facilitate DNA demethylation in the promoter region of Prdm16, thereby inducing the development of fetal brown adipose tissue (Son et al. 2020b). Maternal exercise has also been suggested to enhance mitochondrial biogenesis in offspring skeletal muscle via the apelin-APJ axis, which is closely associated with hypomethylation in the promoter region of the Ppargc1α gene (Son et al. 2020a). Furthermore, reduced apelin levels have been observed in preeclamptic patients (Ivars et al. 2012), while exogenous apelin treatment has been shown to alleviate preeclampsia symptoms (Son et al. 2020b). These findings suggest that apelin may play a critical role in the regulation of pregnancy-related complications.

Adiponectin, a lipid hormone secreted by white adipose tissue and the placenta, is negatively correlated with obesity and insulin resistance. Low adiponectin levels during early pregnancy are strongly associated with increased insulin resistance and a higher risk of GDM. Studies have shown that exercise elevates adiponectin levels, potentially improving maternal-fetal metabolism by enhancing insulin sensitivity

(Blüher et al. 2006). Additionally, adiponectin regulates the differentiation and invasion of placental trophoblast cells, processes that are closely linked to fetal growth and glucose homeostasis (McDonald and Wolfe 2009). Exogenous adiponectin has been demonstrated to ameliorate obesity-related placental dysfunction and provide protective effects on fetal development (Aye et al. 2015).

In summary, research on exerkines provides new insights into the potential links between maternal exercise and maternal-fetal health. Some studies suggest that changes in the levels of these factors may be associated with metabolic regulation during pregnancy and the risk of complications, such as the roles of irisin and apelin in GDM and preeclampsia (Ivars et al. 2012). However, there is currently insufficient evidence to support the practical utility of these factors in personalized exercise interventions or therapeutic applications. Further research is required to elucidate the specific mechanisms and clinical feasibility of leveraging exerkines for pregnancy health management and to clarify their potential value.

3.6 Epigenetic Remodeling of Offspring Tissues Induced by Maternal Exercise

Dynamic epigenetic changes during embryonic development, including DNA methylation, histone modifications, and microRNA regulation, have profound effects on offspring development and long-term health (Perera and Herbstman 2011). Maternal metabolic dysfunction is widely regarded as a key contributor to placental dysfunction, which can disrupt the intrauterine environment—characterized by oxidative stress, inflammation, or impaired nutrient transport—and further alter fetal epigenetic regulation, thereby increasing the risk of metabolic and cardiovascular diseases in offspring (Zhu et al. 2019). Emerging evidence suggests that maternal exercise during critical periods of embryonic development stabilizes the epigenetic landscape of offspring, promoting optimized development and functionality across multiple tissues.

In metabolic tissues, including skeletal muscle, liver, and brown adipose tissue, maternal exercise serves as a key driver of epigenetic remodeling. Studies demonstrate that maternal exercise optimizes metabolic function in offspring by regulating DNA methylation patterns in metabolic genes within the liver and skeletal muscle. Specifically, maternal exercise induces promoter-specific demethylation in key hepatic metabolic genes, such as *Pcx, Cs, Idh3a, Mdh2, Fatp4, and Cpt1a* (Kusuyama et al. 2021). This is characterized by a significant reduction in 5-methylcytosine (5-mC) levels and a concurrent increase in 5-hydroxymethylcytosine (5-hmC) levels. These epigenetic modifications are thought to enhance the transcriptional activity of metabolic genes, thereby improving glucose and lipid metabolism in the offspring liver (Kusuyama et al. 2021, 2022). Maternal exercise also prevents the carbonylation of WD repeat-containing protein 82 (WDR82), restoring H3K4 methyltransferase activity and increasing H3K4me3 levels at the promoters of glucose metabolic genes (Kusuyama et al. 2021). In addition, maternal exercise significantly alters the expression profiles of several microRNAs involved in metabolic regulation, such as miR-204-5p, miR-10b-5p, and miR-139-3p (Zhou et al. 2022). These microRNAs have been closely associated with the optimization of cholesterol metabolism pathways (Zhou et al. 2022). However, the precise targets of these microRNAs and their regulatory roles within the metabolic network require further investigation.

Epigenetic remodeling in skeletal muscle involves not only the regulation of metabolism-related genes but also processes associated with muscle development and function (Mikovic and Lamon 2018). Although direct evidence for the epigenetic regulation of offspring skeletal muscle development by maternal exercise is currently lacking, existing studies provide a strong theoretical foundation for this hypothesis. In exercise training models, *PGC-1α*, a key metabolic regulator, is precisely modulated by DNA methylation and histone modifications (Zechner et al. 2010). This epigenetic regulation plays a central role in mitochondrial biogenesis within muscle fibers, providing a molecular basis for skeletal muscle functional optimization. Research has shown that maternal exercise enhances mitochondrial function and oxidative metabolism in offspring skeletal muscle by promoting DNA demethylation in the *PGC-1α* promoter region (Siti et al. 2019). Given the pivotal role of *PGC-1α* in muscle fiber type switching and mitochondrial biogenesis, epigenetic changes induced by maternal exercise may influence offspring skeletal muscle development through similar mechanisms (Laker et al. 2014). However, further experimental data are required to elucidate the precise molecular pathways involved.

Brown adipose tissue (BAT) enhances energy expenditure and regulates lipid and glucose metabolism through non-shivering thermogenesis mediated by uncoupling protein 1 (*UCP1*). Studies have shown that maternal exercise significantly improves thermogenic capacity and mitochondrial function in offspring BAT (Son et al. 2020b). These improvements are closely associated with reduced risk of obesity and insulin resistance in offspring exposed to a high-fat diet (Son et al. 2020b). Maternal exercise induces DNA demethylation in the promoter region of *Prdm16*, a master regulator of BAT programming, enhancing its transcriptional activity. Activation of *Prdm16* further promotes the expression of thermogenesis-related genes (Son et al. 2020b). This process is regulated by apelin, a maternal exercise-induced exerkine. Apelin activates the AMPK signaling pathway, increasing α-ketoglutarate (α-KG) levels, which facilitates TET enzyme-mediated DNA demethylation (Son et al. 2020b).

In terms of neural development, studies have shown that maternal resistance exercise significantly influences the epigenetic modifications in offspring hippocampal tissue. Regular resistance exercise has been found to alter DNA methylation patterns and histone acetylation levels at specific gene loci in the hippocampus, with these epigenetic changes potentially linked to enhanced neural plasticity (Segabinazi et al. 2019; Meireles et al. 2021). However, further research is needed to elucidate the precise regulatory mechanisms and identify the target genes involved. Additionally, the longevity and functional implications of these epigenetic modifications remain to be fully explored.

Our research demonstrates that maternal exercise provides significant vascular protection for hypertensive offspring through epigenetic regulation. Specifically, maternal exercise induces hypermethylation in the promoter regions of the CaV1.2 channel α1C gene (*CACNA1C*) and Ang II receptor AT1R gene (*Agtr1a*) in offspring vascular smooth muscle cells, mitigating abnormal vascular tension regulation (Li et al. 2021). Additionally, maternal exercise upregulates the expression of DNA methyltransferases (DNMT1 and DNMT3B) and the deacetylase SIRT1, while reducing the expression of NADPH oxidase-4 (Nox4) in endothelial tissue. This reduces oxidative stress and improves endothelium-dependent vasorelaxation in male offspring (Shan et al. 2023). Furthermore, maternal exercise promotes the deacetylation of the *FGF21* gene via SIRT1, facilitating the phenotypic switch of vascular smooth muscle cells from a "synthetic" to a "contractile" state, thereby alleviating vascular remodeling and maintaining vascular elasticity (Shan et al. 2025). These findings indicate that maternal exercise epigenetically reprograms vascular phenotypes in hypertensive offspring, improving vascular function and delaying the progression of elevated blood pressure in SHR offspring. Notably, current studies are limited to hypertensive animal models, and the intergenerational mechanisms of exercise on vascular responses and blood pressure regulation in non-genetic hypertension remain unclear, highlighting an important direction for future research.

In summary, maternal exercise induces functional remodeling across multiple tissues and systems through epigenetic regulation, ranging from metabolic control to vascular function optimization. Its transgenerational effects are particularly pronounced in pathological contexts such as high-fat diets, obesity, and hypertension. This epigenetic remodeling stabilizes metabolic networks, enhances mitochondrial function, and regulates key gene methylation and histone modifications, improving vascular tension regulation, endothelial function, and tissue plasticity. These processes collectively provide a protective barrier for offspring metabolic and cardiovascular health. While current research primarily focuses on pathological models, the epigenetic effects of maternal exercise on normal offspring development remain underexplored, offering a crucial avenue for uncovering the broader transgenerational benefits of exercise.

3.7 Conclusion

This comprehensive review examines the current evidence regarding the effects of maternal exercise on maternal-fetal health outcomes. The literature demonstrates that exercise during pregnancy induces adaptations across multiple physiological systems through well-documented molecular mechanisms. These adaptations manifest at three primary levels: maternal physiological responses, placental remodeling, and offspring developmental programming (Fig. 3.2). Maternal exercise has been shown to optimize metabolic parameters, enhance cardiovascular function, and modulate neuroendocrine responses, contributing to an improved intrauterine

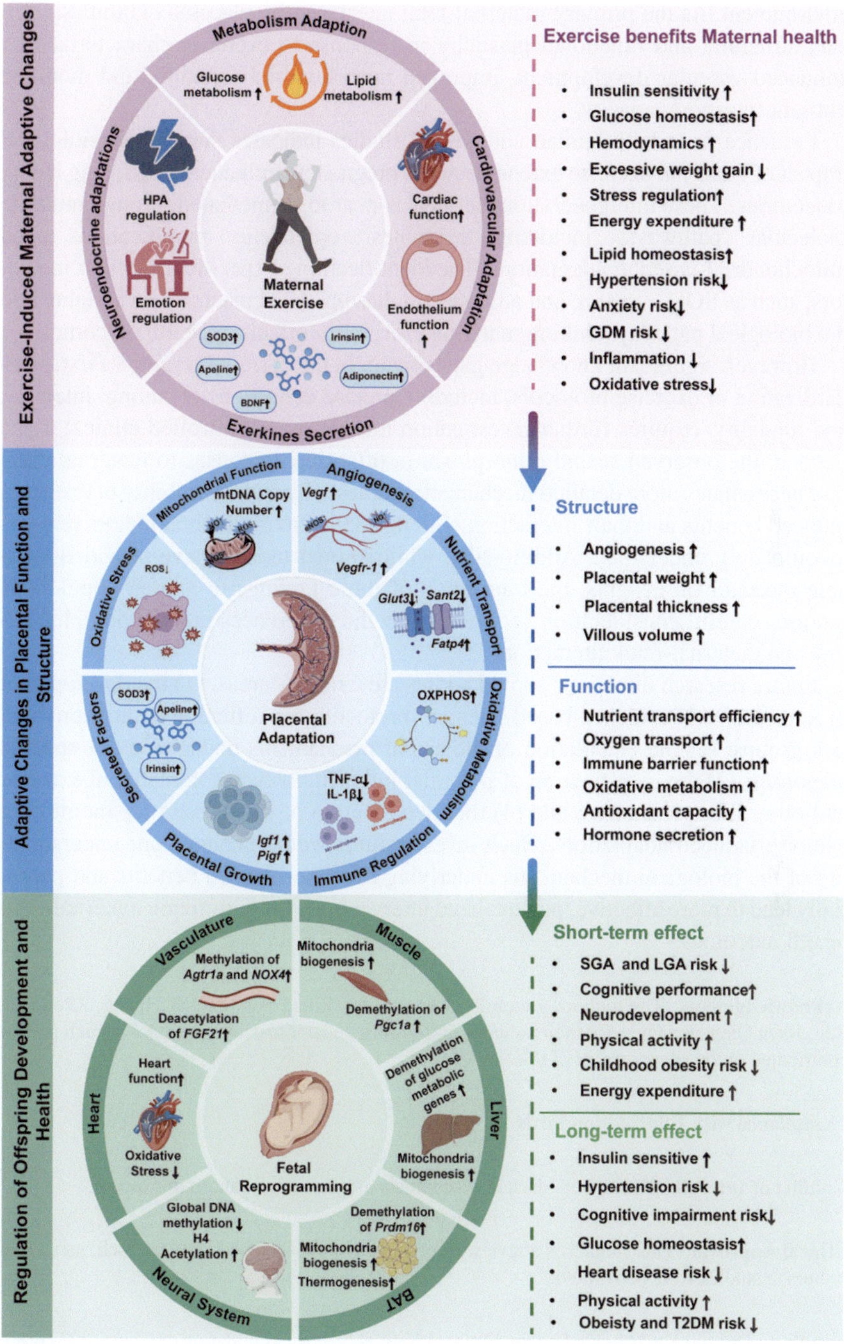

Fig. 3.2 Integrative mechanisms of maternal exercise: placental adaptations and offspring developmental reprogramming

environment. As the primary maternal-fetal interface, the placenta exhibits significant structural and functional plasticity in response to exercise, characterized by enhanced vascular development, improved mitochondrial function, and modified nutrient transport capacity.

Evidence from both human and animal studies indicates that the physiological impact of maternal exercise extends beyond pregnancy, influencing offspring health trajectories. These intergenerational effects appear to be mediated through multiple molecular pathways, including exerkines, epigenetic modifications, and mitochondrial-vascular adaptations. The identification of specific molecular mediators, such as SOD3, apelin, and adiponectin, has provided mechanistic insights into the biological pathways linking maternal exercise to offspring health outcomes.

However, significant knowledge gaps remain in the current literature. First, standardization of exercise protocols, including precise definitions of timing, intensity, and modality, requires further investigation through well-controlled clinical trials. Second, the observed sexual dimorphism in offspring responses to maternal exercise necessitates more detailed mechanistic studies. Third, the durability of exercise-induced benefits and their interaction with postnatal environmental factors remains incompletely understood. Additionally, while animal models have provided valuable mechanistic insights, the translation of these findings to human populations requires careful consideration of species-specific differences in placental physiology and maternal-fetal interactions.

Future research directions should address several key areas: (1) the development of standardized, evidence-based exercise protocols for different populations and risk groups; (2) the elucidation of molecular mechanisms underlying sex-specific responses; (3) the investigation of potential interactions between maternal exercise and other lifestyle factors; and (4) the identification of biomarkers for monitoring exercise-induced adaptations. Such investigations would enhance our understanding of the biological mechanisms underlying exercise-induced benefits and potentially lead to more effective, personalized interventions for optimizing maternal-fetal health outcomes.

Acknowledgments The authors gratefully acknowledge Zihan Fan, Zhihui Huang, Zhongxun Ren, Jiayu Qian, and Qi Li for their valuable discussions and constructive feedback, which greatly contributed to the advancement of this work.

Compliance with Ethical Standards

Conflict of Interest The authors declare no conflicts of interest relevant to this article.

Ethical Approval This chapter is a review of previously published accounts; as such, no animal or human studies were performed.

Funding This work was funded by the National Natural Science Foundation of China (32371183, 32071174, and 32200941), and the Fundamental Research Funds for the Central Universities (2024JCYJ001 and 2024YJSY002).

References

Acosta-Manzano P et al (2020) Association of sedentary time and physical activity levels with immunometabolic markers in early pregnancy: the GESTAFIT project. Scand J Med Sci Sports 30(1):148–158. https://doi.org/10.1111/sms.13547

American College of Sports Medicine (2006) Impact of physical activity during pregnancy and postpartum on chronic disease risk. Med Sci Sports Exerc 38(5):989–1006. https://doi.org/10.1249/01.mss.0000218147.51025.8a

Aye ILMH et al (2015) Adiponectin supplementation in pregnant mice prevents the adverse effects of maternal obesity on placental function and fetal growth. Proc Natl Acad Sci USA 112(41):12858–12863. https://doi.org/10.1073/pnas.1515484112

Barakat R et al (2016) Exercise during pregnancy protects against hypertension and macrosomia: randomized clinical trial. Am J Obstet Gynecol 214(5):649.e1–649.e8. https://doi.org/10.1016/j.ajog.2015.11.039

Bathina S, Das UN (2015) Brain-derived neurotrophic factor and its clinical implications. Arch Med Sci 11(6):1164–1178. https://doi.org/10.5114/aoms.2015.56342

Beeson JH et al (2018) Maternal exercise intervention in obese pregnancy improves the cardiovascular health of the adult male offspring. Mol Metabol 16:35–44. https://doi.org/10.1016/j.molmet.2018.06.009

Beetham KS et al (2019) The effects of vigorous intensity exercise in the third trimester of pregnancy: a systematic review and meta-analysis. BMC Pregnancy Childbirth 19(1):281. https://doi.org/10.1186/s12884-019-2441-1

Bergmann A, Zygmunt M, Clapp JF (2004) Running throughout pregnancy: effect on placental villous vascular volume and cell proliferation. Placenta 25(8–9):694–698. https://doi.org/10.1016/j.placenta.2004.02.005

Bigonnesse E, Sicotte B, Brochu M (2018) Activated NO pathway in uterine arteries during pregnancy in an IUGR rat model. Am J Phys Heart Circ Phys 315(2):H415–H422. https://doi.org/10.1152/ajpheart.00457.2017

Blüher M et al (2006) Circulating adiponectin and expression of adiponectin receptors in human skeletal muscle: associations with metabolic parameters and insulin resistance and regulation by physical training. J Clin Endocrinol Metab 91(6):2310–2316. https://doi.org/10.1210/jc.2005-2556

Boström P et al (2012) A PGC1-α-dependent myokine that drives brown-fat-like development of white fat and thermogenesis. Nature 481(7382):463–468. https://doi.org/10.1038/nature10777

Bousmpoula A et al (2019) Serum and follicular fluid irisin levels in women with polycystic ovaries undergoing ovarian stimulation: correlation with insulin resistance and lipoprotein lipid profiles. Gynecol Endocrinol 35(9):803–806. https://doi.org/10.1080/09513590.2019.1594761

Brombach C, Tong W, Giussani DA (2022) Maternal obesity: new placental paradigms unfolded. Trends Mol Med 28(10):823–835. https://doi.org/10.1016/j.molmed.2022.05.013

Burger RJ et al (2023) Birth-weight centile at term and school performance at 12 years of age: linked cohort study. Ultrasound Obstet Gynecol 61(4):458–465. https://doi.org/10.1002/uog.26160

Carter LG et al (2013) Maternal exercise improves insulin sensitivity in mature rat offspring. Med Sci Sports Exerc 45(5):832–840. https://doi.org/10.1249/MSS.0b013e31827de953

Catalano PM, Shankar K (2017) Obesity and pregnancy: mechanisms of short term and long term adverse consequences for mother and child. BMJ:j1. https://doi.org/10.1136/bmj.j1

Ceylan Hİ et al (2024) Exploring the effect of acute and regular physical exercise on circulating brain-derived neurotrophic factor levels in individuals with obesity: a comprehensive systematic review and meta-analysis. Biology 13(5):323. https://doi.org/10.3390/biology13050323

Chae SA, Son JS, Du M (2022) Prenatal exercise in fetal development: a placental perspective. FEBS J 289(11):3058–3071. https://doi.org/10.1111/febs.16173

Chen Y et al (2021) Effects of maternal exercise during pregnancy on perinatal growth and childhood obesity outcomes: a meta-analysis and meta-regression. Sports Med 51(11):2329–2347. https://doi.org/10.1007/s40279-021-01499-6

Clapp JF (1996) Morphometric and neurodevelopmental outcome at age five years of the offspring of women who continued to exercise regularly throughout pregnancy. J Pediatr 129(6):856–863. https://doi.org/10.1016/S0022-3476(96)70029-X

Cordero Y et al (2015) Exercise is associated with a reduction in gestational diabetes mellitus. Med Sci Sports Exerc 47(7):1328–1333. https://doi.org/10.1249/MSS.0000000000000547

Correia Y et al (2021) Placental mitochondrial function as a driver of angiogenesis and placental dysfunction. Biol Chem 402(8):887–909. https://doi.org/10.1515/hsz-2021-0121

Damm P et al (2016) Gestational diabetes mellitus and long-term consequences for mother and offspring: a view from Denmark. Diabetologia 59(7):1396–1399. https://doi.org/10.1007/s00125-016-3985-5

Davenport MH et al (2018) Prenatal exercise for the prevention of gestational diabetes mellitus and hypertensive disorders of pregnancy: a systematic review and meta-analysis. Br J Sports Med 52(21):1367–1375. https://doi.org/10.1136/bjsports-2018-099355

Davidson S, Duchen M (2006) Effects of NO on mitochondrial function in cardiomyocytes: pathophysiological relevance. Cardiovasc Res 71(1):10–21. https://doi.org/10.1016/j.cardiores.2006.01.019

De Blasio MJ et al (2007) Maternal exposure to dexamethasone or cortisol in early pregnancy differentially alters insulin secretion and glucose homeostasis in adult male sheep offspring. Am J Physiol Endocrinol Metabol 293(1):E75–E82. https://doi.org/10.1152/ajpendo.00689.2006

Delhaes F et al (2018) Altered maternal and placental lipid metabolism and fetal fat development in obesity: current knowledge and advances in non-invasive assessment. Placenta 69:118–124. https://doi.org/10.1016/j.placenta.2018.05.011

Dingsdale H et al (2021) The placenta protects the fetal circulation from anxiety-driven elevations in maternal serum levels of brain-derived neurotrophic factor. Transl Psychiatry 11(1):62. https://doi.org/10.1038/s41398-020-01176-8

Donovan EL (2015) Getting mothers moving: maternal exercise during pregnancy and long-term offspring health benefits. Exerc Sport Sci Rev 43(4):179–180. https://doi.org/10.1249/JES.0000000000000064

Downs DS et al (2012) Physical activity and pregnancy: past and present evidence and future recommendations. Res Q Exerc Sport 83(4):485–502. https://doi.org/10.1080/02701367.2012.10599138

Drewlo S et al (2020) Irisin induces trophoblast differentiation via AMPK activation in the human placenta. J Cell Physiol 235(10):7146–7158. https://doi.org/10.1002/jcp.29613

Eastman AJ et al (2021) The influence of obesity and associated fatty acids on placental inflammation. Clin Ther 43(2):265–278. https://doi.org/10.1016/j.clinthera.2020.12.018

Ebert T et al (2014) Serum levels of irisin in gestational diabetes mellitus during pregnancy and after delivery. Cytokine 65(2):153–158. https://doi.org/10.1016/j.cyto.2013.11.009

Eclarinal JD et al (2016) Maternal exercise during pregnancy promotes physical activity in adult offspring. FASEB J 30(7):2541–2548. https://doi.org/10.1096/fj.201500018R

Ekelund U et al (2015) Physical activity and all-cause mortality across levels of overall and abdominal adiposity in European men and women: the European prospective investigation into cancer and nutrition study (EPIC). Am J Clin Nutr 101(3):613–621. https://doi.org/10.3945/ajcn.114.100065

Fisher J (2020) Mitochondrial adaptions in the placenta during pregnancy. 10.25904/1912/735

Fragoso J et al (2020) Maternal physical activity-induced adaptive transcriptional response in brain and placenta of mothers and rat offspring. J Dev Orig Health Dis 11(2):108–117. https://doi.org/10.1017/S2040174419000333

Fritsche KL (2015) The science of fatty acids and inflammation. Adv Nutr 6(3):293S–301S. https://doi.org/10.3945/an.114.006940

Garcés MF et al (2014) Irisin levels during pregnancy and changes associated with the development of preeclampsia. J Clin Endocrinol Metabol 99(6):2113–2119. https://doi.org/10.1210/jc.2013-4127
Gomes Da Silva S et al (2016) Maternal exercise during pregnancy increases BDNF levels and cell numbers in the hippocampal formation but not in the cerebral cortex of adult rat offspring. PLoS One 11(1):e0147200. https://doi.org/10.1371/journal.pone.0147200
Gomez-Pinilla F et al (2011) Exercise impacts brain-derived neurotrophic factor plasticity by engaging mechanisms of epigenetic regulation. Eur J Neurosci 33(3):383–390. https://doi.org/10.1111/j.1460-9568.2010.07508.x
Gorgij E et al (2021) Treadmill exercise during pregnancy decreased vulnerability to neonatal hypoxia-ischemia through reducing inflammation and increasing Antiapoptotic gene expressions and antioxidant capacity in rats. Stroke Res Treatment 2021:1–10. https://doi.org/10.1155/2021/5512745
Goudreau AD et al (2023) Characterization of Hofbauer cell polarization and VEGF localization in human term placenta from active and inactive pregnant individuals. Physiol Rep 11(11):e15741. https://doi.org/10.14814/phy2.15741
Harris JE, Baer LA, Stanford KI (2018) Maternal exercise improves the metabolic health of adult offspring. Trends Endocrinol Metab 29(3):164–177. https://doi.org/10.1016/j.tem.2018.01.003
Herring A et al (2012) Exercise during pregnancy mitigates Alzheimer-like pathology in mouse offspring. FASEB J 26(1):117–128. https://doi.org/10.1096/fj.11-193193
Huang Z et al (2021) Placental angiogenesis in mammals: a review of the regulatory effects of Signaling pathways and functional nutrients. Adv Nutr 12(6):2415–2434. https://doi.org/10.1093/advances/nmab070
Huertas JR et al (2017) Antioxidant effect of exercise: exploring the role of the mitochondrial complex I superassembly. Redox Biol 13:477–481. https://doi.org/10.1016/j.redox.2017.07.009
Hutchinson KA et al (2020) Physical activity during pregnancy is associated with increased placental FATP4 protein expression. Reprod Sci 27(10):1909–1919. https://doi.org/10.1007/s43032-020-00210-w
Ilekis JV et al (2016) Placental origins of adverse pregnancy outcomes: potential molecular targets: an executive workshop summary of the Eunice Kennedy Shriver National Institute of Child Health and Human Development. Am J Obstet Gynecol 215(1):S1–S46. https://doi.org/10.1016/j.ajog.2016.03.001
Isaksson J et al (2015) High maternal cortisol levels during pregnancy are associated with more psychiatric symptoms in offspring at age of nine – a prospective study from Nicaragua. J Psychiatr Res 71:97–102. https://doi.org/10.1016/j.jpsychires.2015.09.016
Ivars J et al (2012) 767: role of apelinergic system in the pathogenesis of preeclampsia in a pregnant rat model. Am J Obstet Gynecol 206(1):S339. https://doi.org/10.1016/j.ajog.2011.10.785
Iwata S et al (2019) Intrauterine growth and the maturation process of adrenal function. PeerJ 7:e6368. https://doi.org/10.7717/peerj.6368
Jackson MR et al (1995) The effects of maternal aerobic exercise on human placental development: placental volumetric composition and surface areas. Placenta 16(2):179–191. https://doi.org/10.1016/0143-4004(95)90007-1
Jang Y et al (2018) Involuntary swimming exercise in pregnant rats disturbs ERK1/2 signaling in embryonic neurons through increased cortisol in the amniotic fluid. Biochem Biophys Res Commun 495(1):1208–1213. https://doi.org/10.1016/j.bbrc.2017.11.153
Jevtovic F et al (2023) Effects of maternal exercise modes on glucose and lipid metabolism in offspring stem cells. J Clin Endocrinol Metabol 108(7):e360–e370. https://doi.org/10.1210/clinem/dgad059
Kodomari I et al (2009) Maternal supply of BDNF to mouse fetal brain through the placenta. Neurochem Int 54(2):95–98. https://doi.org/10.1016/j.neuint.2008.11.005
Kumorojati R, Alfie A, Warseno A (2021) Effect of yoga exercise on cortisol hormone levels in pregnant women in Kinik Pratama Asih Waluyojati Banguntapan Bantul Yogyakarta. Jurnal Ners dan Kebidanan Indonesia 8(4):292. https://doi.org/10.21927/jnki.2020.8(4).292-297

Kusuyama J et al (2020) Effects of maternal and paternal exercise on offspring metabolism. Nat Metab 2(9):858–872. https://doi.org/10.1038/s42255-020-00274-7
Kusuyama J et al (2021) Placental superoxide dismutase 3 mediates benefits of maternal exercise on offspring health. Cell Metab 33(5):939–956.e8. https://doi.org/10.1016/j.cmet.2021.03.004
Kusuyama J et al (2022) Maternal exercise-induced SOD3 reverses the deleterious effects of maternal high-fat diet on offspring metabolism through stabilization of H3K4me3 and protection against WDR82 Carbonylation. Diabetes 71(6):1170–1181. https://doi.org/10.2337/db21-0706
Laker RC et al (2014) Exercise prevents maternal high-fat diet–induced Hypermethylation of the *Pgc-1α* gene and age-dependent metabolic dysfunction in the offspring. Diabetes 63(5):1605–1611. https://doi.org/10.2337/db13-1614
Li S et al (2021) Prenatal exercise reprograms the development of hypertension progress and improves vascular health in SHR offspring. Vasc Pharmacol 139:106885. https://doi.org/10.1016/j.vph.2021.106885
Lotgering FK (2014) 30+ years of exercise in pregnancy. In: Zhang L, Ducsay CA (eds) Advances in fetal and neonatal physiology. Springer, New York, pp 109–116. https://doi.org/10.1007/978-1-4939-1031-1_10
Lovelady CA et al (1995) Effects of exercise on plasma lipids and metabolism of lactating women. Med Sci Sports Exerc 27(1). https://doi.org/10.1249/00005768-199501000-00005
Mangwiro YTM et al (2018) Maternal exercise in rats upregulates the placental insulin-like growth factor system with diet- and sex-specific responses: minimal effects in mothers born growth restricted. J Physiol 596(23):5947–5964. https://doi.org/10.1113/JP275758
Mangwiro YTM et al (2019) Exercise initiated during pregnancy in rats born growth restricted alters placental mTOR and nutrient transporter expression. J Physiol 597(7):1905–1918. https://doi.org/10.1113/JP277227
McDonald EA, Wolfe MW (2009) Adiponectin attenuation of endocrine function within human term trophoblast cells. Endocrinology 150(9):4358–4365. https://doi.org/10.1210/en.2009-0058
Meireles ALF et al (2021) Maternal resistance exercise promotes changes in neuroplastic and epigenetic marks of offspring's hippocampus during adult life. Physiol Behav 230:113306. https://doi.org/10.1016/j.physbeh.2020.113306
Mikovic J, Lamon S (2018) The effect of maternal metabolic status on offspring health: a role for skeletal muscle? J Physiol 596(21):5079–5080. https://doi.org/10.1113/JP276929
Miller SL, Huppi PS, Mallard C (2016) The consequences of fetal growth restriction on brain structure and neurodevelopmental outcome. J Physiol 594(4):807–823. https://doi.org/10.1113/JP271402
Moholdt T, Hawley JA (2020) Maternal lifestyle interventions: targeting preconception health. Trends Endocrinol Metabol 31(8):561–569. https://doi.org/10.1016/j.tem.2020.03.002
Morita M et al (2015) mTOR coordinates protein synthesis, mitochondrial activity and proliferation. Cell Cycle (Georgetown, Tex) 14(4):473–480. https://doi.org/10.4161/15384101.2014.991572
Mourtakos SP et al (2015) Maternal lifestyle characteristics during pregnancy, and the risk of obesity in the offspring: a study of 5,125 children. BMC Pregnancy Childbirth 15(1):66. https://doi.org/10.1186/s12884-015-0498-z
Nüsken E et al (2016) Increased rat placental fatty acid, but decreased amino acid and glucose transporters potentially modify intrauterine programming. J Cell Biochem 117(7):1594–1603. https://doi.org/10.1002/jcb.25450
Osol G, Ko NL, Mandalà M (2019) Plasticity of the maternal vasculature during pregnancy. Annu Rev Physiol 81(1):89–111. https://doi.org/10.1146/annurev-physiol-020518-114435
Perera F, Herbstman J (2011) Prenatal environmental exposures, epigenetics, and disease. Reprod Toxicol 31(3):363–373. https://doi.org/10.1016/j.reprotox.2010.12.055
Plows JF et al (2018) The pathophysiology of gestational diabetes mellitus. Int J Mol Sci 19(11):3342. https://doi.org/10.3390/ijms19113342

Poston L et al (2016) Preconceptional and maternal obesity: epidemiology and health consequences. Lancet Diabet Endocrinol 4(12):1025–1036. https://doi.org/10.1016/S2213-8587(16)30217-0
Pritchett RV, Daley AJ, Jolly K (2017) Does aerobic exercise reduce postpartum depressive symptoms? A systematic review and meta-analysis. Br J Gen Pract 67(663):e684–e691. https://doi.org/10.3399/bjgp17X692525
Qu H, Khalil RA (2020) Vascular mechanisms and molecular targets in hypertensive pregnancy and preeclampsia. Am J Phys Heart Circ Phys 319(3):H661–H681. https://doi.org/10.1152/ajpheart.00202.2020
Ramírez-Vélez R et al (2013) Effect of exercise training on Enos expression, NO production and oxygen metabolism in human placenta. PLoS One 8(11):e80225. https://doi.org/10.1371/journal.pone.0080225
Ramírez-Vélez R et al (2017) Exercise during pregnancy on maternal lipids: a secondary analysis of randomized controlled trial. BMC Pregnancy Childbirth 17(1):396. https://doi.org/10.1186/s12884-017-1571-6
Renaud SJ et al (2011) Spontaneous pregnancy loss mediated by abnormal maternal inflammation in rats is linked to deficient Uteroplacental perfusion. J Immunol 186(3):1799–1808. https://doi.org/10.4049/jimmunol.1002679
Rojas Vega S et al (2011) Responses of serum neurotrophic factors to exercise in pregnant and postpartum women. Psychoneuroendocrinology 36(2):220–227. https://doi.org/10.1016/j.psyneuen.2010.07.012
Saiyin T et al (2019) Maternal voluntary exercise mitigates oxidative stress and incidence of congenital heart defects in pre-gestational diabetes. J Cell Mol Med 23(8):5553–5565. https://doi.org/10.1111/jcmm.14439
Schulkey CE et al (2015) The maternal-age-associated risk of congenital heart disease is modifiable. Nature 520(7546):230–233. https://doi.org/10.1038/nature14361
Segabinazi E et al (2019) Effects of maternal physical exercise on global DNA methylation and hippocampal plasticity of rat male offspring. Neuroscience 418:218–230. https://doi.org/10.1016/j.neuroscience.2019.08.036
Seneviratne SN et al (2014) Antenatal exercise in overweight and obese women and its effects on offspring and maternal health: design and rationale of the IMPROVE (improving maternal and progeny obesity via exercise) randomised controlled trial. BMC Pregnancy Childbirth 14(1):148. https://doi.org/10.1186/1471-2393-14-148
Sferruzzi-Perri AN (2021) Placental mitochondria central to gestational diabetes pathogenesis? J Physiol 599(4):1019–1020. https://doi.org/10.1113/JP281041
Shan M et al (2023) Maternal exercise upregulates the DNA methylation of Agtr1a to enhance vascular function in offspring of hypertensive rats. Hypertens Res 46(3):654–666. https://doi.org/10.1038/s41440-022-01124-7
Shan M et al (2025) Maternal exercise represses FGF21 via SIRT1 to improve the phenotypic transformation of vascular smooth muscle in hypertensive offspring. Hypertens Res 48(1):353–365. https://doi.org/10.1038/s41440-024-01991-2
Siti F et al (2019) Maternal exercise before and during gestation modifies liver and muscle mitochondria in rat offspring. J Exp Biol:jeb.194969. https://doi.org/10.1242/jeb.194969
Son JS, Liu X, Chen Y et al (2019a) Exercise enhances placental angiogenesis of obese maternal mice. FASEB J 33(S1). https://doi.org/10.1096/fasebj.2019.33.1_supplement.578.1
Son JS, Liu X, Tian Q et al (2019b) Exercise prevents the adverse effects of maternal obesity on placental vascularization and fetal growth. J Physiol 597(13):3333–3347. https://doi.org/10.1113/JP277698
Son JS, Zhao L et al (2020a) Maternal exercise via exerkine apelin enhances brown adipogenesis and prevents metabolic dysfunction in offspring mice. Sci Adv 6(16):eaaz0359. https://doi.org/10.1126/sciadv.aaz0359
Son JS, Chae SA et al (2020b) Maternal inactivity programs skeletal muscle dysfunction in offspring mice by attenuating Apelin Signaling and mitochondrial biogenesis. Cell Rep 33(9):108461. https://doi.org/10.1016/j.celrep.2020.108461

Song L et al (2020) Prenatal exercise reverses high-fat-diet-induced placental alterations and alters male fetal hypothalamus during late gestation in rats†. Biol Reprod 102(3):705–716. https://doi.org/10.1093/biolre/ioz213

Song Z et al (2022) Effects of obesity indices/GDM on the pregnancy outcomes in Chinese women: a retrospective cohort study. Front Endocrinol 13:1029978. https://doi.org/10.3389/fendo.2022.1029978

Sonmez EO et al (2019) Effect of maternal depression on brain-derived neurotrophic factor levels in Fetal cord blood. Clin Psychopharmacol Neurosci 17(2):308–313. https://doi.org/10.9758/cpn.2019.17.2.308

Srilatha S, Sujatha B (2019) Effects of aerobic exercise on blood pressure during pregnancy: a case study. Rcs J Pharm Technol 12(12):5749. https://doi.org/10.5958/0974-360X.2019.00995.8

Stanford KI et al (2017) Maternal exercise improves glucose tolerance in female offspring. Diabetes 66(8):2124–2136. https://doi.org/10.2337/db17-0098

Steckle V et al (2021) Low-intensity physical activity may protect pregnant women against spontaneous preterm labour: a prospective case-control study. Appl Physiol Nutr Metab 46(4):337–345. https://doi.org/10.1139/apnm-2019-0911

Stevanović-Silva J et al (2021) Maternal high-fat high-sucrose diet and gestational exercise modulate hepatic fat accumulation and liver mitochondrial respiratory capacity in mothers and male offspring. Metabolism 116:154704. https://doi.org/10.1016/j.metabol.2021.154704

Suksai M et al (2024) A mitochondrial regulator protein, MNRR1, is elevated in the maternal blood of women with preeclampsia. J Matern Fetal Neonatal Med 37(1):2297158. https://doi.org/10.1080/14767058.2023.2297158

Tinius RA et al (2016) Maternal inflammation during late pregnancy is lower in physically active compared with inactive obese women. Appl Physiol Nutr Metab 41(2):191–198. https://doi.org/10.1139/apnm-2015-0316

Toledano JM et al (2024) Pregnancy disorders: a potential role for mitochondrial altered homeostasis. Antioxidants 13(8):979. https://doi.org/10.3390/antiox13080979

Tomlinson MS et al (2017) Microorganisms in the human placenta are associated with altered CpG methylation of immune and inflammation-related genes. PLoS One 12(12):e0188664. https://doi.org/10.1371/journal.pone.0188664

Wang X et al (2018) Association of maternal serum lipids at late gestation with the risk of neonatal macrosomia in women without diabetes mellitus. Lipids Health Dis 17(1):78. https://doi.org/10.1186/s12944-018-0707-7

Wilczyńska D et al (2024) Stress is not so bad – cortisol level and psychological functioning after 8-week HIIT program during pregnancy: a randomized controlled trial. Front Public Health 11:1307998. https://doi.org/10.3389/fpubh.2023.1307998

Wu Z et al (2018) Effects of Sevoflurane exposure during mid-pregnancy on learning and memory in offspring rats: beneficial effects of maternal exercise. Front Cell Neurosci 12:122. https://doi.org/10.3389/fncel.2018.00122

Xie E et al (2022) The effect of exercise on the prevention of gestational hypertension in obese and overweight pregnant women: an updated meta-analysis. Front Public Health 10:923161. https://doi.org/10.3389/fpubh.2022.923161

Xu H, Liu R (2024) Comprehensive management of gestational diabetes mellitus: practical efficacy of exercise therapy and sustained intervention strategies. Front Endocrinol 15:1347754. https://doi.org/10.3389/fendo.2024.1347754

Xu Y-Y et al (2021) Hypoxic effects on the mitochondrial content and functions of the placenta in fetal growth restriction. Placenta 114:100–107. https://doi.org/10.1016/j.placenta.2021.09.003

Yang Y et al (2021) Beneficial intergenerational effects of exercise on brain and cognition: a multilevel meta-analysis of mean and variance. Biol Rev 96(4):1504–1527. https://doi.org/10.1111/brv.12712

Yang Y et al (2022) Reactive oxygen species are essential for placental angiogenesis during early gestation. Oxidative Med Cell Longev 2022:1–14. https://doi.org/10.1155/2022/4290922

Yoo E-S et al (2009) Reactive oxygen species might have important roles on angiogenesis of endothelial progenitor cells derived from HCB(human cord blood) which are stimulated by VEGF. Blood 114(22):1445–1445. https://doi.org/10.1182/blood.V114.22.1445.1445

Zachary I (2003) VEGF signalling: integration and multi-tasking in endothelial cell biology. Biochem Soc Trans 31(6):1171–1177. https://doi.org/10.1042/bst0311171

Zechner C et al (2010) Total skeletal muscle PGC-1 deficiency uncouples mitochondrial derangements from Fiber type determination and insulin sensitivity. Cell Metab 12(6):633–642. https://doi.org/10.1016/j.cmet.2010.11.008

Zeng Z, Liu F, Li S (2017) Metabolic adaptations in pregnancy: a review. Ann Nutr Metab 70(1):59–65. https://doi.org/10.1159/000459633

Zhang Y et al (2023) Maternal exercise represses Nox4 via SIRT1 to prevent vascular oxidative stress and endothelial dysfunction in SHR offspring. Front Endocrinol 14:1219194. https://doi.org/10.3389/fendo.2023.1219194

Zhou L et al (2022) Maternal exercise programs glucose and lipid metabolism and modulates hepatic miRNAs in adult male offspring. Front Nutr 9:853197. https://doi.org/10.3389/fnut.2022.853197

Zhu Z, Cao F, Li X (2019) Epigenetic programming and fetal metabolic programming. Front Endocrinol 10:764. https://doi.org/10.3389/fendo.2019.00764

Zijlmans MAC, Riksen-Walraven JM, De Weerth C (2015) Associations between maternal prenatal cortisol concentrations and child outcomes: a systematic review. Neurosci Biobehav Rev 53:1–24. https://doi.org/10.1016/j.neubiorev.2015.02.015

Chapter 4
Epigenetic Modifications in Exercise-Exerted Cardiovascular Protection

Yongbo Li and Dan Meng

Abstract Physical inactivity is an important risk factor for many cardiovascular diseases, and regular physical activity can effectively improve cardiovascular function, the specific molecular mechanism of regular physical activity is not well understood. Epigenetics refers to biochemical changes that affect gene expression without relying on changes in DNA sequences. Its high variability in response to environmental and lifestyle interventions is its most predominant characteristic. Specific epigenetic features are strongly associated with healthy cardiovascular function. Recently, a growing body of research has been conducted on epigenetics as a bridge for physical exercise to improve cardiovascular function and reduce the incidence of cardiovascular disease. Physical activity alters numerous epigenetic markers in the cardiovascular system, including DNA methylation, histone modifications, noncoding RNAs, and RNA modifications. In addition, due to the heritability of epigenetics, studies have shown that parental participation in exercise can also reduce the probability of cardiovascular disease in offspring. This review mainly summarizes exercise-induced epigenetic changes associated with cardiovascular health.

Keywords Cardiovascular disease · Epigenetics · Exercise · DNA methylation · Histone posttranslational modifications · Noncoding RNAs · RNA modification · Epigenetics inheritance

---

Y. Li · D. Meng (✉)
Department of Physiology and Pathophysiology, School of Basic Medical Sciences, Fudan University, Shanghai, China
e-mail: dmeng@fudan.edu.cn

F. Gao, X. Zhang (eds.), *Exercise, Epigenetics and Human Health*, Epigenetics and Human Health 15, https://doi.org/10.1007/978-3-032-08045-5_4

Abbreviations

GBD	Global burden of disease
ASC	Apoptosis-associated speck-like protein
RCT	Randomized clinical trials
ACEI	Angiotensin-converting enzyme inhibitors
ARB	Angiotensin II receptor blockers
CCB	Calcium channel blockers
eNOS	endothelial nitric oxide synthase
HF	Heart failure
HIIT	High-intensity interval training
HFrEF	Heart failure with reduced ejection fraction
LVEF	Left ventricular injection fraction
PTMs	Posttranslational modifications
TAC	Transverse aortic constriction
MI	Myocardial infarction
CPhar	Cardiac physiological hypertrophy associated regulator
SHR	Spontaneously hypertensive rats

4.1 Introduction

Cardiovascular disease refers to a range of diseases that affect the heart and blood vessels, including high blood pressure, coronary heart disease, cerebrovascular disease, heart failure, and other heart diseases. According to the Global Burden of Disease (GBD) study, it is estimated that 523 million people worldwide suffer from cardiovascular disease and 18.6 million died from cardiovascular disease in 2019 (Roth et al. 2020). Even more worrying is the fact that the prevalence of cardiovascular disease in young people under the age of 40 has continued to increase over the past decade (Sun et al. 2023). This indicates that more people will be accompanied by cardiovascular disease until death, and a reasonable lifestyle is an effective way to improve the prognosis of patients with cardiovascular disease. During a 30-year follow-up, 300 min of aerobic exercise per week was associated with significantly lower mortality from cardiovascular disease (Lee et al. 2022). Exercise participation in patients with coronary heart disease treated with modern medications reduces cardiovascular mortality, recurrent cardiac events, and hospitalizations, and improves the quality of life (Dibben et al. 2023). Exercise also lowers systolic blood pressure (Loaiza-Betancur et al. 2020), glycemia (Jansson et al. 2022), and lipids (Costa et al. 2019) to reduce the risk of cardiovascular disease. As a result, the American Cardiovascular Association recommends 8–10 different types of resistance training involving major muscle groups at least twice a week to effectively reduce cardiovascular risk (Paluch et al. 2024).

Epigenetic modifications play an important role in the onset and development of cardiovascular disease (Maleszewska et al. 2015; Kurozumi et al. 2019; Hall et al. 2021; Shi et al. 2022). Epigenetics is a regulatory mechanism that perpetuates replacement gene function/expression/activity without altering the content of the DNA sequence. It is considered to be a major response regulatory mechanism of cellular responses to environmental changes. Epigenetics mainly regulate the function and expression level of cardiovascular disease-related genes through DNA methylation, histone modification, and noncoding RNA regulation, thus affecting the progression of cardiovascular diseases. Epigenetic markers are important molecular markers of cardiovascular disease because they occur early in the disease and involve key cardiovascular pathology-related pathways (Palou-Márquez et al. 2021). Most importantly, it can be used as a cardiovascular disease biomarker for cardiovascular. Several recent studies have indicated the involvement of epigenetic modifications in exercise-exerted cardiovascular protection. These findings suggest an important role of epigenetics in the cardiovascular protection of exercise. This review will focus on the epigenetic-mediated cardiovascular protection of exercise.

4.2 DNA Methylation

DNA methylation typically refers to the methylation of the fifth carbon of cytosine residues throughout the genome, DNA methylation is most common in CpG islands. DNA methylation in the promoter regions is associated with transcriptional suppression, while the effects of methylation on gene expression in introns and intergenic regions are variable. DNA methylation can change in response to exercise stimuli. Long-term exercise can lower the methylation level in the promoter regions of smooth muscle contraction genes MYH7 and MYL3 increase their transcription, and increase the methylation in the promoter regions of FOXO3, CREB5, and PGC-1α genes to improve inflammation and metabolism. DNA methylation is closely related to cardiovascular health (Geiger et al. 2024). Evidence generated on blood-based DNA methylation profiles supports the existence of substantial DNA methylation signals linked to blood pressure and cardiovascular disease. So far, DNA methylation predictors have been developed for the diagnosis of large-artery atherosclerosis stroke, hypertension, and risk of cardiovascular disease. DNA methylation is an important epigenetic change during cardiac development, and DNA methylation alterations are associated with a variety of congenital heart diseases such as aortic stenosis, tetralogy of Fallot, and ventricular septal defect. In addition, DNA methylation is also strongly associated with cardiomyopathy, hypertension, and atherosclerosis. ABCA1, TCF7, NFATC1, PRKCZ, and PDGFA DNA methylation as potential epigenetic-sensitive targets in acute coronary syndrome via network analysis (Infante et al. 2022). There is no direct evidence that exercise regulates cardiac DNA methylation, but exercise can reduce the incidence of cardiovascular disease through DNA methylation. In a 45-week follow-up study of middle-aged and older women who had not participated in physical activity for 6 months, it was found that 14 weeks of strength and aerobic exercise led to a reduction in blood pressure and triglyceride levels and altered the methylation of 725 genomic sites, which were primarily located in the promoter regions (da Silva Rodrigues et al. 2023). Of note, these identified hypermethylated sites were located mainly in the genomic islands. On the other hand, the hypomethylated sites were more distributed in other regions that should not, in principle, affect the initiation of the transcription process. These findings suggest that exercise training promotes both hypomethylation and hypermethylation, but the hypermethylation process is more concentrated in transcription start sites and it may affect the transcription process. After conducting enrichment analysis on the genes with altered methylation, they found several metabolic pathways, including AMPK signaling, TGF-β signaling, and insulin signaling. The study also indicated that 7 weeks of exercise had no significant impact on the researchers' physiological markers or DNA methylation. This suggests that exercise is a long-term and gradual process for improving genetic methylation and cardiovascular physiological function. In a 6-month study of middle-aged and older adults aged 40–87 years, it was found that moderate exercise participants had increased methylation of apoptosis-associated speck-like protein (ASC) compared with non-exercise people (Nishida et al. 2019). The higher the level of ASC methylation, the lower the level of inflammation (Huang et al. 2021). In addition, hypermethylation levels of ASC were positively associated with heart failure outcomes (Butts et al. 2016). Another study found that KLF14 methylation in athletes' peripheral blood was significantly elevated, which was also associated with anti-inflammatory levels (Spólnicka et al. 2018). Long-term exercise can also induce DNA methylation of src homologous 2-domain transforming protein C1 in monocytes, reduce microvascular senescence, and reduce the probability of cardiovascular disease (Andreato et al. 2019).

Hypertension is a clinical syndrome characterized by elevated arterial pressure of systemic circulation. Hypertension is an important risk factor leading to cardiovascular and cerebrovascular diseases, which can damage the structure and function of the heart, brain, and kidneys. Remodeling of large and small arteries is an important mechanism in the development of hypertension. The process of hypertension is accompanied by the thickening of the blood vessel wall and the increase of wall/cavity ratio, which leads to the reduction of vascular diameter and the increase of resistance. Exercise has even more encouraging results in the prevention and treatment of hypertension (Kumagai et al. 2022). Long-term, regular endurance and resistance training can lower the compliance of the large arteries (Cornelissen and Fagard 2005). Regular aerobic exercise for 14 weeks reduced blood pressure in normal subjects and increased the methylation levels of Endothelin-1 (encoded by the EDN1 gene) and inducible NOS (NOS2) in blood cells (Ferrari et al. 2019). A comprehensive review of randomized controlled trials (RCTs) investigating the impact of physical activity and drug-based treatments on high blood pressure revealed that exercise is as effective in reducing blood pressure as medications like ACE inhibitors, ARBs, beta-blockers, calcium channel blockers, and diuretics. Research also indicates that retired athletes have a notably lower risk of developing hypertension compared to the general public (Naci et al. 2019). A potential underlying mechanism involves DNA methylation influencing various biological pathways, including vascular smooth muscle contraction, the apelin signaling cascade, cholinergic synaptic transmission, oxytocin signaling, TRP channel modulation by inflammatory mediators, as well as the HIF-1, Notch, aldosterone-mediated sodium reabsorption, APK, Ras, Wnt, AMPK, and VEGF signaling pathways. Consistent physical activity enhances nitric oxide (NO) synthesis in endothelial cells (ECs) via shear stress-mediated upregulation and posttranscriptional modification of endothelial nitric oxide synthase (eNOS). Additionally, exercise modulates calcium channel activity in smooth muscle cells, which plays a critical role in blood pressure reduction. Specifically, exercise counteracts hypertension-induced increases in LTCC and BKCa channel activity in mesenteric arteries by downregulating the expression of the LTCC α1c and BKCa β1 subunits (Zhang et al. 2020). Consistent with these findings, physical exercise-induced hypermethylation of the α1c and β1 genes in spontaneously hypertensive rats (SHR), accompanied by a reduction in TET1 expression and an upregulation of DNMT1 and DNMT3b in mesenteric arteries. Additionally, exercise modulated the peripheral levels of α-ketoglutarate (α-KG) and altered the S-adenosylmethionine to S-adenosylhomocysteine ratio. These combined functional and molecular improvements, driven by exercise, synergistically contribute to mitigating hypertension-induced increases in arterial blood pressure. Specifically, exercise downregulates LTCC and BKCa channel activity through hypermethylation of their α1c and β1 subunits, thereby aiding in the restoration of mesenteric arterial function and promoting vasodilation in hypertensive conditions. Furthermore, exercise enhances nitric oxide (NO) levels in endothelial cells, supporting their vasodilatory capacity and maintaining endothelial diastolic function.

Another cardiovascular disease that has been shown to improve significantly after exercise is heart failure. Heart failure (HF) is a clinical syndrome consisting of sitting breathing, peripheral edema, and fatigue, caused by structural and/or functional abnormalities of the heart. HF is the final stage of cardiovascular disease, HF mortality is very high, 1 year mortality is about 20%, and 5 years mortality is as high as 50–60%. Cardiac remodeling is the basic mechanism of the occurrence and development of heart failure, which mainly includes pathological hypertrophy of cardiomyocytes, cardiomyocyte death, excessive fibrosis, and increased degradation of the cardiomyocyte extracellular matrix. High-intensity interval training (HIIT) is characterized by exercising at ≥80% of peak oxygen consumption (O_{2peak}), interspersed with 40–50% of O_{2peak} for recovery (Ellingsen et al. 2017). HIIT-induced alternating aerobic and anaerobic metabolic responses have been shown to increase aerobic capacity, improve red blood cell deformability, reduce heart cavity size, reduce cardiac fibrosis, and increase 8-month survival in patients with HF. Therefore, HIIT has been advocated as a beneficial therapy for HF patients. Recent studies have shown that HIIT is associated with better 10-year survival in patients with HF. Increased aerobic capacity and reduced ventilatory response to CO_2 production after HIIT and reduced end-diastolic diameter of the left ventricle are promising protective factors against death in patients with heart failure. A significant reversal of cardiac remodeling in patients with heart failure with reduced ejection fraction (HFrEF) after HIIT was observed immediately after 12 weeks of supervised aerobic exercise training. HIIT-related left ventricular remodeling reverses in HF patients with left ventricular injection fraction (LVEF) <50% were similar to those observed in patients treated with angiotensin-converting enzyme inhibitors or beta blockers. Exercise-mediated DNA methylation improves cardiac remodeling in patients with heart failure. In a study involving 12 heart failure patients, it was proven that continuous 12-week HIIT can improve cardiorespiratory function and reduce myocardial fibrosis, improving cardiac remodeling. Further research found that the expression of DNMT1 (cytosine-5)-methyltransferase in the serum of heart failure patients after HIIT increased significantly, accompanied by markedly high methylation of the ACADVL gene transcription start site (Hsu et al. 2023). In cardiomyocytes, knocking down ACADVL increases cardiomyocyte apoptosis and leads to extensive disassembly of actin filaments. ACADVL is located on chromosome 17p13.1 and encodes VLCAD, which is distributed on the inner mitochondrial membrane and is a key enzyme in mitochondrial energy metabolism. High methylation of the ACADVL gene leads to a deficiency of VLCAD in the mitochondria, causing marked mitochondrial swelling. Subsequently, cytochrome c is released from the mitochondria and activates the subsequent caspase-mediated disassembly of actin filaments and nuclear laminin B1. Apoptotic responses and disassembly of actin filaments were also observed in cardiomyocytes incubated with serum from HIIT-treated patients.

Exercise-induced DNA methylation also reduces the probability of cardiovascular disease by inhibiting the occurrence of metabolic syndrome. Metabolic syndrome is a cluster of metabolic risk factors that are closely associated with the incidence and mortality risk of arteriosclerotic cardiovascular diseases and type 2 diabetes. Its main components include impaired glucose tolerance or diabetes, central obesity (abdominal obesity), dyslipidemia, hypertension, low-level inflammation, abnormalities in coagulation and fibrinolysis, etc. The main causes of metabolic syndrome include obesity and insulin resistance. In recent years, the incidence of metabolic syndrome has been increasing rapidly. The methylation of multiple genes is closely related to the progression of metabolic syndrome in patients. Acute exercise decreases genome-wide methylation in skeletal muscle, as well as significantly reduced methylation in the promoter regions of metabolism-related PGC-1α, PDK4, and PPAR-δ genes, as well as in mouse studies. These genes play an important role in glycolipid metabolism and insulin sensitivity (Barrès et al. 2012; Kärkkäinen et al. 2019). In people with diabetes, long-term exercise decreases NRF1 methylation and increases mitochondrial density (Rowlands et al. 2014). In addition, short-term high-intensity exercise reduced the DNA methylation of the whole genome of leukocytes, and the genes with the most drastic changes were mainly enriched in focal adhesions, calcium signaling, mitogen-activated protein kinase (MAPK) signaling, and important pathways related to cardiovascular pathophysiologies such as ischemic cardiomyopathy and coronary atherosclerosis. Among them, exercise training-induced methylation of cg12093976 reduced the mRNA expression of EGF, which is closely related to inflammation and atherosclerosis (Denham et al. 2015). At the same time, short-term high-intensity exercise decreased DNA methylation of miR21 and miR210 genes and affected mature miR-21 and miR-210 expression (Denham et al. 2015). Both miR-21 and miR-210 are induced by hypoxia and are involved in the occurrence and development of a variety of cardiovascular diseases (Bei et al. 2024). miR-21 is upregulated as a protective factor in atherosclerotic plaques and acute coronary syndromes. Elevated miR-21 downregulates programmed cell death 4 (PDCD4) and inhibits ischemia-induced myocardial apoptosis. In addition, miR-21 can inhibit PTEN-induced apoptosis of endothelial cells in response to the stimulation of endothelium, promote eNOS phosphorylation, and maintain normal endothelial cell function in response to the stimulation of endothelium (Xiong et al. 2022). miR-21 can also be upregulated in postexercise cardiac tissue, but its mechanism has not been reported (Raitoharju et al. 2011; Eken et al. 2017; Li et al. 2017), and reactive oxygen species are upregulated by miR-210 through the AKT-ERK pathway. In conclusion, although there is a lack of direct evidence of exercise-induced cardiac DNA methylation, the above studies suggest that exercise-induced DNA methylation is an important mechanism for cardiovascular protection.

4.3 Histone Modifications

DNA binds to proteins (mainly conserved histones) to form complex macromolecular structures called chromatin to assemble large-sized DNA into the nucleus of the cell. Chromatinized DNA ultimately folds into stable and condensed chromosome structures. To transcribe the DNA's genetic information into messenger RNA (mRNA), it must be denatured into linear transcription templates. The basic unit of chromatin is the nucleosome, which consists of a central histone octamer (two copies of histones H2A, H2B, H3, and H4) surrounded by DNA. These histones have a large number of posttranslational modifications (PTMs), commonly referred to as epigenetic markers, which regulate chromatin structure and thus regulate DNA condensation and denaturation. The histone PTM landscape is established, maintained, and reset by a network of interconnected signaling pathways involving enzymes that catalyze the formation of specific types of PTMs (writing proteins), proteins that recognize specific PTMs through specific domains (readers), and enzymes that remove PTMs (erasers). These enzymes primarily affect the N-terminal tails of histones, with a focus on amino acids lysine, arginine, serine, threonine, and tyrosine. Exercise promotes histone turnover and changes histone H3 trimethylation at lysine 4 (H3K4me3), histone H3 trimethylation at lysine 27 (H3K27me3), histone H3 lactylation at lysine 18 (H3K18la), and other histone modifications.

Histone modifications are involved in the pathophysiological processes of a variety of cardiovascular diseases, such as myocardial infarction, hypertension, and atherosclerosis. Exercise-induced cardiovascular protection also requires the involvement of histone modifications, which require histone modification enzymes, and current research focuses on the HDACs family. Intermittent exercise can induce the production of an N-terminal proteolytically derived fragment of HDAC4

(HDAC4-NT) in the heart of mice, and overexpression of HDAC-NT can protect cardiac function and inhibit cardiac fibrosis in mice after transverse aortic constriction (TAC) (Lehmann et al. 2018). HDAC4-NT inhibits the expression of NR4A1 through MEF2, thereby improving glucose metabolism in cardiomyocytes. Another study showed that diabetic patients and obese mice had high levels of HDAC4-NT in peripheral blood mononuclear cells and that overexpression of HDAC4-NT in the heart of diabetic mice inhibited cardiac systolic dysfunction, suggesting that HDAC4-NT is necessary for maintaining diabetic cardiac function (Kronlage et al. 2019). HDAC2 activity in cardiac tissue and HDAC4/5 nuclear export increased in skeletal muscle tissue after 3 days of forced swimming in mice (Richter and Hargreaves 2013). Mammalian switch-independent 3A (mSin3A) and HDAC1 and 2 regulate hypertrophy genes by binding to DNA-binding protein repressor element-1 silencing transcription factor (REST) and O-linked β-N-acetylglucosamine transferase (OGT), and exercise promotes mSin3A/HDAC1/HDAC2 complex with OGT in mouse hearts, reversing the pathological process of the diabetic heart (Cox and Marsh 2013). Interestingly, another study showed that exercise inhibited the interaction between the mammalian switch-independent 3A (mSin3A)/HDAC1/HDAC2 complex and OGT in the mouse heart (Medford et al. 2013). This opposite result may be related to the type of exercise, the length of exercise, and the intensity of exercise in the mice. These results surface a highly dynamic and variable regulatory relationship between exercise and epigenetics. Obesity causes epigenetic memory in fat tissue. Promoters that remained active in obese mice (that is, high H3K4me3 and/or H3K27ac and low H3K27me3) were related to genes involved in extracellular matrix remodeling and inflammatory signaling (for example, Icam1, Lyz2, or Tyrobp).

Exercise-induced epigenetic changes inhibit the inflammation. Compared to sedentary mice that were 4 months old, mice that had been swimming for 3 months since 1 month old showed a reduction in white blood cell (WBC) count, including neutrophils, lymphocytes, and monocytes, and a reduction in pro-inflammatory cytokines TNF-α and IL-1β, but an increase in anti-inflammatory cytokine IL-1Ra (Zaldivar et al. 2006). Pipecolic acid was the only metabolite that was continuously upregulated in the serum of the exercised mice at 4 months, compared to sedentary mice. Pipecolic acid is a metabolic product of lysine, primarily catalyzed by CRYM/ketimine reductase in the liver. SETD1A-mediated H3K4me3 epigenomic modification in the Crym promoter in hepatocytes contributes to the elevation of Crym expression induced by early-life exercise (Zhang et al. 2024).

4.4 Noncoding RNA

Noncoding RNAs include long noncoding RNAs and microRNAs. Long noncoding RNAs are RNAs that are greater than 200 nucleotides in length and lack the potential to encode proteins, and most noncoding RNAs fall into this category. MicroRNAs, on the other hand, are a class of single-stranded nucleotides that are about 20–24 nt in length. Noncoding RNAs play an important role in the epigenetic regulation of genes. A variety of noncoding RNAs in cardiovascular tissues have been found to improve cardiovascular function in response to motor stimuli. A large number of noncoding RNAs are involved in a variety of pathophysiological processes such as myocardial physiological hypertrophy, pathological hypertrophy, angiogenesis, and myocardial death in response to motor stimuli. Due to the use of different exercise models and exercise cycles, the noncoding RNAs reported in heart tissue in response to exercise stimuli are also different in different kinds of literature. miRNAs can be used as novel therapeutic targets for the treatment of human cardiovascular disease, and various strategies for modulating miRNA activity to treat cardiovascular disease are being developed based on preclinical studies. In a study comparing a sedentary control group with a group that received 10 weeks of exercise, 87 differentially expressed miRNAs were identified out of 349 cardiac miRNAs, including 48 upregulated miRNAs and 39 downregulated miRNAs (Soci et al. 2011). It has also been reported that 35 miRNAs have been differentially expressed after short-term exercise. Another study using two exercise models, swimming and roller running, found that 124 miRNAs were differentially expressed in the hearts of roller-running mice and 55 miRNAs in the hearts of swimming mice compared to sedentary controls, with 16 miRNAs differentially expressed in both exercise models (Liu et al. 2015). Among these differentially expressed cardiac miRNAs, several miRNAs are key regulators of cardiac physiological and pathological processes after exercise.

miRNA-222 and miRNA-17-3p are upregulated after exercise training and attenuate cardiac remodeling caused by ischemia-reperfusion injury. miRNA-222 expression is upregulated in mouse models of running and swimming while increasing in the peripheral blood of patients with heart failure after exercise. Serum miR-222 levels were significantly reduced in myocardial infarction (MI) patients, as well as in MI patients who were readmitted to the hospital and/or died compared to those not. MiR-222 plays a role in regulating cardiomyocyte apoptosis by targeting p27, HIPK1, HIPK2, and Hmbox-1, thereby attenuating cardiac remodeling after infarction (Liu et al. 2015). More importantly, miR-222 is required for exercise-induced myocardial growth. MiR-222 also increased in pathological hypertrophy induced by pressure overload. After TAC, miR-222 inhibition accelerated the development of pathological hypertrophy, cardiac dysfunction, and heart failure. P53-upregulated modulator of apoptosis, a pro-apoptotic Bcl-2 family member, and the transcription factors, Hmbox1 and nuclear factor of activated T-cells 3, as direct miR-222 targets contributing to its roles in pathological hypertrophy (Liu et al. 2024). MiR-17-3p expression was significantly increased in mouse models of running and swimming and was also increased in the serum of patients with heart failure after exercise. MiR-17-3p inhibits TIMP3 and attenuates cardiomyocyte apoptosis by activating EGFR/JNK/SP-1 signaling. In addition, PTEN and Akt have been identified as indirectly regulated by miR-17-3p in cardiomyocytes, and PTEN and Akt regulate PTEN/Akt signaling in the motor heart, which is also responsible for miR-17-3p's inhibition of cardiomyocyte apoptosis. At the same time, miR-17-3p also helps exercise to induce heart growth in mice (Shi et al. 2017). Recently, miR-486 is significantly downregulated in myocardial tissue after ischemia-reperfusion (Li et al. 2021; Bei et al. 2022). In addition, miR-486-5p overexpression reduced glucose-deprivation-driven cardiomyocyte apoptosis. It targets PTEN and FOXO1 and activates AKT/mammalian target of rapamycin (mTOR) signaling to protect mice from cardiac dysfunction and myocardial apoptosis after ischemia-reperfusion. The depletion of miR-486-5p abolishes the protective effects of exercise, suggesting that it is required for cardioprotection. Different studies on miRNA-222, miRNA-17-3p, and miR-486-5p have shown that exercise-induced miRNAs are important performers of exercise in improving cardiac function. In addition to cardiac miRNAs, studies have shown that exercise-induced circulating miRNAs also exert cardioprotective effects. Twelve miRNAs were differentially expressed in serum exosomes of rats after exercise, and miR-342-5p attenuated myocardial injury caused by ischemia-reperfusion by targeting Caspase 9 and JNK2 (Hou et al. 2019). MiR-125b-5p, miR-128-3p, and miR-30d-5p secreted by brown adipose tissue target the MAPK pathway to mediate exercise-induced cardioprotective effects (Zhao et al. 2022).

Long noncoding RNAs also play an important role in improving cardiac function with exercise. Recent studies have found that exercise can improve atherosclerosis by altering lncRNA expression. For example, 204 differentially expressed lncRNA were identified after high-intensity interval training (12 weeks) and 43 differentially expressed lncRNA were identified after resistance training. Fifteen differentially expressed lncRNA were identified after combined exercise training. In exercise-induced cardiac physiological hypertrophy in mice, the expression of long noncoding RNA FR236730 increased, which the researchers named CPhar (cardiac physiological hypertrophy associated regulator), which is required for exercise-induced physiological hypertrophy and protects against cardiac fibrosis caused by ischemia-reperfusion injury. CPhar mainly isolates C/EBPβ by binding to the RNA helicase DDX17 of the DEAD-Box family and regulates the expression of transcription factor ATF17, which plays a role in promoting cardiomyocyte proliferation, hypertrophy, and inhibiting apoptosis (Gao et al. 2021). Another study found that the long noncoding myosin heavy chain-associated RNA transcript Mhrt779 was one of the long noncoding RNAs significantly upregulated in exercise mice (Lin et al. 2021). Silencing of Mhrt779 attenuates the antihypertrophic effect of exercise on TAC mice and angiotensin II-treated cardiomyocytes, and overexpression of Mhrt779 enhances the antihypertrophic effect of exercise. It was found that exercise increased H3K4me3 and H3K36me3 modifications in the promoter region of Mhrt779, and Mhrt779 could bind to Brg1 to inhibit the activation of the Hdac2/Akt/GSK3β pathway induced by stress overload. In conclusion, exercise-induced myocardial hypertrophy pretreatment enhances resistance to pathological stress through antihypertrophic effects mediated by Mhrt779/Brg1/HDAC2/p-Akt/p--GSK3β signaling pathways. lncRNA has been validated as a therapeutic target for cardiovascular diseases. Utilizing lentiviral shRNAs that target lncRNA H19 has been shown to significantly reduce the progression of atherosclerosis and enhance plaque stability in vivo.

Circular RNAs are a special class of noncoding RNAs that do not have a 5′-cap and 3′-poly(A) that also play an important role in exercise-induced cardioprotection. CircRNAs lack the typical 5′ caps and 3′ poly(A) tails and, consequently, form stable covalent closed-loop molecules that are resistant to the ubiquitous 3′-exonuclease RNase R. CircRNA can be produced from one or more exons, introns, or two exons and introns by retaining internal introns. Three main types of cyclization mechanisms that lead to reverse splicing have been described: (1) lasso-driven cyclization, also known as exon jump models; (2) intron pairing-driven cyclization; (3) cyclization driven by RNA binding protein (RBP). Most circRNAs have been shown to act as miRNA sponges or decoys, binding to miRNA response elements and subsequently blocking miRNA activity, thereby upregulating the expression of mRNA regulated by disrupted miRNAs. Through this mechanism, circRNA protects mRNAs from miRNA-regulated degradation and enhances their translation to proteins. CircRNAs participate in the regulation of the pathophysiological mechanisms underpinning cardiac remodeling. CircUtrn is a circular RNA that is highly expressed in cardiomyocytes of long-term exercise mice, highly conserved between humans and mice and is predominantly localized to the cytoplasm (Wang et al. 2023). The serum circUtrn level of long-term exercise athletes was significantly increased, indicating that the serum circUtrn level was positively correlated with endurance training. In vitro, overexpression of circUtrn promotes cardiomyocyte hypertrophy and alleviates glucose deprivation-induced apoptosis by binding to and modulating PP5. In vivo, exercise-induced increase in circUtrn is essential for the protective effect of exercise in the ischemia-reperfusion remodeling environment. Overexpression of circUtrn in mouse hearts reduces acute myocardial ischemia-reperfusion injury and prevents ischemia-reperfusion-induced myocardial remodeling.

4.5 RNA Modification

RNA modification is a complex regulatory mechanism that affects gene expression at the posttranscriptional level. The key participants in this process are RNA modifying proteins, which act as "writers," "erasers," and "readers"—adding, deleting, or interpreting chemical groups on RNA to control its function, stability, and interactions with other molecules. Specific RNA modifying enzymes, such as METTL3 (Methyltransferase 3, N6-Adenosine-Methyltransferase Complex Catalytic Subunit) and METTL14 (Methyltransferase 4, N6-Adenosine), act as "writers" catalyzing the addition of m6A markers, which are crucial for RNA stability, splicing, and translation. Erasers like FTO (FTO Alpha-Ketoglutarate Dependent Dioxygenase) and ALKBH5 (AlkB Homolog 5, RNA Demethylase) remove these modifications, dynamically regulating the fate of RNA, while readers like YTHDF (YTH N6-Methyladenosine RNA Binding Protein F3) family and YTHDC1 (containing YTH domain 1) interpret these markers, affecting RNA processing and degradation.

Modifications such as N6 methyladenosine (m6A), N6.2′-O-dimethyladenosine (m6Am), N1 methyladenosine (m1A), 5-methylcytosine (m5C), N4 acetylcytosine (ac4C), N7 methylguanosine (m7G), pseudouridine (PSI), adenosine to inosine (A-to-I) editing.

N6-methyladenosine (m6A) represents the most ubiquitous RNA epigenetic modification in eukaryotic systems, critically involved in cellular homeostasis and cardiovascular pathophysiology. This reversible methylation mark constitutes over 80% of total RNA methylation events, serving as a regulatory nexus for posttranscriptional control. Cardiac-specific dysregulation of m6A dynamics induces cardiomyocyte functional aberrations and disrupts myocardial equilibrium. Chronic endurance exercise protocols induce global m6A hypomethylation in cardiac tissue, as evidenced by high-resolution methylome mapping. Comparative epitranscriptomic profiling revealed 165 significantly elevated m6A peaks and 310 diminished methylation loci (FDR <0.05) in exercised versus sedentary myocardial samples, indicating exercise-driven reprogramming of the cardiac epitranscriptome (Wang et al. 2022). Exercise triggers bidirectional regulation of METTL14 expression, manifesting as transcriptional suppression (1.8-fold mRNA reduction) and proteasomal degradation (62% protein depletion) in cardiac tissue. Functional studies in transgenic models establish a requirement for METTL14's catalytic activity. Adenoviral delivery of the wild-type protein abolishes exercise-induced cardioprotective hypertrophy. In contrast, the catalytically dead mutant (D395A/D399A) is ineffective. Conversely, pharmacological METTL14 inhibition (via CRISPR-Cas9 knockout or small-molecule antagonists) induces: Physiological cardiomyocyte enlargement (1.4-fold cross-sectional area increase); Enhanced mitotic activity (BrdU+ nuclei: 12.3% vs 3.1% in controls); Cell cycle reentry (Cyclin D1 upregulation: 2.1-fold). Cardiac-specific METTL14 ablation in conditional knockout murine models confers protection against acute ischemia-reperfusion injury (infarct size reduction: 38%) and attenuates maladaptive remodeling (fibrosis area: 15% vs 42% in WT). Mechanistically, METTL14 silencing mediates epitranscriptomic reprogramming through: Phlpp2 m6A demethylation (YTHDF2-dependent decay inhibition); Akt-Ser473 phosphorylation axis activation (3.5-fold vs baseline). This METTL14-m6A-Akt signaling axis orchestrates exercise-adaptive cardiac plasticity by balancing hypertrophic growth with ischemic resilience. METTL14 plays an important role in maintaining cardiac homeostasis. The abundance of m6A methylation exhibits significant elevation across multiple vascular cell types affected by atherosclerosis, demonstrating that this epigenetic modification contributes to the pathomechanisms underlying atherosclerotic progression. Exercise-modulated long noncoding RNA NEAT1 (nuclear-enriched assembly transcript 1) influences atherogenesis following N6-adenosine methylation alterations (Yang et al. 2023). The transcriptional activity of NEAT1 is suppressed following exercise regimens, and this suppression constitutes a critical determinant in exercise-induced atherosclerotic amelioration. Exercise-driven functional attenuation of NEAT1 impedes atherogenesis progression. At the molecular level, physical activity stimulates m6A methylation and marked reduction of METTL14, which interacts with NEAT1's m6A motif to facilitate NEAT1 transcription via YTHDC1-mediated recognition, thereby exacerbating pyroptotic processes in endothelial cells. Furthermore, NEAT1 triggers endothelial pyroptosis by associating with KLF4 (Krüppel-like factor 4) to enhance transcriptional initiation of the pivotal pyroptotic mediator NLRP3 (NOD-like receptor pyrin domain-containing protein 3). Exercise interventions counteract NEAT1-dependent endothelial pyroptotic pathways and mitigate atherosclerotic pathogenesis.

Adenosine to inosine is another RNA modification. Adenosine deaminase (ADAR) acting on RNA is an enzyme that acts on double-stranded RNA (dsRNA), converting adenosine to inosine (A to I). There are three known members of the ADAR gene family (ADAR1–3) in vertebrates. ADAR1 and ADAR2 are commonly expressed, while ADAR3 is only expressed in the brain. ADAR-mediated A-to-I editing is the most common type of RNA editing in humans, and inosine is explained by cellular mechanisms as guanosine. A-to-I modifications of RNA at the posttranscriptional level typically result in alternative splicing, RNA folding, and RNA sequencing. The myocardial-specific deficiency of ADAR1 can lead to severe cardiac dysfunction and ultimately result in embryonic or postnatal death. The expression of ADAR2 increases in the exercise heart (Wu et al. 2022). In vitro, ADAR2 promotes cardiomyocyte (CM) proliferation and inhibits DOX-induced CM cell apoptosis. AAV9 (adeno-associated virus serotype 9) carrying the heart-specific troponin T (cTnT) promoter-driven overexpression of ADAR2 (AAV9-cTnT-ADAR2) in vivo can alleviate acute myocardial infarction (AMI), MI remodeling, and DOX-induced cardiac toxicity. ADAR2 negatively regulates mature miR-34a and ADAR2's RNA editing activity contributes to the ADAR2/miR-34a regulatory axis in CM. In addition, ADAR2 regulates CM proliferation and cell apoptosis through miR-34a and Sirt1, Cyclin D1, and Bcl2. Overexpression of ADAR2 or posttranscriptional RNA editing through ADAR2 may be a promising strategy for treating heart disease.

4.6 Exercise and Offspring Cardiovascular Health

The maternal lifestyle during pregnancy can affect the health of the fetus, and epigenetic inheritance is an important driver of these changes. There may be a window period of epigenetic instability during fetal development, during which life events that occur are permanently "imprinted" on an individual's phenotype and the individual's risk of developing cardiovascular disease. Due to its phenotypic plasticity, the fetus is highly susceptible and produces lasting effects. Negative maternal risk factors during pregnancy, such as poor nutrition, smoking, and lack of oxygen, have been reported to increase the risk of cardiovascular disease in adulthood. Maternal exercise during pregnancy has emerged as a potentially promising way to protect offspring from cardiovascular disease, including high blood pressure. Swimming training induced the expression of SIRT1 in blood vessels of spontaneously hypertensive rats (SHRs) and Wistar-Kyoto rats, reduced the level of histone H3 acetylation at lysine 9 (H3K9ac), and inhibited the transcription of Nox4 in the mesenteric artery of hypertensive fetuses (Zhang et al. 2023). This change induced the vasodilation injury induced by acetylcholine. In addition, another study demonstrated that maternal exercise in spontaneously hypertensive rats and Wistar-Kyoto rats upregulated SIRT1 expression, resulting in decreased H3K9ac levels in the FGF21 gene promoter region (Shan et al. 2024). This epigenetic modification leads to transcriptional downregulation of FGF21 in hypertensive fetuses, improving vascular remodeling and hypertension. DNA methylation is also involved in the protective effect of exercise against fetal hypertension. Exercise during pregnancy may reduce the mRNA and protein expression of the CaV1.2 channel α1C subunit by regulating the DNA methylation pattern in the offspring of hypertensive rats, thereby inhibiting the upregulation of CaV1.2 channel function during hypertension (Li et al. 2019). Exercise during pregnancy improves cardiovascular function in the adult offspring of SHR. Maternal exercise during pregnancy triggers hypermethylation and transcriptional inhibition of the Agtr1a gene by increasing the expression of DNMT1 and DNMT3B in the mesenteric arteries of the offspring of SHR (Shan et al. 2023). Downregulation of AT1R expression reduces the contribution of angiotensin-2 to vascular tone and ultimately improves vascular structure and function. Paternal exercise may also have a beneficial effect on the cardiovascular health of offspring. Offspring of trained male mice exhibited improved glucose homeostasis and insulin sensitivity (Costa-Júnior et al. 2021).

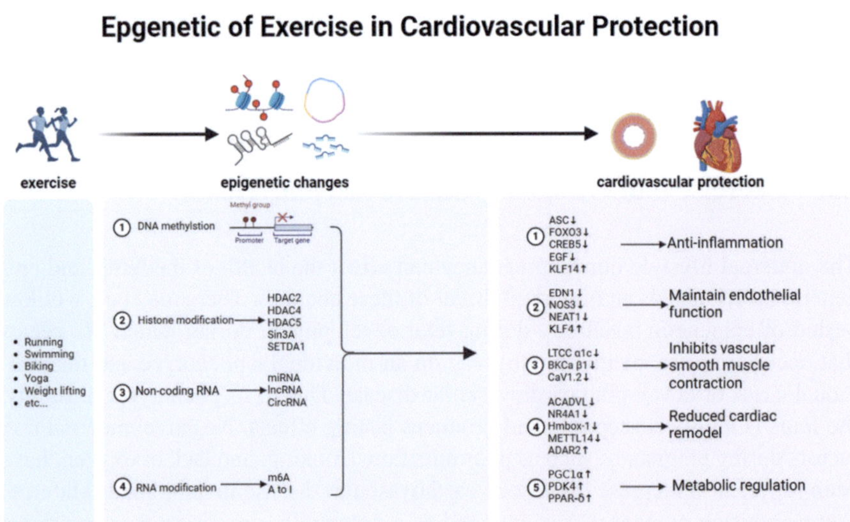

Fig. 4.1 Exercise induces alterations in DNA methylation, histone modification, noncoding RNA expression and RNA modification, which contribute to cardiovascular protection through anti-inflammation, maintain endothelial and vascular smooth muscle function, reduced cardiac remodel and metabolic remodeling

4.7 Conclusion

The protective role of exercise on cardiovascular health has gradually become a consensus, but less is known about the role of epigenetics in this process. This review summarizes the recent reports on the role of epigenetics in the protection of cardiovascular health by exercise, including DNA methylation, histone modification, and noncoding RNA (Fig. 4.1). In recent decades, along with economic development and the wide promotion of high-sugar and high-fat foods, cardiovascular diseases have become the greatest disease burden worldwide, and the affected population of cardiovascular diseases has also shown a trend of younger age. Secondary preventive measures for cardiovascular diseases mainly based on drugs require good compliance from patients to play a role in improving patient prognosis. Exercise is a more acceptable way of managing cardiovascular diseases for the population. It has more significant application prospects among populations with medication contraindications, such as pregnant women. For certain specific diseases such as hypertension, the improvement effect of exercise is no less than that of existing clinical drugs. Revealing the mechanism by which exercise improves cardiovascular function is of great significance for the prevention and control of cardiovascular diseases through exercise, and epigenetics is an irreplaceable mechanism for exercise to improve cardiovascular conditions. In conclusion, more and more evidence indicates that epigenetics, the epigenetic modification changes brought about by exercise, can improve cardiovascular function. In-depth research in this field helps us discover new biomarkers of cardiovascular diseases and may lead to more scientific and effective exercise prescriptions.

Compliance with Ethical Standards

Conflict of Interest The authors declare that there is no conflict of interest.

Ethical Approval This chapter does not contain any studies with human participants or animals performed by any of the authors.

References

Andreato LV, Machado UF, Esteves JV (2019) Distracting factors involving the high-intensity interval training as modulator of retinal microvascular phenotype and DNA methylation of SHC1 gene. Eur Heart J 40(48):3947. https://doi.org/10.1093/eurheartj/ehz899

Barrès R, Yan J, Egan B, Treebak JT, Rasmussen M, Fritz T, Caidahl K, Krook A, O'Gorman DJ, Zierath JR (2012) Acute exercise remodels promoter methylation in human skeletal muscle. Cell Metab 15(3):405–411. https://doi.org/10.1016/j.cmet.2012.01.001

Bei Y, Lu D, Bär C, Chatterjee S, Costa A, Riedel I, Mooren FC, Zhu Y, Huang Z, Wei M, Hu M, Liu S, Yu P, Wang K, Thum T, Xiao J (2022) miR-486 attenuates cardiac ischemia/reperfusion injury and mediates the beneficial effect of exercise for myocardial protection. Mol Ther 30(4):1675–1691. https://doi.org/10.1016/j.ymthe.2022.01.031

Bei Y, Wang H, Liu Y, Su Z, Li X, Zhu Y, Zhang Z, Yin M, Chen C, Li L, Wei M, Meng X, Liang X, Huang Z, Cao RY, Wang L, Li G, Cretoiu D, Xiao J (2024) Exercise-induced miR-210 promotes Cardiomyocyte proliferation and survival and mediates exercise-induced cardiac protection against ischemia/reperfusion injury. Research (Wash D C) 7:0327. https://doi.org/10.34133/research.0327

Butts B, Gary RA, Dunbar SB, Butler J (2016) Methylation of apoptosis-associated speck-like protein with a caspase recruitment domain and outcomes in heart failure. J Card Fail 22(5):340–346. https://doi.org/10.1016/j.cardfail.2015.12.004

Cornelissen VA, Fagard RH (2005) Effects of endurance training on blood pressure, blood pressure-regulating mechanisms, and cardiovascular risk factors. Hypertension 46(4):667–675. https://doi.org/10.1161/01.Hyp.0000184225.05629.51

Costa RR, Buttelli ACK, Vieira AF, Coconcelli L, Magalhães RL, Delevatti RS, Kruel LFM (2019) Effect of strength training on lipid and inflammatory outcomes: systematic review with meta-analysis and meta-regression. J Phys Act Health 16(6):477–491. https://doi.org/10.1123/jpah.2018-0317

Costa-Júnior JM, Ferreira SM, Kurauti MA, Bernstein DL, Ruano EG, Kameswaran V, Schug J, Freitas-Dias R, Zoppi CC, Boschero AC, Oliveira CAM, Santos GJ, Carneiro EM, Kaestner KH (2021) Paternal exercise improves the metabolic health of offspring via epigenetic modulation of the germline. Int J Mol Sci 23(1). https://doi.org/10.3390/ijms23010001

Cox EJ, Marsh SA (2013) Exercise and diabetes have opposite effects on the assembly and O-GlcNAc modification of the mSin3A/HDAC1/2 complex in the heart. Cardiovasc Diabetol 12:101. https://doi.org/10.1186/1475-2840-12-101

da Silva Rodrigues G, Noronha NY, Almeida ML, Sobrinho A, Watanabe LM, Pinhel MAS, de Lima JGR, Zhang R, Nonino CB, Alves CRR, Bueno Júnior CR (2023) Exercise training modifies the whole blood DNA methylation profile in middle-aged and older women. J Appl Physiol 134(3):610–621. https://doi.org/10.1152/japplphysiol.00237.2022

Denham J, O'Brien BJ, Marques FZ, Charchar FJ (2015) Changes in the leukocyte methylome and its effect on cardiovascular-related genes after exercise. J Appl Physiol 118(4):475–488. https://doi.org/10.1152/japplphysiol.00878.2014

Dibben GO, Faulkner J, Oldridge N, Rees K, Thompson DR, Zwisler AD, Taylor RS (2023) Exercise-based cardiac rehabilitation for coronary heart disease: a meta-analysis. Eur Heart J 44(6):452–469. https://doi.org/10.1093/eurheartj/ehac747

Eken SM, Jin H, Chernogubova E, Li Y, Simon N, Sun C, Korzunowicz G, Busch A, Bäcklund A, Österholm C, Razuvaev A, Renné T, Eckstein HH, Pelisek J, Eriksson P, González Díez M, Perisic Matic L, Schellinger IN, Raaz U, Leeper NJ, Hansson GK, Paulsson-Berne G, Hedin U, Maegdefessel L (2017) MicroRNA-210 enhances fibrous cap stability in advanced atherosclerotic lesions. Circ Res 120(4):633–644. https://doi.org/10.1161/circresaha.116.309318

Ellingsen Ø, Halle M, Conraads V, Støylen A, Dalen H, Delagardelle C, Larsen AI, Hole T, Mezzani A, Van Craenenbroeck EM, Videm V, Beckers P, Christle JW, Winzer E, Mangner N, Woitek F, Höllriegel R, Pressler A, Monk-Hansen T, Snoer M, Feiereisen P, Valborgland T, Kjekshus J, Hambrecht R, Gielen S, Karlsen T, Prescott E, Linke A (2017) High-intensity interval training in patients with heart failure with reduced ejection fraction. Circulation 135(9):839–849. https://doi.org/10.1161/circulationaha.116.022924

Ferrari L, Vicenzi M, Tarantini L, Barretta F, Sironi S, Baccarelli AA, Guazzi M, Bollati V (2019) Effects of physical exercise on endothelial function and DNA methylation. Int J Environ Res Public Health 16(14). https://doi.org/10.3390/ijerph16142530

Gao R, Wang L, Bei Y, Wu X, Wang J, Zhou Q, Tao L, Das S, Li X, Xiao J (2021) Long noncoding RNA cardiac physiological hypertrophy-associated regulator induces cardiac physiological hypertrophy and promotes functional recovery after myocardial ischemia-reperfusion injury. Circulation 144(4):303–317. https://doi.org/10.1161/circulationaha.120.050446

Geiger C, Needhamsen M, Emanuelsson EB, Norrbom J, Steindorf K, Sundberg CJ, Reitzner SM, Lindholm ME (2024) DNA methylation of exercise-responsive genes differs between trained and untrained men. BMC Biol 22(1):147. https://doi.org/10.1186/s12915-024-01938-6

Hall IF, Climent M, Viviani Anselmi C, Papa L, Tragante V, Lambroia L, Farina FM, Kleber ME, März W, Biguori C, Condorelli G, Elia L (2021) rs41291957 controls miR-143 and miR-145 expression and impacts coronary artery disease risk. EMBO Mol Med 13(10):e14060. https://doi.org/10.15252/emmm.202114060

Hou Z, Qin X, Hu Y, Zhang X, Li G, Wu J, Li J, Sha J, Chen J, Xia J, Wang L, Gao F (2019) Longterm exercise-derived Exosomal miR-342-5p: a novel Exerkine for Cardioprotection. Circ Res 124(9):1386–1400. https://doi.org/10.1161/circresaha.118.314635

Hsu CC, Wang JS, Shyu YC, Fu TC, Juan YH, Yuan SS, Wang CH, Yeh CH, Liao PC, Wu HY, Hsu PH (2023) Hypermethylation of ACADVL is involved in the high-intensity interval training-associated reduction of cardiac fibrosis in heart failure patients. J Transl Med 21(1):187. https://doi.org/10.1186/s12967-023-04032-7

Huang Y, Xu W, Zhou R (2021) NLRP3 inflammasome activation and cell death. Cell Mol Immunol 18(9):2114–2127. https://doi.org/10.1038/s41423-021-00740-6

Infante T, Franzese M, Ruocco A, Schiano C, Affinito O, Pane K, Memoli D, Rizzo F, Weisz A, Bontempo P, Grimaldi V, Berrino L, Soricelli A, Mauro C, Napoli C (2022) ABCA1, TCF7, NFATC1, PRKCZ, and PDGFA DNA methylation as potential epigenetic-sensitive targets in acute coronary syndrome via network analysis. Epigenetics 17(5):547–563. https://doi.org/10.1080/15592294.2021.1939481

Jansson AK, Chan LX, Lubans DR, Duncan MJ, Plotnikoff RC (2022) Effect of resistance training on HbA1c in adults with type 2 diabetes mellitus and the moderating effect of changes in muscular strength: a systematic review and meta-analysis. BMJ Open Diabetes Res Care 10(2). https://doi.org/10.1136/bmjdrc-2021-002595

Kärkkäinen O, Tuomainen T, Mutikainen M, Lehtonen M, Ruas JL, Hanhineva K, Tavi P (2019) Heart specific PGC-1α deletion identifies metabolome of cardiac restricted metabolic heart failure. Cardiovasc Res 115(1):107–118. https://doi.org/10.1093/cvr/cvy155

Kronlage M, Dewenter M, Grosso J, Fleming T, Oehl U, Lehmann LH, Falcão-Pires I, Leite-Moreira AF, Volk N, Gröne HJ, Müller OJ, Sickmann A, Katus HA, Backs J (2019) O-GlcNAcylation of histone deacetylase 4 protects the diabetic heart from failure. Circulation 140(7):580–594. https://doi.org/10.1161/circulationaha.117.031942

Kumagai H, Miyamoto-Mikami E, Someya Y, Kidokoro T, Miller B, Kumagai ME, Yoshioka M, Choi Y, Tagawa K, Maeda S, Kohmura Y, Suzuki K, Machida S, Naito H, Fuku N (2022) Sports activities at a young age decrease hypertension risk-the J-fit(+) study. Physiol Rep 10(12):e15364. https://doi.org/10.14814/phy2.15364

Kurozumi A, Nakano K, Yamagata K, Okada Y, Nakayamada S, Tanaka Y (2019) IL-6 and sIL-6R induces STAT3-dependent differentiation of human VSMCs into osteoblast-like cells through JMJD2B-mediated histone demethylation of RUNX2. Bone 124:53–61. https://doi.org/10.1016/j.bone.2019.04.006

Lee DH, Rezende LFM, Joh HK, Keum N, Ferrari G, Rey-Lopez JP, Rimm EB, Tabung FK, Giovannucci EL (2022) Long-term leisure-time physical activity intensity and all-cause and cause-specific mortality: a prospective cohort of US adults. Circulation 146(7):523–534. https://doi.org/10.1161/circulationaha.121.058162

Lehmann LH, Jebessa ZH, Kreusser MM, Horsch A, He T, Kronlage M, Dewenter M, Sramek V, Oehl U, Krebs-Haupenthal J, von der Lieth AH, Schmidt A, Sun Q, Ritterhoff J, Finke D, Völkers M, Jungmann A, Sauer SW, Thiel C, Nickel A, Kohlhaas M, Schäfer M, Sticht C, Maack C, Gretz N, Wagner M, El-Armouche A, Maier LS, Londoño JEC, Meder B, Freichel M, Gröne HJ, Most P, Müller OJ, Herzig S, Furlong EEM, Katus HA, Backs J (2018) A proteolytic fragment of histone deacetylase 4 protects the heart from failure by regulating the hexosamine biosynthetic pathway. Nat Med 24(1):62–72. https://doi.org/10.1038/nm.4452

Li Y, Yang C, Zhang L, Yang P (2017) MicroRNA-210 induces endothelial cell apoptosis by directly targeting PDK1 in the setting of atherosclerosis. Cell Mol Biol Lett 22:3. https://doi.org/10.1186/s11658-017-0033-5

Li S, Chen Y, Zhang Y, Zhang H, Wu Y, He H, Gong L, Zeng F, Shi L (2019) Exercise during pregnancy enhances vascular function via epigenetic repression of Ca(V)1.2 channel in offspring of hypertensive rats. Life Sci 231:116576. https://doi.org/10.1016/j.lfs.2019.116576

Li Q, Xu Y, Lv K, Wang Y, Zhong Z, Xiao C, Zhu K, Ni C, Wang K, Kong M, Li X, Fan Y, Zhang F, Chen Q, Li Y, Li Q, Liu C, Zhu J, Zhong S, Wang J, Chen Y, Zhao J, Zhu D, Wu R, Chen J, Zhu W, Yu H, Ardehali R, Zhang JJ, Wang J, Hu X (2021) Small extracellular vesicles containing miR-486-5p promote angiogenesis after myocardial infarction in mice and nonhuman primates. Sci Transl Med 13(584). https://doi.org/10.1126/scitranslmed.abb0202

Lin H, Zhu Y, Zheng C, Hu D, Ma S, Chen L, Wang Q, Chen Z, Xie J, Yan Y, Huang X, Liao W, Kitakaze M, Bin J, Liao Y (2021) Antihypertrophic memory after regression of exercise-induced physiological myocardial hypertrophy is mediated by the long noncoding RNA Mhrt779. Circulation 143(23):2277–2292. https://doi.org/10.1161/circulationaha.120.047000

Liu X, Xiao J, Zhu H, Wei X, Platt C, Damilano F, Xiao C, Bezzerides V, Boström P, Che L, Zhang C, Spiegelman BM, Rosenzweig A (2015) miR-222 is necessary for exercise-induced cardiac growth and protects against pathological cardiac remodeling. Cell Metab 21(4):584–595. https://doi.org/10.1016/j.cmet.2015.02.014

Liu X, Li H, Hastings MH, Xiao C, Damilano F, Platt C, Lerchenmüller C, Zhu H, Wei XP, Yeri A, Most P, Rosenzweig A (2024) miR-222 inhibits pathological cardiac hypertrophy and heart failure. Cardiovasc Res 120(3):262–272. https://doi.org/10.1093/cvr/cvad184

Loaiza-Betancur AF, Pérez Bedoya E, Montoya Dávila J, Chulvi-Medrano I (2020) Effect of isometric resistance training on blood pressure values in a group of normotensive participants: a systematic review and meta-analysis. Sports Health 12(3):256–262. https://doi.org/10.1177/1941738120908070

Maleszewska M, Gjaltema RA, Krenning G, Harmsen MC (2015) Enhancer of zeste homolog-2 (EZH2) methyltransferase regulates transgelin/smooth muscle-22α expression in endothelial cells in response to interleukin-1β and transforming growth factor-β2. Cell Signal 27(8):1589–1596. https://doi.org/10.1016/j.cellsig.2015.04.008

Medford HM, Porter K, Marsh SA (2013) Immediate effects of a single exercise bout on protein O-GlcNAcylation and chromatin regulation of cardiac hypertrophy. Am J Physiol Heart Circ Physiol 305(1):H114–H123. https://doi.org/10.1152/ajpheart.00135.2013

Naci H, Salcher-Konrad M, Dias S, Blum MR, Sahoo SA, Nunan D, Ioannidis JPA (2019) How does exercise treatment compare with antihypertensive medications? A network meta-analysis of 391 randomised controlled trials assessing exercise and medication effects on systolic blood pressure. Br J Sports Med 53(14):859–869. https://doi.org/10.1136/bjsports-2018-099921

Nishida Y, Hara M, Higaki Y, Taguchi N, Nakamura K, Nanri H, Horita M, Shimanoe C, Yasukata J, Miyoshi N, Yamada Y, Higashimoto K, Soejima H, Tanaka K (2019) Habitual light-intensity physical activity and ASC methylation in a middle-aged population. Int J Sports Med 40(10):670–677. https://doi.org/10.1055/a-0965-1374

Palou-Márquez G, Subirana I, Nonell L, Fernández-Sanlés A, Elosua R (2021) DNA methylation and gene expression integration in cardiovascular disease. Clin Epigenetics 13(1):75. https://doi.org/10.1186/s13148-021-01064-y

Paluch AE, Boyer WR, Franklin BA, Laddu D, Lobelo F, Lee DC, McDermott MM, Swift DL, Webel AR, Lane A (2024) Resistance exercise training in individuals with and without cardiovascular disease: 2023 update: a scientific statement from the American Heart Association. Circulation 149(3):e217–e231. https://doi.org/10.1161/cir.0000000000001189

Raitoharju E, Lyytikäinen LP, Levula M, Oksala N, Mennander A, Tarkka M, Klopp N, Illig T, Kähönen M, Karhunen PJ, Laaksonen R, Lehtimäki T (2011) miR-21, miR-210, miR-34a, and miR-146a/b are up-regulated in human atherosclerotic plaques in the Tampere vascular study. Atherosclerosis 219(1):211–217. https://doi.org/10.1016/j.atherosclerosis.2011.07.020

Richter EA, Hargreaves M (2013) Exercise, GLUT4, and skeletal muscle glucose uptake. Physiol Rev 93(3):993–1017. https://doi.org/10.1152/physrev.00038.2012

Roth GA, Mensah GA, Johnson CO, Addolorato G, Ammirati E, Baddour LM, Barengo NC, Beaton AZ, Benjamin EJ, Benziger CP, Bonny A, Brauer M, Brodmann M, Cahill TJ, Carapetis J, Catapano AL, Chugh SS, Cooper LT, Coresh J, Criqui M, DeCleene N, Eagle KA, Emmons-Bell S, Feigin VL, Fernández-Solà J, Fowkes G, Gakidou E, Grundy SM, He FJ, Howard G, Hu F, Inker L, Karthikeyan G, Kassebaum N, Koroshetz W, Lavie C, Lloyd-Jones D, Lu HS,

Mirijello A, Temesgen AM, Mokdad A, Moran AE, Muntner P, Narula J, Neal B, Ntsekhe M, Moraes de Oliveira G, Otto C, Owolabi M, Pratt M, Rajagopalan S, Reitsma M, Ribeiro ALP, Rigotti N, Rodgers A, Sable C, Shakil S, Sliwa-Hahnle K, Stark B, Sundström J, Timpel P, Tleyjeh IM, Valgimigli M, Vos T, Whelton PK, Yacoub M, Zuhlke L, Murray C, Fuster V (2020) Global burden of cardiovascular diseases and risk factors, 1990-2019: update from the GBD 2019 study. J Am Coll Cardiol 76(25):2982–3021. https://doi.org/10.1016/j.jacc.2020.11.010

Rowlands DS, Page RA, Sukala WR, Giri M, Ghimbovschi SD, Hayat I, Cheema BS, Lys I, Leikis M, Sheard PW, Wakefield SJ, Breier B, Hathout Y, Brown K, Marathi R, Orkunoglu-Suer FE, Devaney JM, Leiken B, Many G, Krebs J, Hopkins WG, Hoffman EP (2014) Multi-omic integrated networks connect DNA methylation and miRNA with skeletal muscle plasticity to chronic exercise in type 2 diabetic obesity. Physiol Genomics 46(20):747–765. https://doi.org/10.1152/physiolgenomics.00024.2014

Shan M, Li S, Zhang Y, Chen Y, Zhou Y, Shi L (2023) Maternal exercise upregulates the DNA methylation of Agtr1a to enhance vascular function in offspring of hypertensive rats. Hypertens Res 46(3):654–666. https://doi.org/10.1038/s41440-022-01124-7

Shan M, Qiu F, Li P, Zhang Y, Shi L (2024) Maternal exercise represses FGF21 via SIRT1 to improve the phenotypic transformation of vascular smooth muscle in hypertensive offspring. Hypertens Res. https://doi.org/10.1038/s41440-024-01991-2

Shi J, Bei Y, Kong X, Liu X, Lei Z, Xu T, Wang H, Xuan Q, Chen P, Xu J, Che L, Liu H, Zhong J, Sluijter JP, Li X, Rosenzweig A, Xiao J (2017) miR-17-3p contributes to exercise-induced cardiac growth and protects against myocardial ischemia-reperfusion injury. Theranostics 7(3):664–676. https://doi.org/10.7150/thno.15162

Shi Y, Zhang H, Huang S, Yin L, Wang F, Luo P, Huang H (2022) Epigenetic regulation in cardiovascular disease: mechanisms and advances in clinical trials. Signal Transduct Target Ther 7(1):200. https://doi.org/10.1038/s41392-022-01055-2

Soci UP, Fernandes T, Hashimoto NY, Mota GF, Amadeu MA, Rosa KT, Irigoyen MC, Phillips MI, Oliveira EM (2011) MicroRNAs 29 are involved in the improvement of ventricular compliance promoted by aerobic exercise training in rats. Physiol Genomics 43(11):665–673. https://doi.org/10.1152/physiolgenomics.00145.2010

Spólnicka M, Pośpiech E, Adamczyk JG, Freire-Aradas A, Pepłońska B, Zbieć-Piekarska R, Makowska Ż, Pięta A, Lareu MV, Phillips C, Płoski R, Żekanowski C, Branicki W (2018) Modified aging of elite athletes revealed by analysis of epigenetic age markers. Aging (Albany NY) 10(2):241–252. https://doi.org/10.18632/aging.101385

Sun J, Qiao Y, Zhao M, Magnussen CG, Xi B (2023) Global, regional, and national burden of cardiovascular diseases in youths and young adults aged 15-39 years in 204 countries/territories, 1990-2019: a systematic analysis of global burden of disease study 2019. BMC Med 21(1):222. https://doi.org/10.1186/s12916-023-02925-4

Wang L, Wang J, Yu P, Feng J, Xu GE, Zhao X, Wang T, Lehmann HI, Li G, Sluijter JPG, Xiao J (2022) METTL14 is required for exercise-induced cardiac hypertrophy and protects against myocardial ischemia-reperfusion injury. Nat Commun 13(1):6762. https://doi.org/10.1038/s41467-022-34434-y

Wang L, Feng J, Feng X, Meng D, Zhao X, Wang J, Yu P, Xu GE, Hu M, Wang T, Lehmann HI, Li G, Sluijter JPG, Xiao J (2023) Exercise-induced circular RNA circUtrn is required for cardiac physiological hypertrophy and prevents myocardial ischaemia-reperfusion injury. Cardiovasc Res 119(16):2638–2652. https://doi.org/10.1093/cvr/cvad161

Wu X, Wang L, Wang K, Li J, Chen R, Wu X, Ni G, Liu C, Das S, Sluijter JPG, Li X, Xiao J (2022) ADAR2 increases in exercised heart and protects against myocardial infarction and doxorubicin-induced cardiotoxicity. Mol Ther 30(1):400–414. https://doi.org/10.1016/j.ymthe.2021.07.004

Xiong W, Huang XL, Li CY, Wang XL, Lan HW, Wang TT, Chen ZL, Yang QP, Hu AL, Xia YF, Zhu FR, Zhou ZZ (2022) Astragaloside IV – mediated endothelial progenitor cell exosomes promote autophagy and inhibit apoptosis in hyperglycemic damaged endothelial cells via miR-21/PTEN axis. Folia Histochem Cytobiol 60(4):323–334. https://doi.org/10.5603/FHC.a2022.0030

Yang Q, Chen S, Wang X, Yang X, Chen L, Huang T, Zheng Y, Zheng X, Wu X, Sun Y, Wu J (2023) Exercise mitigates endothelial Pyroptosis and atherosclerosis by downregulating NEAT1 through N6-Methyladenosine modifications. Arterioscler Thromb Vasc Biol 43(6):910–926. https://doi.org/10.1161/atvbaha.123.319251

Zaldivar F, Wang-Rodriguez J, Nemet D, Schwindt C, Galassetti P, Mills PJ, Wilson LD, Cooper DM (2006) Constitutive pro- and anti-inflammatory cytokine and growth factor response to exercise in leukocytes. J Appl Physiol 100(4):1124–1133. https://doi.org/10.1152/japplphysiol.00562.2005

Zhang Y, Chen Y, Xu Z, Wu Y, Zhang Y, Shi L (2020) Chronic exercise mediates epigenetic suppression of L-type Ca2+ channel and BKCa channel in mesenteric arteries of hypertensive rats. J Hypertens 38(9):1763–1776. https://doi.org/10.1097/hjh.0000000000002457

Zhang Y, Shan M, Ding X, Sun H, Qiu F, Shi L (2023) Maternal exercise represses Nox4 via SIRT1 to prevent vascular oxidative stress and endothelial dysfunction in SHR offspring. Front Endocrinol (Lausanne) 14:1219194. https://doi.org/10.3389/fendo.2023.1219194

Zhang N, Wang X, Feng M, Li M, Wang J, Yang H, He S, Xia Z, Shang L, Jiang X, Sun M, Wu Y, Ren C, Zhang X, Li J, Gao F (2024) Early-life exercise induces immunometabolic epigenetic modification enhancing anti-inflammatory immunity in middle-aged male mice. Nat Commun 15(1):3103. https://doi.org/10.1038/s41467-024-47458-3

Zhao H, Chen X, Hu G, Li C, Guo L, Zhang L, Sun F, Xia Y, Yan W, Cui Z, Guo Y, Guo X, Huang C, Fan M, Wang S, Zhang F, Tao L (2022) Small extracellular vesicles from Brown adipose tissue mediate exercise Cardioprotection. Circ Res 130(10):1490–1506. https://doi.org/10.1161/circresaha.121.320458

Chapter 5
Exercise, Epigenetics, and Mental Health: Unveiling the Connection

Xinyu Liao, Ziang Zhang, and Xuyun Liu

Abstract Physical exercise has long been recognized for its numerous benefits to physical health, yet its profound effects on mental health are equally significant. This chapter delves into the intricate relationship between exercise, mental health, and epigenetics, highlighting how recent advances in epigenetics have started to uncover the molecular mechanisms through which exercise enhances mental well-being. This chapter reviews evidence demonstrating that regular physical activity can alleviate symptoms of depression and anxiety, enhance stress resilience, improve cognitive function, and promote better sleep quality. Moreover, it explores the epigenetic molecular mechanisms underlying these mental health benefits, such as histone modifications, DNA methylation, and microRNA regulation. These mechanisms play a crucial role in mediating the positive effects of exercise on mental health by influencing gene expression without altering the DNA sequence. By elucidating these epigenetic changes, the work aims to establish a robust theoretical foundation for future research and to suggest potential directions for developing more effective and personalized interventions for mental health disorders. Ultimately, this exploration of the interplay between exercise, mental health, and epigenetics underscores the importance of continued scientific inquiry in this area. It positions epigenetics as a critical field with significant implications for advancing both prevention and treatment strategies in mental healthcare, paving the way for innovative approaches to enhancing mental well-being through physical activity.

Keywords Exercise · Epigenetics · Mental health · Histone modifications · DNA methylation · MicroRNA regulation

X. Liao · X. Liu (✉)
Center for Mitochondrial Biology and Medicine, The Key Laboratory of Biomedical Information Engineering of Ministry of Education, School of Life Science and Technology, Xi'an Jiaotong University, Xi'an, China
e-mail: xuyunliu@xjtu.edu.cn

Z. Zhang
Key Laboratory of Aerospace Medicine of the Ministry of Education, School of Aerospace Medicine, Fourth Military Medical University, Xi'an, China

F. Gao, X. Zhang (eds.), *Exercise, Epigenetics and Human Health*, Epigenetics and Human Health 15, https://doi.org/10.1007/978-3-032-08045-5_5

Abbreviations

PTSD	Post-traumatic stress disorder
5-HT(1B)	5-Hydroxytryptamine receptor 1B
5-HTT	5-Hydroxytryptamine transporter
5-HT	5-Hydroxytryptamine
PD	Parkinson's disease
PSQI	Pittsburgh Sleep Quality Index
TET1	Ten-eleven translocation 1
AD	Alzheimer's disease
HDAC3	Histone deacetylase 3

5.1 Introduction

Physical exercise has long been recognized for its numerous benefits to physical health, but its profound effects on mental health are equally noteworthy. Despite extensive research into the benefits of exercise on mental health, our understanding of the underlying mechanisms remains limited. However, insights into the mechanisms by which exercise promotes mental health have significant implications for the application and translation of exercise as a therapeutic tool. Recent advances in the field of epigenetics have begun to shed light on how exercise influences mental well-being at the molecular level, including the promotion of neurotransmitter and neurotrophic factor release, reduction of inflammatory responses, modulation of the microbiome-gut-brain axis, and especially changes in epigenetic regulation.

This chapter aims to summarize and elucidate the relationship between exercise, mental health, and epigenetic mechanisms, with a particular emphasis on the epigenetic mechanisms that underlie the alleviation of mental health disorders through exercise. By exploring these connections, this chapter seeks to establish a theoretical foundation and suggest potential directions for future research into the mechanisms by which exercise benefits mental health. The integration of epigenetic perspectives into the study of exercise and mental health not only enriches our understanding of the biological processes involved but also opens new avenues for developing more effective and personalized interventions for mental health disorders.

In essence, this chapter highlights the importance of continued investigation into the interplay between exercise, mental health, and epigenetics, positioning it as a critical area of scientific inquiry that holds promise for advancing both prevention and treatment strategies in mental healthcare. Through rigorous exploration of these relationships, we can better leverage the therapeutic potential of physical activity to promote mental well-being and contribute to the development of novel treatments that address the complex nature of mental health disorders.

5.2 Mental Health Benefits of Exercise

5.2.1 *Stress Reduction*

Regular exercise supports the body's stress response system. A large meta-analysis involving over 1000 participants examined various mental health disorders, including depression, post-traumatic stress disorder (PTSD), schizophrenia, anxiety disorders, alcohol dependence, and bipolar disorder. The results indicated that yoga is more effective than conventional treatments and attention control in alleviating depressive symptoms. Furthermore, a higher frequency of weekly yoga practice correlates with a more pronounced reduction in depression symptoms (Brinsley et al. 2021). A brief yoga-based lifestyle intervention can lead to significant reductions in stress (as indicated by lower plasma cortisol and β-endorphin levels) and inflammation (as indicated by interleukin-6 and tumor necrosis factor-α) from Day 0 to Day 10 of the intervention (Yadav et al. 2012). In a study involving six experienced runners, plasma levels of β-endorphin/β-lipotropin were assessed before and after 30 min of treadmill running at self-selected paces of 60% and 80% VO_2 max, along with a control experiment consisting of 30 min of rest. The findings demonstrated that treadmill running promotes the release of β-endorphin and β-lipotropin, hormones commonly referred to as "feel-good" hormones (Farrell et al. 1982). This hormonal change induces a sense of relaxation and calmness, facilitating the individual's ability to manage daily stressors. Engaging in brisk jogging or long walks can significantly lower levels of stress hormones such as cortisol in the body (Carneiro et al. 2017; Zhao et al. 2024). Research indicates that wheel running produces a rapid and lasting reduction in 5-Hydroxytryptamine Receptor 1B (5-HT(1B)) mRNA in the ventral dorsal area. Additionally, three weeks of wheel running leads to decreased 5-Hydroxytryptamine Transporter (5-HTT) mRNA in both the dorsal and median raphe nuclei. After 6 weeks of wheel running, 5-HTT mRNA levels remain reduced, suggesting increased 5-Hydroxytryptamine (5-HT) levels, which can enhance mood and alleviate stress (Greenwood et al. 2005).

5.2.2 *Mood Improvement*

Individuals who engage in regular exercise are more likely to experience an enhanced sense of well-being. This improvement can be attributed to the fact that exercise not only influences the body's biochemical processes but also fosters a sense of accomplishment. A systematic review of randomized controlled trials examining the effects of physical exercise on body esteem among females will be conducted using databases such as PubMed, Embase, Web of Science, Cochrane Library, and Scopus. The search will encompass all available studies from the inception of these databases up to June 20, 2024. Meta-analysis will be performed using Stata 17.0 software, and preliminary findings suggest that physical exercise has a significant positive effect on enhancing body esteem in females (Zhang et al. 2024). In a human study involving 39 previously sedentary healthy participants, subjects were randomly assigned to one of two groups: a control group maintaining their sedentary habits and an experimental group engaging in a standardized progressive aerobic exercise program for 16 weeks. Psychological health assessments, including emotional state evaluations and magnetic resonance imaging of hippocampal subfields, were conducted at baseline and every 4 weeks during the intervention. Results indicated that aerobic exercise significantly improved mood and increased hippocampal subfield volumes compared to the control group (Frodl et al. 2020). Additionally, a pilot study involving 11 older adults assessed the effects of 6 min of self-paced walking, monitoring fatigue, tension, depression, confusion, mental and physical energy, and motivation to perform cognitive tasks over a three-day period. Findings revealed that 6 min of self-paced walking significantly improved the mood of older adults, providing scientific support for physical therapists in prescribing exercise to this population (Boolani et al. 2021).

5.2.3 *Cognitive Function Enhancement*

Research involving both humans and non-human animals has demonstrated that aerobic exercise can enhance various aspects of cognition and performance. The decline in physical activity, particularly among children in developed countries, is a significant contributor to the rising rates of obesity. Engaging in regular exercise may not only improve physical health but also enhance academic performance (Hillman et al. 2008). Moreover, exercise has been shown to protect against age-related cognitive decline. A study involving 60 patients with Parkinson's Disease (PD) utilized questionnaires to assess health-related quality of life. The findings revealed that individuals with PD who engaged in physical activity prior to their diagnosis reported higher scores in quality of life, suggesting that exercise can indeed enhance cognitive function (Koutsouras et al. 2020). Physical exercise has proven effective in mitigating cognitive deficits in older adults and neurological patients with glioma. In a study, 34 clinically stable patients diagnosed with World Health Organization grades II/III glioma were randomized into either a home-based, remotely coached exercise group or an active control group. Participants in the exercise group engaged in moderate to vigorous physical activity three times per week for 20–45 min over a duration of 6 months. Cognitive performance and patient-reported outcomes were evaluated at baseline and immediately following the intervention through neuropsychological assessments and questionnaires. The exercise group exhibited small to medium improvements in self-reported cognitive symptoms, fatigue, sleep quality, mood, and mental health-related quality of life. This exploratory randomized controlled trial in glioma patients provides preliminary evidence supporting the potential for aerobic exercise to enhance cognitive functioning and patient-reported outcomes, thereby justifying the need for larger exercise trials in brain tumor populations (Gehring et al. 2020).

5.2.4 *Sleep Quality Improvement*

Good mental health is often associated with proper sleep, and exercise can help with that. For example, those who suffer from insomnia may find that regular exercise during the day helps them fall asleep faster and stay asleep longer. A comprehensive literature search was conducted from 2004 to October 2024 across multiple databases, including PubMed, Embase, Web of Science, Google Scholar, Cochrane Library, and Chinese databases (CNKI and Wanfang), to identify randomized controlled trials published in both Chinese and English that investigate the effects of Tai Chi on sleep quality in older adults. A total of nine randomized controlled trials, comprising 1166 older participants, were included in the analysis. The results indicated that, compared to the control group, Tai Chi significantly reduced overall scores on the Pittsburgh Sleep Quality Index (PSQI) among older adults. Specifically, a total weekly duration of Tai Chi practice between 150 and 300 min was associated with a significant decrease in PSQI scores. Furthermore, durations of Tai Chi exercise programs ranging from 8 to 12 weeks and 16 to 24 weeks were both effective in reducing PSQI scores significantly. Additionally, single session durations of Tai Chi practice lasting 25–45 min and 60 min also led to significant reductions in PSQI scores. These findings suggest that older individuals engaging in at least 150 min of Tai Chi practice per week can experience improved sleep quality (Lei et al. 2024). Additionally, a systematic search of MEDLINE, Embase, the Cochrane Library, and Web of Science was conducted from the initiation of database indexing until March 2024 to identify randomized controlled trials examining the effects of yoga on specific outcomes in breast cancer survivors. Paired and dose-response meta-analyses were conducted using Bayesian hierarchical random-effects models to analyze the gathered data. A total of 2423 breast cancer survivors were included across 32 studies, with 1252 participants in the yoga intervention group and 1171 in the control group. The results demonstrated that the yoga intervention significantly improved fatigue, sleep quality, and quality of life among breast cancer survivors (Liu et al. 2024).

In summary, exercise has proven to be a powerful means of enhancing mental well-being, offering advantages in various aspects such as stress relief, mood elevation, cognitive enhancement, and improved sleep quality (Fig. 5.1).

5.3 Epigenetic Mechanisms of Exercise Benefiting Mental Health

5.3.1 *Histone Modifications*

Exercise has been demonstrated in rodent studies to enhance synaptic integrity and neuroplasticity in the brain through environmental enrichment that incorporates voluntary physical activity, thereby improving memory, learning, and stress

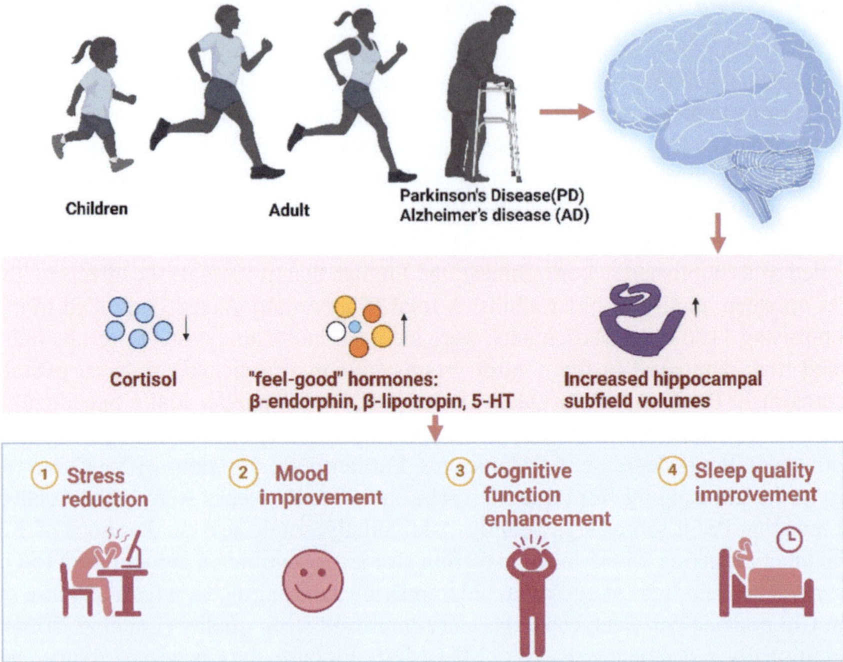

Fig. 5.1 Mental health benefits of exercise. Exercise has established itself as a potent tool for promoting mental health, providing benefits across multiple domains, including stress reduction, mood enhancement, cognitive function improvement, and better sleep quality. This is primarily achieved through the inhibition of cortisol production, the release of "feel-good" hormones such as β-endorphin, β-lipotropin, and serotonin (5-HT), as well as the increased volumes of hippocampal subfields

responses. These beneficial outcomes have been correlated with increased histone-tail acetylation in the hippocampus and cerebral cortex. Furthermore, the elevation of histone acetylation facilitated by inhibitors of histone deacetylases has been shown to promote dendritic sprouting, augment the number of synapses, and restore learning behaviors alongside access to long-term memories. These findings suggest that the inhibition of histone deacetylases may represent a promising therapeutic strategy for neurodegenerative diseases characterized by learning and memory impairments and also raises the potential for rehabilitating long-term memories in patients with dementia (Fischer et al. 2007).

Research conducted by the Reul team in the UK has revealed that regular physical exercise induces epigenetic modifications in the dentate gyrus, which may regulate gene expression involved in neuroplastic and cognitive responses to stressful events. Their findings indicate that exercised rats exhibit lower levels of anxiety than sedentary rats when exposed to a novel environment. After 30 min of exposure, the exercised rats tended to lie down and sleep, whereas the sedentary controls engaged in continuous exploratory behavior indicative of an anxious and vigilant state. In a forced swim test, both sedentary and exercised rats displayed indistinguishable behavior, such as struggling and swimming. However, in the forced swim re-test, the exercised rats exhibited significantly greater immobility behavior, which appears to be an adaptive response that may enhance survival. These behavioral changes in response to exercise were found to correlate with alterations in histone H3 acetylation at lysine 14 (H3K14Ac) and phosphorylation at serine 10 (H3S10ph), which are implicated in the immobility response (Chandramohan et al. 2008; Collins et al. 2009).

5.3.2 DNA Methylation

The most widely recognized epigenetic modification induced by physical exercise is DNA methylation. Research indicates that the effects of exercise are highly contingent upon its characteristics, including type, intensity, duration, and the specific tissue or cell type involved. Furthermore, it has been demonstrated that regular exercise promotes the hypomethylation of CpG sites in the promoter region of the Brain-Derived Neurotrophic Factor (BDNF), leading to a sustained increase in BDNF expression. This enhancement, in turn, improves hippocampal synaptic plasticity, learning, and memory (Gomez-Pinilla et al. 2011).

Additionally, one study revealed that physical exercise, specifically swimming, resulted in a decreased expression of the DNA methyltransferase 1 (Dnmt1) gene in the hippocampus and hypothalamus of animals engaged in physical activity, which consequently led to a significant increase in methylation within the hypothalamus of these animals (Kashimoto et al. 2016).

In another investigation, the authors employed whole-genome bisulfite sequencing at single-base resolution to determine that each exercise model induces specific alterations in the epigenome, primarily characterized by a loss of DNA methylation. Exercise is preferentially associated with DNA hypomethylation across various training models. Notably, among all exercise models, there was significant enrichment of CpG sites exhibiting a loss of DNA methylation at the binding sites of the demethylase ten-eleven translocation 1 (TET1). This finding indicates that the identified global DNA hypomethylation could be, in part, influenced by this particular epigenetic transcription factor. Additionally, alterations in DNA methylation at the binding sites of TET1 exhibited the strongest correlation with changes in gene expression across all exercise models. This suggests that, in conjunction with global DNA hypomethylation, this epigenetic transcription factor may be critical in modulating the expression of various target genes (Urdinguio et al. 2021).

Intriguingly, exercise appears to produce intergenerational epigenetic effects, as evidenced by research indicating that maternal pregestational treadmill exercise can influence the hippocampal development of male offspring. This influence occurs through the modulation of their epigenetic mechanisms, resulting in enhanced cognitive function and neuroplasticity. Notably, pregestational exercise alone significantly decreased global DNA methylation levels in the hippocampus of the offspring.

Consequently, this study provides pioneering evidence that maternal pregestational exercise impacts the epigenome of the progeny's somatic cells, contributing to mechanisms of environmental inheritance (Segabinazi et al. 2019). Moreover, it is crucial to investigate the modulation of the methylation status of specific genes that play significant roles in synaptic plasticity and memory, including activity-regulated cytoskeleton-associated protein, synaptophysin, synapsin-1, and postsynaptic density protein 95, as a result of maternal exercise conducted at any point during gestation.

5.3.3 *MicroRNAs*

MicroRNAs (miRNAs) represent an additional mechanism of epigenetic regulation by inhibiting or silencing protein translation (Gong 2017). A study involving healthy young men (aged 19–30), who engaged in ten 2-min bouts of cycle ergometer exercise with 1-min rest intervals, demonstrated that exercise significantly altered the expression of 38 miRNAs in neutrophils (Radom-Aizik et al. 2010). In addition, miRNAs play a crucial role in the benefits of exercise for ameliorating degenerative diseases, with an increasing body of research indicating that physical activity enhances mental health by modulating miRNA expression (Kuznetsova et al. 2020). One of the few genome-wide studies investigating miRNAs identified 34 exercise-responsive miRNAs following high-intensity intermittent swimming training (Zhao et al. 2019). In murine models, aerobic exercise significantly elevated the expression of miR-223 in the hippocampus and inhibited the downstream TLR4/MyD88-NF-κB signaling pathway. This modulation is posited to contribute to improved hippocampal function by reducing inflammatory responses (Qu et al. 2020). Furthermore, physical training has been shown to increase the levels of miR-129-5p in transgenic mice with Alzheimer's disease (AD) and in AD patients. This increase correlates with enhanced cognitive function and diminished inflammatory responses (Li et al. 2020).

In a recent study, 200 differentially expressed miRNAs were identified in exercising wild-type (WT) mice compared to their sedentary counterparts. Notably, miR-223-3p and miR-129-5p were upregulated in the exercising WT mice. The elevated expression of these miRNAs was associated with a decrease in calcium-binding protein levels, which may serve as a marker for the calcium-dependent anti-inflammatory effects of exercise on hippocampal function. This finding was further supported by KEGG pathway analysis, revealing several enriched pathways, including the calcium signaling pathway, Hippo signaling pathway, and Tumor Necrosis Factor (TNF) signaling pathway, in WT mice postexercise (Kuznetsova et al. 2024).

Additional studies have demonstrated that exercise upregulates miR-132 in cultured neurons treated with beta-amyloid oligomers, resulting in the inhibition of the histone deacetylase 3 (HDAC3) signaling pathway. Consequently, both artificial

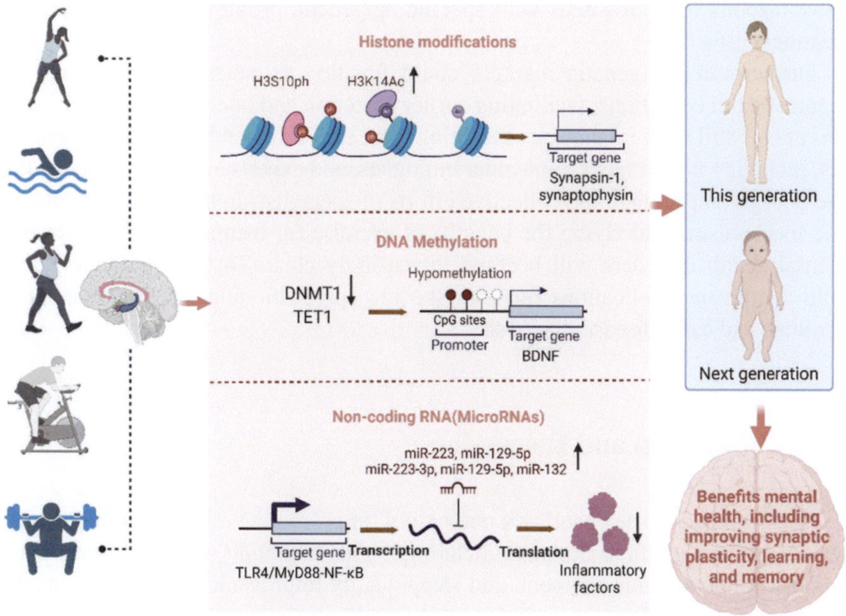

Fig. 5.2 Epigenetic mechanisms underlying exercise-mediated mental health benefits. The epigenetic molecular mechanisms, including histone modifications, DNA methylation, and microRNA regulation, play a significant role in mediating the beneficial effects of exercise on mental health

overexpression of miR-132 and inhibition of HDAC3 in vivo yielded similar enhancements in synaptic plasticity measures (Wei et al. 2020).

Moreover, oxidative stress is recognized as a significant mechanism underlying neurodegenerative diseases. Research indicates that swimming mitigates d-galactose-induced brain aging by suppressing miR-34a-mediated autophagy impairment and abnormal mitochondrial dynamics, highlighting that the suppression of miR-34a expression is a core mechanism through which exercise counters brain aging (Kou et al. 2017).

To sum up, the epigenetic molecular mechanisms, such as histone modifications, DNA methylation and microRNA regulation, play a crucial part in mediating the positive impacts of exercise on mental health (Fig. 5.2).

5.4 Clinical Implications and Future Directions

Understanding the epigenetic basis of exercise's effects on mental health opens new avenues for therapeutic interventions. Tailoring exercise programs to maximize these epigenetic benefits could serve as a complementary approach in the treatment of mental health disorders. Personalized exercise regimens may be particularly advantageous for individuals with specific epigenetic profiles, thereby enhancing treatment efficacy.

Furthermore, epigenetic markers could function as potential biomarkers for mental health conditions, facilitating earlier detection and intervention. Research in this area is still in its infancy, necessitating increased collaboration among researchers, including neurologists, molecular biologists, and experts in exercise and health. We believe that through the collective efforts of specialists in this field, the epigenetic mechanisms underlying the benefits of exercise for mental health and various mental health disorders will become increasingly clear. This understanding will hold significant implications for both specific epigenetic-related molecular interventions and early disease diagnosis.

5.5 Conclusion and Discussion

Exercise has established itself as a potent tool for promoting mental health, offering benefits across multiple domains including stress reduction, mood improvement, cognitive function enhancement, and sleep quality improvement. The mechanisms underlying these effects are multifaceted, involving the promotion of neurotransmitter and neurotrophic factor release, reduction of inflammatory responses, modulation via the microbiome-gut-brain axis, and epigenetic regulation. Despite this promising evidence, several limitations in current research highlight areas requiring further investigation.

There is significant variability in individual responses to exercise interventions, influenced by factors such as age, gender, type of exercise, and individual psychological conditions. This variability underscores the need for personalized approaches to exercise prescription, recognizing that one-size-fits-all solutions may not be effective for all individuals. Moreover, while some neurobiological mechanisms have been identified, the specific pathways through which exercise exerts its effects on mental health remain incompletely understood. Further research is essential to unravel these complex interactions, deepening our theoretical knowledge and informing more effective intervention strategies.

Barriers to engaging in regular physical activity—such as socioeconomic status, physical limitations, and lack of motivation or support—can significantly impede individuals from benefiting from exercise's positive impact on mental health. Addressing these barriers requires a multidisciplinary approach that considers social determinants alongside individual factors. Strategies aimed at improving access and adherence could substantially enhance the overall impact of exercise on mental health outcomes. The lack of standardized exercise protocols in research studies complicates efforts to compare findings and draw definitive conclusions about the optimal parameters for exercise to improve mental health. Establishing consensus on key variables such as type, duration, and intensity of exercise is crucial for advancing the field. Standardization would facilitate more robust meta-analyses and systematic reviews, thereby strengthening the evidence base for exercise as a mental health intervention.

In conclusion, while exercise has demonstrated benefits for mental health across multiple dimensions, addressing these limitations through further research is essential to enhance the applicability of exercise as an effective intervention for mental health improvement. Investigating and understanding the epigenetic molecular mechanisms through which exercise benefits mental health could help overcome these challenges.

The interplay between exercise, mental health, and epigenetics represents a promising area of scientific inquiry with significant implications for our understanding of mental health disorders. By elucidating the molecular mechanisms through which exercise engenders positive mental health outcomes, we can better appreciate the importance of physical activity as a tool for prevention and intervention. Continued research in this field will not only advance our knowledge of epigenetic regulation but will also pave the way for more effective and targeted mental health treatments, helping individuals achieve improved mental well-being through exercise.

Compliance with Ethical Standards

Conflict of Interest All authors declare that he/she has no conflict of interest.

Ethical Approval This chapter is a review of previously published accounts, as such, no animal or human studies were performed.

Funding This work was supported by grants from the National Natural Science Foundation of China (32100917).

References

Boolani A, Sur S, Yang D, Avolio A, Goodwin A, Mondal S, Fulk G, Towler C, Smith ML (2021) Six minutes of physical activity improves mood in older adults: a pilot study. J Geriatr Phys Ther 44(1):18–24. https://doi.org/10.1519/Jpt.0000000000000233

Brinsley J, Schuch F, Lederman O, Girard D, Smout M, Immink MA, Stubbs B, Firth J, Davison K, Rosenbaum S (2021) Effects of yoga on depressive symptoms in people with mental disorders: a systematic review and meta-analysis. Br J Sports Med 55(17):992–1000. https://doi.org/10.1136/bjsports-2019-101242

Carneiro LSF, Mota MP, Vieira-Coelho MA, Alves RC, Fonseca AM, Vasconcelos-Raposo J (2017) Monoamines and cortisol as potential mediators of the relationship between exercise and depressive symptoms. Eur Arch Psychiatry Clin Neurosci 267(2):117–121. https://doi.org/10.1007/s00406-016-0719-0

Chandramohan Y, Droste SK, Arthur JS, Reul JM (2008) The forced swimming-induced behavioural immobility response involves histone H3 phospho-acetylation and c-Fos induction in dentate gyrus granule neurons via activation of the N-methyl-D-aspartate/extracellular signal-regulated kinase/mitogen- and stress-activated kinase signalling pathway. Eur J Neurosci 27(10):2701–2713. https://doi.org/10.1111/j.1460-9568.2008.06230.x

Collins A, Hill LE, Chandramohan Y, Whitcomb D, Droste SK, Reul JM (2009) Exercise improves cognitive responses to psychological stress through enhancement of epigenetic mechanisms and gene expression in the dentate gyrus. PLoS One 4(1):e4330. https://doi.org/10.1371/journal.pone.0004330

Farrell PA, Gates WK, Maksud MG, Morgan WP (1982) Increases in plasma beta-endorphin/beta-lipotropin immunoreactivity after treadmill running in humans. J Appl Physiol Respir Environ Exerc Physiol 52(5):1245–1249. https://doi.org/10.1152/jappl.1982.52.5.1245

Fischer A, Sananbenesi F, Wang XY, Dobbin M, Tsai LH (2007) Recovery of learning and memory is associated with chromatin remodelling. Nature 447(7141):178–1U2. https://doi.org/10.1038/nature05772

Frodl T, Strehl K, Carballedo A, Tozzi L, Doyle M, Amico F, Gormley J, Lavelle G, O'Keane V (2020) Aerobic exercise increases hippocampal subfield volumes in younger adults and prevents volume decline in the elderly. Brain Imaging Behav 14(5):1577–1587. https://doi.org/10.1007/s11682-019-00088-6

Gehring K, Stuiver MM, Visser E, Kloek C, van den Bent M, Hanse M, Tijssen C, Rutten GJ, Taphoorn MJB, Aaronson NK, Sitskoorn MM (2020) A pilot randomized controlled trial of exercise to improve cognitive performance in patients with stable glioma: a proof of concept. Neuro-Oncology 22(1):103–115. https://doi.org/10.1093/neuonc/noz178

Gomez-Pinilla F, Zhuang Y, Feng J, Ying Z, Fan G (2011) Exercise impacts brain-derived neurotrophic factor plasticity by engaging mechanisms of epigenetic regulation. Eur J Neurosci 33(3):383–390. https://doi.org/10.1111/j.1460-9568.2010.07508.x

Gong H (2017) Chapter 8—The epigenetics of brain aging and psychiatric disorders A2—Yasui, Dag H. Available at: https://www.sciencedirect.com/science/article/abs/pii/B9780128002261000083#:~:text=In%20this%20chapter%2C%20we%20attempt%20to%20review%20the,on%20brain%20aging%20as%20well%20as%20psychiatric%20disorders

Greenwood BN, Foley TE, Day HE, Burhans D, Brooks L, Campeau S, Fleshner M (2005) Wheel running alters serotonin (5-HT) transporter, 5-HT1A, 5-HT1B, and alpha 1b-adrenergic receptor mRNA in the rat raphe nuclei. Biol Psychiatry 57(5):559–568. https://doi.org/10.1016/j.biopsych.2004.11.025

Hillman CH, Erickson KI, Kramer AF (2008) Be smart, exercise your heart: exercise effects on brain and cognition. Nat Rev Neurosci 9(1):58–65. https://doi.org/10.1038/nrn2298

Kashimoto RK, Toffoli LV, Manfredo MHF, Volpini VL, Martins-Pinge MC, Pelosi GG, Gomes MV (2016) Physical exercise affects the epigenetic programming of rat brain and modulates the adaptive response evoked by repeated restraint stress. Behav Brain Res 296:286–289. https://doi.org/10.1016/j.bbr.2015.08.038

Kou X, Li J, Liu X, Chang J, Zhao Q, Jia S, Fan J, Chen N (2017) Swimming attenuates d-galactose-induced brain aging via suppressing miR-34a-mediated autophagy impairment and abnormal mitochondrial dynamics. J Appl Physiol 122(6):1462–1469

Koutsouras GW, Levine K, Duroseau N, Ciraco C, Chan V, Pergament K, Chan T, Mancini JD, Leder AN, Krishnamachari B (2020) Effects of depression and exercise on health-related quality of life in patients with Parkinson's disease. Chronic Illn 16(3):190–200. https://doi.org/10.1177/1742395318796166

Kuznetsova M, Wilson C, Hannan AJ, Renoir T (2020) How the enriched get richer? Experience-dependent modulation of microRNAs and the therapeutic effects of environmental enrichment. Pharmacol Biochem Behav 195:172940. https://doi.org/10.1016/j.pbb.2020.172940

Kuznetsova M, Wilson C, Cheng L, Pang T, Li S, Roberts BR, Lago LC, Tran H, Hill AF, Hannan AJ, Renoir T (2024) Serotonergic-dependent effects of exercise and elevated stress hormone on small non-coding RNA transcriptomics and proteomics in a mouse model of affective disorders. Neuropharmacology 265:110240. https://doi.org/10.1016/j.neuropharm.2024.110240

Lei L, Chang S, Cheng L (2024) At least 150 min per week of tai chi practice improves sleep quality in the older people: evidence from a meta-analysis. Eur Geriatr Med. https://doi.org/10.1007/s41999-024-01125-4

Li Z, Chen Q, Liu J, Du Y (2020) Physical exercise ameliorates the cognitive function and attenuates the Neuroinflammation of Alzheimer's disease via miR-129-5p. Dement Geriatr Cogn Disord 49(2):163–169. https://doi.org/10.1159/000507285

Liu W, Li W, Wu B, Chen H, Zhao X, Wangwan H (2024) Optimal yoga dose for improving cancer-related fatigue, quality of life, and sleep quality in breast cancer survivors: a Bayesian dose-response meta-analysis. J Cancer Surviv. https://doi.org/10.1007/s11764-024-01699-9

Qu H, Liu R, Chen J, Zheng L, Chen R (2020) Aerobic exercise inhibits CUMS-depressed mice hippocampal inflammatory response via activating hippocampal miR-223/TLR4/MyD88-NF-kappaB pathway. Int J Environ Res Public Health 17(8). https://doi.org/10.3390/ijerph17082676

Radom-Aizik S, Zaldivar F, Oliver S, Galassetti P, Cooper DM (2010) Evidence for microRNA involvement in exercise-associated neutrophil gene expression changes. J Appl Physiol 109(1):252–261. https://doi.org/10.1152/japplphysiol.01291.2009

Segabinazi E, Spindler C, Meireles ALF, Piazza FV, Mega F, Salvalaggio GDS, Achaval M, Marcuzzo S (2019) Effects of maternal physical exercise on global DNA methylation and hippocampal plasticity of rat male offspring. Neuroscience 418:218–230. https://doi.org/10.1016/j.neuroscience.2019.08.036

Urdinguio RG, Tejedor JR, Fernandez-Sanjurjo M, Perez RF, Penarroya A, Ferrero C, Codina-Martinez H, Diez-Planelles C, Pinto-Hernandez P, Castilla-Silgado J, Coto-Vilcapoma A, Diez-Robles S, Blanco-Agudin N, Tomas-Zapico C, Iglesias-Gutierrez E, Fernandez-Garcia B, Fernandez AF, Fraga MF (2021) Physical exercise shapes the mouse brain epigenome. Mol Metab 54:101398. https://doi.org/10.1016/j.molmet.2021.101398

Wei Z, Meng X, El Fatimy R, Sun B, Mai D, Zhang J, Arora R, Zeng A, Xu P, Qu S, Krichevsky AM, Selkoe DJ, Li S (2020) Environmental enrichment prevents Abeta oligomer-induced synaptic dysfunction through mirna-132 and hdac3 signaling pathways. Neurobiol Dis 134:104617. https://doi.org/10.1016/j.nbd.2019.104617

Yadav RK, Magan D, Mehta N, Sharma R, Mahapatra SC (2012) Efficacy of a short-term yoga-based lifestyle intervention in reducing stress and inflammation: preliminary results. J Altern Complement Med 18(7):662–667. https://doi.org/10.1089/acm.2011.0265

Zhang Y, Wen ZH, Zhu YB, Guan GF (2024) Effects of physical exercise on body esteem among females: a meta-analysis. BMC Public Health 24(1):ARTN 3387 10.1186/s12889-024-20861-7

Zhao Y, Zhang A, Wang Y, Hu S, Zhang R, Qian S (2019) Genome-wide identification of brain miRNAs in response to high-intensity intermittent swimming training in Rattus norvegicus by deep sequencing. BMC Mol Biol 20(1):3. https://doi.org/10.1186/s12867-019-0120-4

Zhao YL, Sun SY, Qin HC, Zhu YL, Luo ZW, Qian Y, Chen S (2024) Research progress on the mechanism of exercise against depression. World J Psychiatry 14(11):1611–1617. https://doi.org/10.5498/wjp.v14.i11.1611

Chapter 6
Histone Acetylation/Deacetylation and Exercise: Epigenetic Mechanisms Underlying Skeletal Muscle Health and Aging

Shuang Hao, Yumeng Hu, Lining Zhang, and Jia Li

Abstract Skeletal muscle aging, often accompanied by sarcopenia and metabolic dysfunction, is closely associated with an imbalance in epigenetic regulation involving histone acetylation (HATs) and deacetylation (HDACs). HATs activate muscle metabolic genes (such as PGC-1α and GLUT4) by enhancing chromatin accessibility, thereby promoting mitochondrial biogenesis and glucose metabolism. Conversely, HDACs contribute to muscle atrophy by inhibiting transcription. Studies have revealed that endurance exercise activates HATs through the AMPK/CaMK signaling pathway, while resistance training enhances transcriptional activity by reducing HDAC4/5 levels. Both types of exercise can delay age-related muscle function decline. During the aging process, a decrease in histone acetylation levels coupled with an increase in HDACs activity suppresses the expression of muscle repair genes, exacerbating mitochondrial dysfunction. Targeted interventions, such as the use of HDAC inhibitors, have been shown to reverse pathological gene networks and promote regeneration in animal models. Future research should integrate single-cell epigenomics and multi-omics approaches to elucidate the mechanisms underlying exercise-induced chromatin remodeling, aiming to develop precision treatment strategies. This article systematically elucidates the central role of the HATs/HDACs balance in muscle health and aging, providing a theoretical basis for delaying muscle decline through exercise and epigenetic interventions.

Keywords Skeletal muscle · Aging · HATs · HDACs

S. Hao · Y. Hu · J. Li (✉)
School of Aerospace Medicine, Fourth Military Medical University, Xi'an, China
e-mail: jiali816@fmmu.edu.cn

L. Zhang
Department of Rehabilitation, First Medical Center, Chinese PLA General Hospital, Beijing, China

F. Gao, X. Zhang (eds.), *Exercise, Epigenetics and Human Health*, Epigenetics and Human Health 15, https://doi.org/10.1007/978-3-032-08045-5_6

Abbreviations

AMPK	Adenosine monophosphate-activated protein kinase
ATAC-seq	Assay for transposase-accessible chromatin using sequencing
CRE	Cyclic-AMP response element
CRISPR	Clustered regularly interspaced short palindromic repeats
dCas9	dead Cas9
DMD	Duchenne muscular dystrophy
EVs	Extracellular vesicles
FOXO	Forkhead Box O
GLUT4	Glucose transporter type 4
HATs	Histone acetyltransferases
HDACs	Histone deacetylases
IL-6	Interleukin-6
MEF2	Myocyte enhancer factor 2
NAD+	Nicotinamide adenine dinucleotide.
NAMPT	Nicotinamide phosphoribosyltransferase
OGG1	8-Oxoguanine DNA glycosylase
PGC-1α	Peroxisome proliferator-activated receptor gamma coactivator 1-alpha
PPARGC1A	Peroxisome proliferator-activated receptor gamma coactivator 1-alpha gene
qPCR	quantitative polymerase chain reaction
SIRT1–7	Sirtuins 1–7
TSA	Trichostatin A

6.1 Introduction

In mammals, skeletal muscle stands as the primary metabolic organ, constituting 40–60% of an animal's body weight. This tissue demonstrates remarkable capabilities in substrate oxidation and energy storage, exerting a pivotal influence on basal metabolic rate, systemic lipid homeostasis, and blood glucose regulation (Egan and Zierath 2013; Mallett 2025). Emerging evidence from multiple investigations highlights the involvement of histone acetylation modifications in various physiological mechanisms of skeletal muscle. Such mechanisms encompass the skeletal muscle cell cycle, myofiber phenotypic transition, muscle atrophy processes, exercise performance, endurance capacity, and muscle aging phenomena (Wang et al. 2022; Wu et al. 2024; Mallett 2025; Murach and Bagley 2025).

HATs catalyze histone acetylation to enhance chromatin accessibility for transcriptional activation, whereas HDACs remove acetyl groups to promote chromatin condensation and gene silencing (Meynial-Denis 2019). The balance between HATs and HDACs is crucial for maintaining the proper function and adaptation of skeletal muscle (Xu et al. 2023). Exercise has been recognized as a powerful modulator of epigenetic marks in skeletal muscle. It can influence histone acetylation and deacetylation, thereby affecting gene expression and muscle function (Mallett 2025). For example, endurance exercise increases HAT activity, promoting histone acetylation at the promoters of genes involved in muscle metabolism and contractile function, such as PGC-1α and GLUT4 (McGee and Hargreaves 2019), leading to enhanced mitochondrial biogenesis and improved insulin sensitivity. Resistance exercise regulates histone acetylation, influencing muscle fiber type conversion and muscle hypertrophy (Egan and Sharples 2023; Roberts et al. 2023). By modulating these epigenetic mechanisms, exercise helps maintain muscle health and delay age-related muscle decline.

This review aims to synthesize current knowledge on histone acetylation/deacetylation in skeletal muscle health, explore exercise-mediated epigenetic adaptations, and discuss therapeutic strategies targeting these pathways.

6.2 Histone Acetylation and Deacetylation: An Overview

Histone modifications, encompassing acetylation, methylation, phosphorylation, and ubiquitination, govern the interaction between histones and DNA, thereby influencing gene expression (Zhou et al. 2011). Histone acetylation, facilitated by histone acetyltransferases (HATs), adds acetyl groups to lysine residues on histones, neutralizing their positive charge. This reduces histone-DNA interaction, leading to a relaxed chromatin state that enhances transcription factor accessibility and gene transcription (Bannister and Kouzarides 2011). Conversely, histone deacetylation mediated by histone deacetylases (HDACs) removes acetyl groups, thereby reinstating the positive charge and giving rise to a compact chromatin configuration that represses gene transcription. These processes are vital for cellular functions such as differentiation, development, and environmental response (Kim et al. 2022). Recent research highlights the pivotal role of histone acetylation/deacetylation in skeletal muscle health and aging, particularly in response to exercise (Mallett 2025). Targeting histone-modifying enzymes and associated signaling pathways offers promising therapeutic potential for improving muscle health.

6.3 Epigenetic Regulation of Skeletal Muscle Health

The balance between HATs and HDACs is essential during myogenesis (Xu et al. 2023). HATs, such as CBP/p300 and PCAF/GCN5, acetylate lysine residues on histone tails, creating binding sites for bromodomain-containing proteins. This facilitates the recruitment and stabilization of transcriptional complexes at promoter regions, enhancing the activation of muscle-specific genes (Wapenaar and Dekker 2016). The MYST family, including MOF (MYST1), Tip60, and HBO1, also exhibits HAT activity, contributing to muscle cell differentiation and function (Xu et al. 2023). HBO1 acetylates histone H3, potentially regulating genes involved in muscle differentiation (Xiao et al. 2021). Tip60 interacts with MyoD to activate muscle-specific gene transcription and plays roles in DNA repair and cell cycle regulation, essential for myoblast proliferation and differentiation (Kim et al. 2011). Collectively, these HATs maintain a crucial balance of histone acetylation during myogenesis, vital for muscle health. In contrast, as a family of protein deacetylases, HDACs have been shown to modulate physiological homeostasis and developmental processes through deacetylation (Bannister and Kouzarides 2011; Ibrahim et al. 2025). Class I HDACs (HDAC1, 2, 3, and 8) are nuclear enzymes that prevent premature differentiation of myoblasts, maintaining their proliferation until differentiation signals are received (Beharry et al. 2014; Joung et al. 2018; Li et al. 2019). Class II HDACs (HDAC4, 5, 7, and 9) shuttle between the nucleus and cytoplasm, deacetylating MEF2 transcription factors to regulate the timing of muscle gene expression (Renzini et al. 2018; Li et al. 2019). Class IIB HDACs, particularly HDAC6, influence cytoskeletal dynamics and myoblast fusion into myotubes by deacetylating non-histone proteins like alpha-tubulin (Kumar and Datta 2022; Osseni et al. 2022). Together, these HDACs ensure proper myogenesis progression and muscle function maintenance.

The role of epigenetic regulation in myogenesis is also gaining attention. Recent studies have identified novel transcription factors, such as Bach2, MafK, and Fosl2, regulated by H3K27 acetylation in the myogenic differentiation of porcine skeletal muscle satellite cells (Zhou et al. 2024). These transcription factors play crucial roles in muscle development and regeneration. Additionally, the study by Chen and Li highlights the involvement of HDAC11 in immune response, metabolic pathways, and cell cycle regulation in skeletal muscle (Chen and Li 2024). The detailed mechanisms by which HDAC11 controls muscle-specific gene expression and myogenesis remain to be fully characterized, but these insights underscore the need for further research to elucidate the specific downstream targets and pathways involved.

Building on this foundation, exercise can increase the activity of HATs, which promotes histone acetylation at the promoters of specific genes, thereby upregulating the expression of genes involved in muscle metabolism and contractile function, such as PGC-1α and GLUT4 (McGee and Hargreaves 2019). Specifically, exercise-induced signaling pathways, such as AMPK (adenosine monophosphate-activated protein kinase) and CaMK (calcium/calmodulin-dependent protein kinase), can activate HATs like p300 and CBP (Smith et al. 2023). These HATs acetylate histones at the promoters of genes such as PGC-1α, promoting their expression and enhancing mitochondrial biogenesis and muscle oxidative metabolism, thus improving exercise endurance and muscle function (Egan et al. 2010). Additionally, HATs can acetylate histones at the GLUT4 gene promoter, increasing its transcription level and enhancing glucose uptake by muscle cells, improving insulin sensitivity, and supporting muscle energy metabolism and function (Niu et al. 2017).

Conversely, during muscle atrophy, the activity of HDACs may be upregulated, leading to histone deacetylation and the suppression of muscle-specific genes, resulting in the loss of muscle mass and function. For example, HDAC1 can activate the FOXO signaling pathway, inhibiting the expression of muscle-specific genes. FOXO transcription factors play a key role in muscle atrophy by upregulating genes such as atrogin-1, promoting muscle protein degradation and leading to muscle fiber atrophy and reduced muscle mass and strength (Beharry et al. 2014). Moreover, HDAC3 is also important in muscle energy metabolism, and its upregulation can lead to the deacetylation of histones, inhibiting the expression of genes related to muscle metabolism and function, such as PGC-1α, reducing mitochondrial biogenesis, and further exacerbating muscle atrophy and functional loss (Hong et al. 2017). Therefore, the balance between HATs and HDACs is crucial for maintaining muscle health and function, and its dysregulation can lead to muscle diseases. Garcinol and Anacardic Acid, two natural inhibitors of histone acetyltransferases, have been reported to suppress the growth and proliferation of RMS cells through targeting P300/CBP (Tomasiak et al. 2023). Additionally, studies indicate that glucocorticoids can enhance P300 expression in skeletal muscle while reducing HDAC6 activity, a process that contributes to the development of muscle atrophy (Alamdari et al. 2010).

6.4 Exercise as a Modulator of Histone Acetylation/Deacetylation

6.4.1 Endurance Exercise

A recent study by Hong and colleagues has shed light on the nexus between histone acetylation and exercise tolerance (Hong et al. 2017). By selectively depleting HDAC3 in mice, they uncovered a paradoxical phenomenon: despite the presence of severe insulin resistance and diminished muscle strength, these mice exhibited enhanced endurance and resistance to muscle fatigue. This was attributed to a metabolic shift from glucose to fat oxidation, underscoring the intricate relationship between epigenetic modifications and metabolic pathways in muscle physiology. Given the circadian regulation of HDAC3's genomic association, this study suggests that the timing of exercise may profoundly influence muscle adaptation and performance, a hypothesis that merits further exploration.

In contrast, Ohsawa et al. reported a decrease in pan-acetylated H3 levels in rat skeletal muscle following 4 weeks of regular treadmill running, with the plantaris muscle showing particularly notable changes (Ohsawa et al. 2018). Through ChIP analysis, they identified significant gene-specific distribution differences of this histone modification. Their findings indicate that exercise not only modulates histone modifications but also influences histone "variant switching," such as the replacement of specific histone H3 isoforms, which may be integral to skeletal muscle adaptation to training.

The transcriptional coactivator PGC-1α, pivotal for muscle adaptation to exercise (Smith et al. 2023; Feng et al. 2025; Naito et al. 2025; Posridee et al. 2025), recruits histone acetyltransferases (HATs) to specific transcription factors (McGee and Hargreaves 2019). The PPARGC1A gene, a key player in this process, has a promoter region that contains conserved sequences for myocyte enhancer factor 2 (MEF2) and cyclic-AMP response element (CRE), which regulate PGC-1α transcription in response to physiological stimuli (Egan et al. 2010). After 3 weeks of endurance training, PPARGC1A and VEGFA were upregulated, leading to a shift in the muscle transcriptome towards an endurance phenotype by Week 8 (Beiter et al. 2024). This observation points to a threshold system in muscle adaptation that balances the costs and benefits of muscle remodeling.

6.4.2 Resistance Exercise

Skeletal muscle exhibits a remarkably malleable phenotype, demonstrating dynamic adaptation to diverse exercise modalities. Resistance training elicits specific adaptive responses, including enhanced muscular strength, skeletal muscle hypertrophy, and metabolic resilience (Egan and Sharples 2023; Roberts et al. 2023). Notably, research by Gehlert and colleagues has illuminated the broad spectrum of skeletal muscle metabolic adaptations following acute unaccustomed resistance exercise and subsequent training periods, encompassing shifts in energy metabolism and tissue remodeling processes (Gehlert et al. 2022). Beyond the observed changes in the metabolome, such extensive metabolic alterations are intrinsically linked to underlying genetic changes.

In the epigenome, many studies have also found an enhancement of transcriptional activity markers after resistance exercise training, with histone acetylation being an important representative (Seaborne et al. 2018; Turner et al. 2019; Lim et al. 2020; Pilotto et al. 2025). This is supported by Lim's study, which showed an increase in H3 acetylation after acute RE (Lim et al. 2020). Actually, research by McGee et al. has previously demonstrated that the acetylation level of histone 3 at lysine 36 rises rapidly following an hour of cycling with a moderate to high intensity. This increase in acetylation correlates with a decrease in the nuclear presence of HDAC4 and HDAC5, with HDAC5 being notably targeted for ubiquitination as a result of the exercise regimen (McGee et al. 2009). While Mattingly et al. revealed that there was a significant decrease in the levels of HDAC6 protein in both the cytoplasm and the nucleus with one bout of resistance exercise. Additionally, the study discovered that resistance exercise led to a decrease in the overall acetylation levels of skeletal muscle proteins, which naturally warrants further investigation (Mattingly et al. 2024). Additionally, Ohsawa and colleagues revealed that exercise training accelerates histone turnover at transcriptionally active sites in rat and mouse skeletal muscle, while inhibiting the accumulation of acetylated histones during long-term training. Notably, their findings showed that although rat skeletal muscle exhibited reduced histone turnover, this led to the accumulation of acetylated histones within the tissue (Ohsawa et al. 2018; Ohsawa and Kawano 2021). These results align with Lim et al.'s proposal that limitations in histone turnover during repeated resistance exercise (RE) stimulation actually promote the accumulation of acetylated histones (Lim et al. 2020). The buildup of acetylated histones may disrupt nucleosome formation, maintain transcriptional activation at these loci, and enhance gene expression following RE training.

The positive role of resistance exercise in improving muscle atrophy can be partly attributed to the regulation of acetylation (Zheng et al. 2024; Li et al. 2025). Specifically, resistance exercise can regulate the activity of key enzymes in sugar metabolism, thereby promoting glucose metabolism in the skeletal muscles of T2D rats. In this process, resistance exercise further regulates Sirt1-mediated histone H4K16ac, thus promoting the proliferation of muscle satellite cells and ultimately achieving the improvement of muscle atrophy.

6.5 Histone Acetylation/Deacetylation in Muscle Aging

One of the key molecular alterations observed in aging muscle is the dysregulation of histone acetylation and deacetylation patterns (Alamdari et al. 2013). Histone acetylation, typically associated with active transcription, tends to decrease with age, leading to a more condensed chromatin structure and reduced gene expression of critical muscle maintenance and repair genes. Conversely, the activity of histone deacetylases (HDACs) often increases, further contributing to the repression of genes involved in muscle growth and regeneration (McGee-Lawrence et al. 2013; Wang et al. 2023). This shift in the balance of acetylation and deacetylation disrupts the transcriptional landscape necessary for maintaining muscle homeostasis and adaptability. Additionally, age-related changes in the expression and activity of specific histone acetyltransferases (HATs) and HDACs have been implicated in the impaired response to anabolic stimuli and reduced mitochondrial function observed in older muscle tissue (Wang et al. 2022). Understanding these epigenetic alterations provides insight into the mechanisms of muscle aging and highlights potential targets for therapeutic intervention to preserve muscle health in the elderly.

6.5.1 Histone Acetylation and Skeletal Muscle Aging

Histone acetyltransferases (HATs) play a significant role in the regulation of skeletal muscle senescence, involving multiple aspects such as gene expression, mitochondrial function, protein homeostasis, and apoptosis. Based on subcellular localization and substrate specificity, HATs are categorized into two types: A and B. Research has indicated that A-type HATs assume a pivotal role in gene expression regulation (Lee and Workman 2007; Parthun 2007). To the present, over 20 HAT types have been identified, classified into distinct families according to their structural domains. Prominent families include the P300/CBP family (such as P300 and CBP), the GNAT family (e.g., Gcn5 and PCAF), and the MYST family (e.g., Tip60, MOZ, MORF, HBO1).

P300 and CBP are essential histone acetyltransferases that regulate muscle cell survival (Kuninger et al. 2006), differentiation (Chen et al. 2000), contraction function (Svensson et al. 2020), and glucose uptake in response to insulin stimulation (LaBarge et al. 2015; Xiong et al. 2020; Martins et al. 2022). Aged skeletal muscle exhibits elevated levels of oxidative DNA damage markers such as 8-oxoG, alongside increased OGG1 activity. Notably, p300/CBP-mediated acetylation of OGG1 counteracts this damage by facilitating DNA repair processes (Radak et al. 2011; Meynial-Denis 2019). In models of muscle atrophy, elevated P300/CBP mRNA levels may inhibit Forkhead Box O (FOXO) transcription factors (Beharry and Judge 2015), potentially mitigating muscle atrophy due to FOXO's role in regulating this process (Senf et al. 2011). Elevated extracellular C1qbp in aged and disuse-induced muscle atrophy suppresses the NFATc1/p300 complex, reducing histone H3 acetylation and impairing myogenic differentiation (Kim et al. 2024a). These findings indicate that the P300/CBP family regulates skeletal muscle aging and atrophy through histone acetylation-mediated DNA repair and transcriptional regulation.

The GNAT family, including GCN5 and PCAF (p300/CBP-associated factor) (Vetting et al. 2005; Abbassi et al. 2020), regulates skeletal muscle metabolism and mitochondrial function (Favrot et al. 2016). GCN5 acetylates Polo-like kinase 4 (PLK4) and T-box protein 5 (TBX5), while PCAF acetylates TBX5, influencing centrosome duplication and muscle cell differentiation (Xia et al. 2020). GCN5 mediates acetylation of the transcription factor YY1, a negative regulator of muscle fibers, which disrupts the interaction between the zinc finger domain of the YY1 protein and DNA. This acetylation event promotes the upregulation of structurally essential proteins in muscle tissue, thereby preserving the integrity of skeletal muscle cells (Addicks et al. 2022). PCAF can compensate for GCN5 deficiency, indicating functional redundancy (Xu et al. 2023). The GNAT family is also involved in age-related histone modifications, particularly the acetylation of histone H3 and H4 N-terminal lysine residues (Yoshihara et al. 2019). Notably, histone H3 acetylation decreases in aged rat gastrocnemius muscle (Vetting et al. 2005). These findings indicate that the GNAT family regulates skeletal muscle metabolism, mitochondrial function, and cellular integrity by acetylating specific targets (e.g., PLK4, TBX5, YY1) and modulating age-related histone modifications.

The MYST family, including MOZ, HBO1, MOF, MORF, and TIP60 (Sapountzi and Côté 2011), influences cell senescence and muscle cell physiology. KAT6A (MOZ) and KAT6B (MORF) inhibit cell senescence (Baell et al. 2018), while MOZ is critical for p53-mediated cell cycle control (Rokudai et al. 2013). HBO1 may play a role in early skeletal muscle regeneration (Ochiai et al. 2016), and TIP60 promotes muscle cell differentiation (Kim et al. 2011). MOF enhances skeletal muscle stem cell self-renewal and differentiation through PAX7 acetylation (Thomas et al. 2008; Sincennes et al. 2021).

Collectively, these HAT families form an intricate epigenetic network that coordinates chromatin remodeling, gene expression, and stress adaptation in skeletal muscle, playing multifaceted roles in regulating skeletal muscle senescence, atrophy, and regeneration through their diverse enzymatic activities and substrate specificities.

6.5.2 Histone Deacetylation and Skeletal Muscle Aging

Histone deacetylases (HDACs) are categorized into four classes, each playing unique roles in skeletal muscle aging. Class I HDACs (HDAC1–3, HDAC8) modulate key pathways in muscle homeostasis, including atrophy signaling and metabolic adaptation (Meynial-Denis 2019). HDAC1 inhibition prevents muscle atrophy induced by nutritional deficiency (Walsh and Van Remmen 2016). HDAC1 regulates FOXO and the atrophy program by deacetylating FOXO3a, modulating the expression of atrogin-1, a key atrophy-related gene (Beharry et al. 2014). HDAC1 expression and activity increase with age, impairing muscle stem cell differentiation and regeneration (Kosan et al. 2018; Ying et al. 2024). HDAC2 collaborates with HDAC1 to regulate autophagy and homeostasis in skeletal muscle, modulating Fgf21 expression (Renzini et al. 2022). Reduced HDAC2 levels in COPD patients and aging muscle trigger apoptosis and dysfunction (To et al. 2017), while HDAC2 overexpression reverses cigarette smoke extract (CSE)-induced senescence (Li et al. 2021). HDAC3 regulates skeletal muscle metabolism. HDAC3 knockout mice exhibit increased lipid utilization, amino acid catabolism, and age-related muscle loss (Molinari et al. 2023). These mice also show reduced glucose uptake and enhanced exercise endurance. HDAC3 inhibition prevents muscle atrophy under nutritional stress (Razidlo et al. 2010; Walsh and Van Remmen 2016). HDAC8 regulates muscle differentiation via the canonical Wnt pathway. HDAC8 inhibition impairs differentiation in zebrafish embryos and C2C12 cells (Ferrari et al. 2019). In Duchenne muscular dystrophy (DMD), HDAC8 overexpression contributes to muscle degeneration, and its inhibition preserves muscle integrity (Spreafico et al. 2021; Fontana et al. 2022).

Class IIA HDACs, including HDAC4 and HDAC5, are involved in regulating muscle atrophy, mitochondrial function, and apoptosis. HDAC4 expression increases during muscle atrophy and aging. HDAC4 deacetylates MyoH, PGC-1α, and Hsc70, downregulating PGC-1α to disrupt mitochondrial function and lead to atrophy (Luo et al. 2019). HDAC4 inhibition under long-term denervation can be detrimental (Pigna et al. 2018), but HDAC4 inhibitors show promise in treating muscle atrophy (Luo et al. 2019). HDAC5 regulates muscle fiber apoptosis and ferroptosis, maintaining muscle mass during aging (Martin et al. 2023). HDAC5 also modulates IL-6 secretion and insulin signaling, enhancing glucose uptake and glycogenesis (Klymenko et al. 2020). During myoblast differentiation, HDAC5 translocates from the nucleus to the cytoplasm, promoting differentiation (McKinsey et al. 2000).

Class IIB HDACs, including HDAC6, are associated with muscle aging. HDAC6 expression correlates with muscle aging. HDAC6 inhibition improves muscle function and reduces atrophy by regulating stress protein misfolding and autophagolysosome fusion (Ratti et al. 2015; Sandonà et al. 2023). HDAC6 is upregulated during atrophy and interacts with MAFbx to degrade MyoD (Walsh et al. 2015). Class III HDACs, consisting of SIRT1–7, are involved in regulating mitochondrial function, metabolism, and DNA repair. SIRT1, an NAD + -dependent deacetylase, activates

PGC-1α, enhancing mitochondrial biogenesis and muscle endurance (Domi et al. 2021). SIRT1 expression declines with age, and its activation improves muscle function and delays aging (Williams and Gurd 2012; Myers et al. 2021). SIRT1 also regulates Akt signaling and DNA repair efficiency (LaBarge et al. 2015). SIRT3 expression decreases with age but increases with exercise, improving mitochondrial function and antioxidant capacity (Zhou et al. 2022). SIRT3 regulates metabolic pathways and mitochondrial ATP production, and its deficiency accelerates muscle aging (Lin et al. 2014; Robin et al. 2020). Class IV HDACs, including HDAC11, whose deficiency can activate the AMPK-ACC pathway to promote muscle regeneration, enhance fatty acid oxidation and muscle endurance (Cheng et al. 2014; Hurtado et al. 2021). HDAC11 deficiency inhibits MyoD-dependent repression of myoblast differentiation, accelerating regeneration (Lee et al. 2021).

In summary, HDACs play multifaceted and class-specific roles in skeletal muscle aging, influencing processes such as atrophy, metabolism, mitochondrial function, and regeneration. Future research should explore isoform-specific modulators and combinatorial approaches to optimize muscle health in aging and disease.

6.6 Exercise Interventions in Aging Muscle: Epigenetic Insights

6.6.1 Effects of Exercise on Histone Acetylation/Deacetylation

Exercise modulates histone acetylation/deacetylation, influencing muscle contraction and energy metabolism (Liang et al. 2022), thereby delaying skeletal muscle aging (Plaza-Diaz et al. 2022). For example, cycling increases H3K36 acetylation and reduces nuclear HDAC4/5 levels (McGee et al. 2009). Endurance training enhances SIRT3 expression, while resistance training increases SIRT3 in elderly muscle (Zhou et al. 2022). Exercise also activates AMPK, phosphorylating Class IIa HDACs and promoting MEF2-dependent transcription (Smith et al. 2008; McGee and Walder 2017).

6.6.2 Exercise Reverses Age-Related Epigenetic Changes

Exercise delays aging by regulating gene expression and epigenetic mechanisms (Murach et al. 2022). It increases histone acetylation, activates HATs, and enhances DNA repair via p300/CBP-mediated OGG1 acetylation (Radak et al. 2011). Exercise also boosts SIRT1 activity, improving mitochondrial function and reducing oxidative stress (Koltai et al. 2010; Musci et al. 2019). SIRT1 regulates muscle repair, hypertrophy, and angiogenesis, counteracting sarcopenia (Radak et al. 2020).

Exercise induces muscle fiber remodeling. Aerobic exercise increases type I and IIa fiber area and contraction speed (Harber et al. 2012), while endurance training promotes slow-twitch fiber conversion and capillary growth (Hemmeter and Ngamsri 2022). Resistance training stimulates hypertrophy and strength gains (Tumkur Anil Kumar et al. 2021). Exercise also modulates sarcomeric protein acetylation and promotes muscle fiber regeneration (Stitt et al. 2004).

6.7 Therapeutic Implications and Future Directions

The potential for targeting histone acetylation and deacetylation pathways to enhance muscle health and mitigate the effects of aging is significant. HDAC inhibitors, such as P7C3 and TSA, have shown promise in preclinical models. For instance, P7C3 has been demonstrated to enhance muscle repair and reduce serum CK activity in a barium chloride-induced muscle injury model (Kim et al. 2024b). Analogously, TSA-induced hyperacetylation promoted myotube fusion and augmented the secretion of miR-873-3p-enriched extracellular vesicles (EVs), indicating that HDAC inhibition can mimic exercise-mediated bone health benefits (Chen et al. 2024). Furthermore, HDACi-mediated restoration of pathogenic gene expression networks in dystrophin-deficient skeletal muscles holds promise as a pharmacological strategy to mitigate the progression of Duchenne Muscular Dystrophy (DMD) (Mozzetta et al. 2024). Overall, these findings highlight the potential of targeting histone acetylation and deacetylation mechanisms as a novel strategy to enhance muscle function, promote repair, and offer protective benefits against age-related muscle decline and specific muscular pathologies.

The role of epigenetic regulation in myogenesis is also gaining attention. Recent studies have identified novel transcription factors, such as Bach2, MafK, and Fosl2, regulated by H3K27 acetylation in the myogenic differentiation of porcine skeletal muscle satellite cells (Zhou et al. 2024). These transcription factors play crucial roles in muscle development and regeneration. Additionally, the study by Chen and Li (2024) highlights the involvement of HDAC11 in immune response, metabolic pathways, and cell cycle regulation in skeletal muscle (Chen and Li 2024). The detailed mechanisms by which HDAC11 controls muscle-specific gene expression and myogenesis remain to be fully characterized, but these insights underscore the need for further research to elucidate the specific downstream targets and pathways involved.

Emerging technologies and methodologies in epigenetics, such as single-cell epigenomics and CRISPR-based epigenome editing, offer powerful tools to advance research in this area. Single-cell epigenomics technologies, such as single-cell ATAC-seq and single-cell ChIP-seq, can provide insights into the epigenetic heterogeneity of muscle cells and the dynamic changes in histone acetylation and deacetylation during muscle development and aging. CRISPR-based epigenome editing tools, such as CRISPR/dCas9 fused with histone acetyltransferases or deacetylases, enable precise manipulation of histone acetylation at specific genomic loci, facilitating the study of epigenetic regulation in muscle cells.

The integration of multi-omics approaches, including transcriptomics, proteomics, and metabolomics, with epigenetic data can provide a more comprehensive understanding of the complex regulatory networks involved in muscle health and aging. For example, the study by Mallett (2025) discusses the effects of exercise on skeletal muscle epigenetics and metabolic adaptations, suggesting that a multi-omics approach could reveal new insights into the molecular mechanisms underlying muscle adaptation to exercise (Mallett 2025). This approach could help identify new therapeutic targets and strategies to enhance muscle health and mitigate aging effects.

Despite the promising findings, several gaps in knowledge remain. The long-term effects of epigenetic therapies on muscle health and aging need to be investigated. While short-term studies have shown promising results, the potential for epigenetic drift and off-target effects in long-term treatment remains a concern. Future research should focus on elucidating the detailed mechanisms of epigenetic regulation in muscle cells, developing targeted epigenetic therapies, and integrating epigenetic data with other omics approaches to gain a more comprehensive understanding of muscle health and aging.

6.8 Conclusions

As shown in Fig. 6.1, by orchestrating chromatin dynamics, histone acetyltransferases (HATs) and deacetylases (HDACs) critically govern muscle adaptation, metabolic homeostasis, and insulin sensitivity. Exercise emerges as a potent epigenetic modulator, enhancing HAT activity to promote mitochondrial biogenesis and glucose metabolism (e.g., via PGC-1α acetylation) while counteracting HDAC-driven atrophy pathways. These findings underscore the therapeutic potential of targeting HAT/HDAC balance to mitigate age-related muscle decline. However, the complexity of epigenetic crosstalk necessitates further exploration of additional regulatory mechanisms and long-term safety of interventions. Future research should prioritize translating these insights into strategies that preserve muscle function in aging populations.

Compliance with Ethical Standards

Conflict of Interest All authors declare no conflict of interest.

Ethical Approval This chapter is a review of previously published accounts; as such, no animal or human studies were performed.

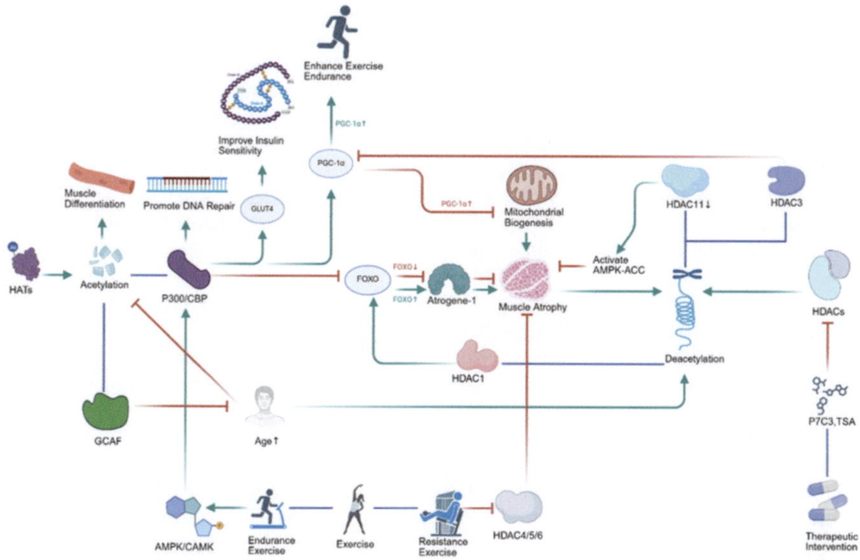

Fig. 6.1 The roles of histone acetyltransferases (HATs) and histone deacetylases (HDACs) induced by exercise in skeletal muscle. Acetylation mediated by HATs promotes muscle differentiation, DNA repair, etc. Deacetylation mediated by HDACs is associated with muscle atrophy, etc. Exercise (endurance exercise, resistance exercise, etc.) can regulate key factors such as PGC-1α and FOXO1 through pathways like activating AMPK/CAMK, influencing mitochondrial biogenesis, insulin sensitivity, etc. Additionally, epigenetic therapeutic interventions (such as drugs like P7C3, TSA) and their effects on HDACs are also involved. (Created with BioRender.com)

References

Abbassi NE, Biela A, Glatt S, Lin TY (2020) How Elongator acetylates tRNA bases. Int J Mol Sci 21(21). https://doi.org/10.3390/ijms21218209

Addicks GC, Zhang H, Ryu D, Vasam G, Green AE, Marshall PL, Patel S, Kang BE, Kim D, Katsyuba E, Williams EG, Renaud JM, Auwerx J, Menzies KJ (2022) GCN5 maintains muscle integrity by acetylating YY1 to promote dystrophin expression. J Cell Biol 221(2). https://doi.org/10.1083/jcb.202104022

Alamdari N, Smith IJ, Aversa Z, Hasselgren PO (2010) Sepsis and glucocorticoids upregulate p300 and downregulate HDAC6 expression and activity in skeletal muscle. Am J Physiol Regul Integr Comp Physiol 299(2):R509–R520. https://doi.org/10.1152/ajpregu.00858.2009

Alamdari N, Aversa Z, Castillero E, Hasselgren PO (2013) Acetylation and deacetylation – novel factors in muscle wasting. Metabolism 62(1):1–11

Baell JB, Leaver DJ, Hermans SJ, Kelly GL, Brennan MS, Downer NL, Nguyen N, Wichmann J, McRae HM, Yang Y, Cleary B, Lagiakos HR, Mieruszynski S, Pacini G, Vanyai HK, Bergamasco MI, May RE, Davey BK, Morgan KJ, Sealey AJ, Wang B, Zamudio N, Wilcox S, Garnham AL, Sheikh BN, Aubrey BJ, Doggett K, Chung MC, de Silva M, Bentley J, Pilling P, Hattarki M, Dolezal O, Dennis ML, Falk H, Ren B, Charman SA, White KL, Rautela J, Newbold A, Hawkins ED, Johnstone RW, Huntington ND, Peat TS, Heath JK, Strasser A, Parker MW, Smyth GK, Street IP, Monahan BJ, Voss AK, Thomas T (2018) Inhibitors of histone acetyltransferases KAT6A/B induce senescence and arrest tumour growth. Nature 560(7717):253–257. https://doi.org/10.1038/s41586-018-0387-5

Bannister AJ, Kouzarides T (2011) Regulation of chromatin by histone modifications. Cell Res 21(3):381–395. https://doi.org/10.1038/cr.2011.22.10.1038/cr.2011.22

Beharry AW, Judge AR (2015) Differential expression of HDAC and HAT genes in atrophying skeletal muscle. Muscle Nerve 52(6):1098–1101. https://doi.org/10.1002/mus.24912

Beharry AW, Sandesara PB, Roberts BM, Ferreira LF, Senf SM, Judge AR (2014) HDAC1 activates FoxO and is both sufficient and required for skeletal muscle atrophy. J Cell Sci 127(Pt 7):1441–1453. https://doi.org/10.1242/jcs.136390

Beiter T, Zügel M, Hudemann J, Schild M, Fragasso A, Burgstahler C, Krüger K, Mooren FC, Steinacker JM, Nieß AM (2024) The acute, short-, and long-term effects of endurance exercise on skeletal muscle transcriptome profiles. Int J Mol Sci 25(5). https://doi.org/10.3390/ijms25052881

Chen J, Li Q (2024) Emerging role of HDAC11 in skeletal muscle biology. Front Cell Dev Biol 12. https://doi.org/10.3389/fcell.2024.1368171

Chen SL, Dowhan DH, Hosking BM, Muscat GE (2000) The steroid receptor coactivator, GRIP-1, is necessary for MEF-2C-dependent gene expression and skeletal muscle differentiation. Genes Dev 14(10):1209–1228

Chen M, Li Y, Zhang M, Ge S, Feng T, Chen R, Shen J, Li R, Wang Z, Xie Y, Wang D, Liu J, Lin Y, Chang F, Chen J, Sun X, Cheng D, Huang X, Wu F, Zhang Q, Cai P, Yin P, Zhang L, Tang P (2024) Histone deacetylase inhibition enhances extracellular vesicles from muscle to promote osteogenesis via miR-873-3p. Signal Transduct Target Ther 9(1):256. https://doi.org/10.1038/s41392-024-01976-0

Cheng F, Lienlaf M, Perez-Villarroel P, Wang HW, Lee C, Woan K, Woods D, Knox T, Bergman J, Pinilla-Ibarz J, Kozikowski A, Seto E, Sotomayor EM, Villagra A (2014) Divergent roles of histone deacetylase 6 (HDAC6) and histone deacetylase 11 (HDAC11) on the transcriptional regulation of IL10 in antigen presenting cells. Mol Immunol 60(1):44–53. https://doi.org/10.1016/j.molimm.2014.02.019

Domi E, Hoxha M, Prendi E, Zappacosta B (2021) A systematic review on the role of SIRT1 in Duchenne muscular dystrophy. Cells 10(6). https://doi.org/10.3390/cells10061380

Egan B, Sharples AP (2023) Molecular responses to acute exercise and their relevance for adaptations in skeletal muscle to exercise training. Physiol Rev 103(3):2057–2170. https://doi.org/10.1152/physrev.00054.2021

Egan B, Zierath JR (2013) Exercise metabolism and the molecular regulation of skeletal muscle adaptation. Cell Metab 17(2):162–184. https://doi.org/10.1016/j.cmet.2012.12.012

Egan B, Carson BP, Garcia-Roves PM, Chibalin AV, Sarsfield FM, Barron N, McCaffrey N, Moyna NM, Zierath JR, O'Gorman DJ (2010) Exercise intensity-dependent regulation of peroxisome proliferator-activated receptor coactivator-1 mRNA abundance is associated with differential activation of upstream signalling kinases in human skeletal muscle. J Physiol 588(Pt 10):1779–1790. https://doi.org/10.1113/jphysiol.2010.188011

Favrot L, Blanchard JS, Vergnolle O (2016) Bacterial GCN5-related N-acetyltransferases: from resistance to regulation. Biochemistry 55(7):989–1002. https://doi.org/10.1021/acs.biochem.5b01269

Feng Y, Rao Z, Tian X, Hu Y, Yue L, Meng Y, Zhong Q, Chen W, Xu W, Li H, Hu Y, Shi R (2025) Endurance training enhances skeletal muscle mitochondrial respiration by promoting MOTS-c secretion. Free Radic Biol Med 227:619–628. https://doi.org/10.1016/j.freeradbiomed.2024.12.038

Ferrari L, Bragato C, Brioschi L, Spreafico M, Esposito S, Pezzotta A, Pizzetti F, Moreno-Fortuny A, Bellipanni G, Giordano A, Riva P, Frabetti F, Viani P, Cossu G, Mora M, Marozzi A, Pistocchi A (2019) HDAC8 regulates canonical Wnt pathway to promote differentiation in skeletal muscles. J Cell Physiol 234(5):6067–6076. https://doi.org/10.1002/jcp.27341

Fontana A, Cursaro I, Carullo G, Gemma S, Butini S, Campiani G (2022) A therapeutic perspective of HDAC8 in different diseases: an overview of selective inhibitors. Int J Mol Sci 23(17). https://doi.org/10.3390/ijms231710014

Gehlert S, Weinisch P, Römisch-Margl W, Jaspers RT, Artati A, Adamski J, Dyar KA, Aussieker T, Jacko D, Bloch W, Wackerhage H, Kastenmüller G (2022) Effects of acute and chronic resistance exercise on the skeletal muscle metabolome. Meta 12(5). https://doi.org/10.3390/metabo12050445

Harber MP, Konopka AR, Undem MK, Hinkley JM, Minchev K, Kaminsky LA, Trappe TA, Trappe S (2012) Aerobic exercise training induces skeletal muscle hypertrophy and age-dependent adaptations in myofiber function in young and older men. J Appl Physiol 113(9):1495–1504. https://doi.org/10.1152/japplphysiol.00786.2012

Hemmeter UM, Ngamsri T (2022) Physical activity and mental health in the elderly. Praxis (Bern 1994) 110(4):193–198. https://doi.org/10.1024/1661-8157/a003853

Hong S, Zhou W, Fang B, Lu W, Loro E, Damle M, Ding G, Jager J, Zhang S, Zhang Y, Feng D, Chu Q, Dill BD, Molina H, Khurana TS, Rabinowitz JD, Lazar MA, Sun Z (2017) Dissociation of muscle insulin sensitivity from exercise endurance in mice by HDAC3 depletion. Nat Med 23(2):223–234. https://doi.org/10.1038/nm.4245

Hurtado E, Núñez-Álvarez Y, Muñoz M, Gutiérrez-Caballero C, Casas J, Pendás AM, Peinado MA, Suelves M (2021) HDAC11 is a novel regulator of fatty acid oxidative metabolism in skeletal muscle. FEBS J 288(3):902–919. https://doi.org/10.1111/febs.15456

Ibrahim SY, Carter J, Bagchi RA (2025) Histone deacetylases in metabolism: the known and the unexplored. Physiology (Bethesda) 40(3). https://doi.org/10.1152/physiol.00044.2024

Joung H, Kwon S, Kim KH, Lee YG, Shin S, Kwon DH, Lee YU, Kook T, Choe N, Kim JC, Kim YK, Eom GH, Kook H (2018) Sumoylation of histone deacetylase 1 regulates MyoD signaling during myogenesis. Exp Mol Med 50(1):e427. https://doi.org/10.1038/emm.2017.236

Kim JW, Jang SM, Kim CH, An JH, Kang EJ, Choi KH (2011) Tip60 regulates myoblast differentiation by enhancing the transcriptional activity of MyoD via their physical interactions. FEBS J 278(22):4394–4404. https://doi.org/10.1111/j.1742-4658.2011.08362.x

Kim DS, Wheeler MT, Ashley EA (2022) The genetics of human performance. Nat Rev Genet 23(1):40–54. https://doi.org/10.1038/s41576-021-00400-5

Kim JM, Kim HK, Cho HJ, Moon SA, Kim Y, Hong JY, Lee SH, Kim K, Koh JM (2024a) Extracellular C1qbp inhibits myogenesis by suppressing NFATc1. Sci Rep 14(1):15678. https://doi.org/10.1038/s41598-024-66549-1

Kim JW, Manickam R, Sinha P, Xuan W, Huang J, Awad K, Brotto M, Tipparaju SM (2024b) P7C3 ameliorates barium chloride-induced skeletal muscle injury activating transcriptomic and epigenetic modulation of myogenic regulatory factors. J Cell Physiol 239(9):e31346. https://doi.org/10.1002/jcp.31346

Klymenko O, Brecklinghaus T, Dille M, Springer C, de Wendt C, Altenhofen D, Binsch C, Knebel B, Scheller J, Hardt C, Herwig R, Chadt A, Pfluger PT, Al-Hasani H, Kabra DG (2020) Histone deacetylase 5 regulates interleukin 6 secretion and insulin action in skeletal muscle. Mol Metab 42:101062. https://doi.org/10.1016/j.molmet.2020.101062

Koltai E, Szabo Z, Atalay M, Boldogh I, Naito H, Goto S, Nyakas C, Radak Z (2010) Exercise alters SIRT1, SIRT6, NAD and NAMPT levels in skeletal muscle of aged rats. Mech Ageing Dev 131(1):21–28. https://doi.org/10.1016/j.mad.2009.11.002

Kosan C, Heidel FH, Godmann M, Bierhoff H (2018) Epigenetic erosion in adult stem cells: drivers and passengers of aging. Cells 7(12). https://doi.org/10.3390/cells7120237

Kumar A, Datta M (2022) H19 inhibition increases HDAC6 and regulates IRS1 levels and insulin signaling in the skeletal muscle during diabetes. Mol Med 28(1):81. https://doi.org/10.1186/s10020-022-00507-3

Kuninger D, Wright A, Rotwein P (2006) Muscle cell survival mediated by the transcriptional coactivators p300 and PCAF displays different requirements for acetyltransferase activity. Am J Physiol Cell Physiol 291(4):C699–C709. https://doi.org/10.1152/ajpcell.00056.2006

LaBarge S, Migdal C, Schenk S (2015) Is acetylation a metabolic rheostat that regulates skeletal muscle insulin action? Mol Cells 38(4):297–303. https://doi.org/10.14348/molcells.2015.0020

Lee KK, Workman JL (2007) Histone acetyltransferase complexes: one size doesn't fit all. Nat Rev Mol Cell Biol 8(4):284–295. https://doi.org/10.1038/nrm2145

Lee Y, Shin MH, Kim MK, Kim YK, Shin HS, Lee DH, Chung JH (2021) Increased histone acetylation and decreased expression of specific histone deacetylases in ultraviolet-irradiated and intrinsically aged human skin in vivo. Int J Mol Sci 22(4). https://doi.org/10.3390/ijms22042032

Li H, Hou L, Zhang Y, Jiang F, Zhu Y, Li QX, Hu CY, Wang C (2019) PFN2a suppresses C2C12 myogenic development by inhibiting proliferation and promoting apoptosis via the p53 pathway. Cells 8(9). https://doi.org/10.3390/cells8090959

Li C, Deng Z, Zheng G, Xie T, Wei X, Huo Z, Bai J (2021) Histone deacetylase 2 suppresses skeletal muscle atrophy and senescence via NF-κB Signaling pathway in cigarette smoke-induced mice with emphysema. Int J Chron Obstruct Pulmon Dis 16:1661–1675. https://doi.org/10.2147/copd.S314640

Li Y, Xiao N, Wang Q, Liu B, Cui Y, Liu Y, Ji Y, Zheng M (2025) Research on the mechanism of resistance exercise in promoting glucose metabolic shift to regulate muscle satellite cell proliferation in type 2 diabetic rats. Biochem Biophys Res Commun 751:151401. https://doi.org/10.1016/j.bbrc.2025.151401

Liang D, Chen C, Huang S, Liu S, Fu L, Niu Y (2022) Alterations of lysine acetylation profile in murine skeletal muscles upon exercise. Front Aging Neurosci 14:859313. https://doi.org/10.3389/fnagi.2022.859313

Lim C, Shimizu J, Kawano F, Kim HJ, Kim CK (2020) Adaptive responses of histone modifications to resistance exercise in human skeletal muscle. PLoS One 15(4):e0231321. https://doi.org/10.1371/journal.pone.0231321

Lin L, Chen K, Abdel Khalek W, Ward JL 3rd, Yang H, Chabi B, Wrutniak-Cabello C, Tong Q (2014) Regulation of skeletal muscle oxidative capacity and muscle mass by SIRT3. PLoS One 9(1):e85636

Luo L, Martin SC, Parkington J, Cadena SM, Zhu J, Ibebunjo C, Summermatter S, Londraville N, Patora-Komisarska K, Widler L, Zhai H, Trendelenburg AU, Glass DJ, Shi J (2019) HDAC4 controls muscle homeostasis through Deacetylation of myosin heavy chain, PGC-1α, and Hsc70. Cell Rep 29(3):749–763.e12. https://doi.org/10.1016/j.celrep.2019.09.023

Mallett G (2025) The effect of exercise and physical activity on skeletal muscle epigenetics and metabolic adaptations. Eur J Appl Physiol. https://doi.org/10.1007/s00421-025-05704-6

Martin SD, Connor T, Sanigorski A, McEwen KA, Henstridge DC, Nijagal B, De Souza D, Tull DL, Meikle PJ, Kowalski GM, Bruce CR, Gregorevic P, Febbraio MA, Collier FM, Walder KR, McGee SL (2023) Class IIa HDACs inhibit cell death pathways and protect muscle integrity in response to lipotoxicity. Cell Death Dis 14(12):787. https://doi.org/10.1038/s41419-023-06319-5

Martins VF, LaBarge SA, Stanley A, Svensson K, Hung CW, Keinan O, Ciaraldi TP, Banoian D, Park JE, Ha C, Hetrick B, Meyer GA, Philp A, David LL, Henry RR, Aslan JE, Saltiel AR, McCurdy CE, Schenk S (2022) p300 or CBP is required for insulin-stimulated glucose uptake in skeletal muscle and adipocytes. JCI Insight 7(1). https://doi.org/10.1172/jci.insight.141344

Mattingly ML, Anglin DA, Ruple BA, Scarpelli MC, Bergamasco JG, Godwin JS, Mobley CB, Frugé AD, Libardi CA, Roberts MD (2024) Acute and chronic resistance training, acute endurance exercise, nor physiologically plausible lactate in vitro affect skeletal muscle Lactylation. Int J Mol Sci 25(22). https://doi.org/10.3390/ijms252212216

McGee SL, Hargreaves M (2019) Epigenetics and exercise. Trends Endocrinol Metabol 30(9):636–645

McGee SL, Walder KR (2017) Exercise and the skeletal muscle Epigenome. Cold Spring Harb Perspect Med 7(9). https://doi.org/10.1101/cshperspect.a029876

McGee SL, Fairlie E, Garnham AP, Hargreaves M (2009) Exercise-induced histone modifications in human skeletal muscle. J Physiol 587(Pt 24):5951–5958. https://doi.org/10.1113/jphysiol.2009.181065

McGee-Lawrence ME, Bradley EW, Dudakovic A, Carlson SW, Ryan ZC, Kumar R, Dadsetan M, Yaszemski MJ, Chen Q, An KN, Westendorf JJ (2013) Histone deacetylase 3 is required for maintenance of bone mass during aging. Bone 52(1):296–307. https://doi.org/10.1016/j.bone.2012.10.015

McKinsey TA, Zhang CL, Lu J, Olson EN (2000) Signal-dependent nuclear export of a histone deacetylase regulates muscle differentiation. Nature 408(6808):106–111. https://doi.org/10.1038/35040593

Meynial-Denis D (ed) (2019) Sarcopenia: molecular, cellular, and nutritional aspects, 1st edn. CRC Press

Molinari S, Imbriano C, Moresi V, Renzini A, Belluti S, Lozanoska-Ochser B, Gigli G, Cedola A (2023) Histone deacetylase functions and therapeutic implications for adult skeletal muscle metabolism. Front Mol Biosci 10:1130183. https://doi.org/10.3389/fmolb.2023.1130183

Mozzetta C, Sartorelli V, Steinkuhler C, Puri PL (2024) HDAC inhibitors as pharmacological treatment for Duchenne muscular dystrophy: a discovery journey from bench to patients. Trends Mol Med 30(3):278–294. https://doi.org/10.1016/j.molmed.2024.01.007

Murach KA, Bagley JR (2025) A primer on global molecular responses to exercise in skeletal muscle: omics in focus. J Sport Health Sci 101029. https://doi.org/10.1016/j.jshs.2025.101029

Murach KA, Dimet-Wiley AL, Wen Y, Brightwell CR, Latham CM, Dungan CM, Fry CS, Watowich SJ (2022) Late-life exercise mitigates skeletal muscle epigenetic aging. Aging Cell 21(1):e13527. https://doi.org/10.1111/acel.13527

Musci RV, Hamilton KL, Linden MA (2019) Exercise-induced Mitohormesis for the maintenance of skeletal muscle and Healthspan extension. Sports (Basel) 7(7). https://doi.org/10.3390/sports7070170

Myers MJ, Shaik F, Shaik F, Alway SE, Mohamed JS (2021) Skeletal muscle gene expression profile in response to caloric restriction and aging: a role for SirT1. Genes (Basel) 12(5). https://doi.org/10.3390/genes12050691

Naito A, Sato T, Kimura I, Tokuda N, Yamauchi N, Fusagawa H, Yamada T (2025) Effects of contraction frequency during high-intensity training on fatigue resistance and aerobic adaptations in mouse skeletal muscle. J Appl Physiol 138(1):107–120. https://doi.org/10.1152/japplphysiol.00530.2024

Niu Y, Wang T, Liu S, Yuan H, Li H, Fu L (2017) Exercise-induced GLUT4 transcription via inactivation of HDAC4/5 in mouse skeletal muscle in an AMPKα2-dependent manner. Biochim Biophys Acta Mol basis Dis 1863(9):2372–2381. https://doi.org/10.1016/j.bbadis.2017.07.001

Ochiai N, Nishizuka M, Osada S, Imagawa M (2016) Fad24, a positive regulator of Adipogenesis, is required for S phase re-entry of C2C12 myoblasts arrested in G0 phase and involved in p27(Kip1) expression at the protein level. Biol Pharm Bull 39(5):807–814. https://doi.org/10.1248/bpb.b15-00954

Ohsawa I, Kawano F (2021) Chronic exercise training activates histone turnover in mouse skeletal muscle fibers. FASEB J 35(4):e21453. https://doi.org/10.1096/fj.202002027RR

Ohsawa I, Konno R, Masuzawa R, Kawano F (2018) Amount of daily exercise is an essential stimulation to alter the epigenome of skeletal muscle in rats. J Appl Physiol 125(4):1097–1104. https://doi.org/10.1152/japplphysiol.00074.2018

Osseni A, Ravel-Chapuis A, Belotti E, Scionti I, Gangloff YG, Moncollin V, Mazelin L, Mounier R, Leblanc P, Jasmin BJ, Schaeffer L (2022) Pharmacological inhibition of HDAC6 improves muscle phenotypes in dystrophin-deficient mice by downregulating TGF-β via Smad3 acetylation. Nat Commun 13(1):7108. https://doi.org/10.1038/s41467-022-34831-3

Parthun MR (2007) Hat1: the emerging cellular roles of a type B histone acetyltransferase. Oncogene 26(37):5319–5328. https://doi.org/10.1038/sj.onc.1210602

Pigna E, Renzini A, Greco E, Simonazzi E, Fulle S, Mancinelli R, Moresi V, Adamo S (2018) HDAC4 preserves skeletal muscle structure following long-term denervation by mediating distinct cellular responses. Skelet Muscle 8(1):6. https://doi.org/10.1186/s13395-018-0153-2

Pilotto AM, Turner DC, Mazzolari R, Crea E, Brocca L, Pellegrino MA, Miotti D, Bottinelli R, Sharples AP, Porcelli S (2025) Human skeletal muscle possesses an epigenetic memory of high-intensity interval training. Am J Physiol Cell Physiol 328(1):C258–c272. https://doi.org/10.1152/ajpcell.00423.2024

Plaza-Diaz J, Izquierdo D, Torres-Martos Á, Baig AT, Aguilera CM, Ruiz-Ojeda FJ (2022) Impact of physical activity and exercise on the Epigenome in skeletal muscle and effects on systemic metabolism. Biomedicine 10(1). https://doi.org/10.3390/biomedicines10010126

Posridee K, Kupittayanant S, Rachnavy P, Oonsivilai A, Oonsivilai R (2025) Role of ginseng and L-carnitine in modulating exercise endurance and oxidative stress in rats. Nutrients 17(3). https://doi.org/10.3390/nu17030568

Radak Z, Bori Z, Koltai E, Fatouros IG, Jamurtas AZ, Douroudos II, Terzis G, Nikolaidis MG, Chatzinikolaou A, Sovatzidis A, Kumagai S, Naito H, Boldogh I (2011) Age-dependent changes in 8-oxoguanine-DNA glycosylase activity are modulated by adaptive responses to physical exercise in human skeletal muscle. Free Radic Biol Med 51(2):417–423. https://doi.org/10.1016/j.freeradbiomed.2011.04.018

Radak Z, Suzuki K, Posa A, Petrovszky Z, Koltai E, Boldogh I (2020) The systemic role of SIRT1 in exercise mediated adaptation. Redox Biol 35:101467. https://doi.org/10.1016/j.redox.2020.101467

Ratti F, Ramond F, Moncollin V, Simonet T, Milan G, Méjat A, Thomas JL, Streichenberger N, Gilquin B, Matthias P, Khochbin S, Sandri M, Schaeffer L (2015) Histone deacetylase 6 is a FoxO transcription factor-dependent effector in skeletal muscle atrophy. J Biol Chem 290(7):4215–4224. https://doi.org/10.1074/jbc.M114.600916

Razidlo DF, Whitney TJ, Casper ME, McGee-Lawrence ME, Stensgard BA, Li X, Secreto FJ, Knutson SK, Hiebert SW, Westendorf JJ (2010) Histone deacetylase 3 depletion in osteo/chondroprogenitor cells decreases bone density and increases marrow fat. PLoS One 5(7):e11492. https://doi.org/10.1371/journal.pone.0011492

Renzini A, Marroncelli N, Noviello C, Moresi V, Adamo S (2018) HDAC4 regulates skeletal muscle regeneration via soluble factors. Front Physiol 9:1387. https://doi.org/10.3389/fphys.2018.01387

Renzini A, D'Onghia M, Coletti D, Moresi V (2022) Histone deacetylases as modulators of the crosstalk between skeletal muscle and other organs. Front Physiol 13:706003. https://doi.org/10.3389/fphys.2022.706003

Roberts MD, McCarthy JJ, Hornberger TA, Phillips SM, Mackey AL, Nader GA, Boppart MD, Kavazis AN, Reidy PT, Ogasawara R, Libardi CA, Ugrinowitsch C, Booth FW, Esser KA (2023) Mechanisms of mechanical overload-induced skeletal muscle hypertrophy: current understanding and future directions. Physiol Rev 103(4):2679–2757. https://doi.org/10.1152/physrev.00039.2022

Robin JD, Jacome Burbano MS, Peng H, Croce O, Thomas JL, Laberthonniere C, Renault V, Lototska L, Pousse M, Tessier F, Bauwens S, Leong W, Sacconi S, Schaeffer L, Magdinier F, Ye J, Gilson E (2020) Mitochondrial function in skeletal myofibers is controlled by a TRF2-SIRT3 axis over lifetime. Aging Cell 19(3):e13097. https://doi.org/10.1111/acel.13097

Rokudai S, Laptenko O, Arnal SM, Taya Y, Kitabayashi I, Prives C (2013) MOZ increases p53 acetylation and premature senescence through its complex formation with PML. Proc Natl Acad Sci USA 110(10):3895–3900. https://doi.org/10.1073/pnas.1300490110

Sandonà M, Cavioli G, Renzini A, Cedola A, Gigli G, Coletti D, McKinsey TA, Moresi V, Saccone V (2023) Histone deacetylases: molecular mechanisms and therapeutic implications for muscular dystrophies. Int J Mol Sci 24(5). https://doi.org/10.3390/ijms24054306

Sapountzi V, Côté J (2011) MYST-family histone acetyltransferases: beyond chromatin. Cell Mol Life Sci 68(7):1147–1156. https://doi.org/10.1007/s00018-010-0599-9

Seaborne RA, Strauss J, Cocks M, Shepherd S, O'Brien TD, van Someren KA, Bell PG, Murgatroyd C, Morton JP, Stewart CE, Sharples AP (2018) Human skeletal muscle possesses an epigenetic memory of hypertrophy. Sci Rep 8(1):1898. https://doi.org/10.1038/s41598-018-20287-3

Senf SM, Sandesara PB, Reed SA, Judge AR (2011) p300 acetyltransferase activity differentially regulates the localization and activity of the FOXO homologues in skeletal muscle. Am J Physiol Cell Physiol 300(6):C1490–C1501. https://doi.org/10.1152/ajpcell.00255.2010

Sincennes MC, Brun CE, Lin AYT, Rosembert T, Datzkiw D, Saber J, Ming H, Kawabe YI, Rudnicki MA (2021) Acetylation of PAX7 controls muscle stem cell self-renewal and differentiation potential in mice. Nat Commun 12(1):3253. https://doi.org/10.1038/s41467-021-23577-z

Smith JA, Kohn TA, Chetty AK, Ojuka EO (2008) CaMK activation during exercise is required for histone hyperacetylation and MEF2A binding at the MEF2 site on the Glut4 gene. Am J Physiol Endocrinol Metab 295(3):E698–E704. https://doi.org/10.1152/ajpendo.00747.2007

Smith JAB, Murach KA, Dyar KA, Zierath JR (2023) Exercise metabolism and adaptation in skeletal muscle. Nat Rev Mol Cell Biol 24(9):607–632

Spreafico M, Cafora M, Bragato C, Capitanio D, Marasca F, Bodega B, De Palma C, Mora M, Gelfi C, Marozzi A, Pistocchi A (2021) Targeting HDAC8 to ameliorate skeletal muscle differentiation in Duchenne muscular dystrophy. Pharmacol Res 170:105750. https://doi.org/10.1016/j.phrs.2021.105750

Stitt TN, Drujan D, Clarke BA, Panaro F, Timofeyva Y, Kline WO, Gonzalez M, Yancopoulos GD, Glass DJ (2004) The IGF-1/PI3K/Akt pathway prevents expression of muscle atrophy-induced ubiquitin ligases by inhibiting FOXO transcription factors. Mol Cell 14(3):395–403. https://doi.org/10.1016/s1097-2765(04)00211-4

Svensson K, LaBarge SA, Sathe A, Martins VF, Tahvilian S, Cunliffe JM, Sasik R, Mahata SK, Meyer GA, Philp A, David LL, Ward SR, McCurdy CE, Aslan JE, Schenk S (2020) p300 and cAMP response element-binding protein-binding protein in skeletal muscle homeostasis, contractile function, and survival. J Cachexia Sarcopenia Muscle 11(2):464–477. https://doi.org/10.1002/jcsm.12522

Thomas T, Dixon MP, Kueh AJ, Voss AK (2008) Mof (MYST1 or KAT8) is essential for progression of embryonic development past the blastocyst stage and required for normal chromatin architecture. Mol Cell Biol 28(16):5093–5105. https://doi.org/10.1128/mcb.02202-07

To M, Swallow EB, Akashi K, Haruki K, Natanek SA, Polkey MI, Ito K, Barnes PJ (2017) Reduced HDAC2 in skeletal muscle of COPD patients. Respir Res 18(1):99. https://doi.org/10.1186/s12931-017-0588-8

Tomasiak P, Janisiak J, Rogińska D, Perużyńska M, Machaliński B, Tarnowski M (2023) Garcinol and Anacardic acid, natural inhibitors of histone acetyltransferases, inhibit rhabdomyosarcoma growth and proliferation. Molecules 28(14). https://doi.org/10.3390/molecules28145292

Tumkur Anil Kumar N, Oliver JL, Lloyd RS, Pedley JS, Radnor JM (2021) The influence of growth, maturation and resistance training on muscle-tendon and neuromuscular adaptations: a narrative review. Sports (Basel) 9(5). https://doi.org/10.3390/sports9050059

Turner DC, Seaborne RA, Sharples AP (2019) Comparative transcriptome and Methylome analysis in human skeletal muscle anabolism. Hypertrophy and epigenetic memory. Sci Rep 9(1):4251. https://doi.org/10.1038/s41598-019-40787-0

Vetting MW et al (2005) Structure and functions of the GNAT superfamily of acetyltransferases. Arch Biochem Biophys 433(1):212–226. https://doi.org/10.1016/j.abb.2004.09.003

Walsh ME, Van Remmen H (2016) Emerging roles for histone deacetylases in age-related muscle atrophy. Nutr Healthy Aging 4(1):17–30. https://doi.org/10.3233/nha-160005

Walsh ME, Bhattacharya A, Sataranatarajan K, Qaisar R, Sloane L, Rahman MM, Kinter M, Van Remmen H (2015) The histone deacetylase inhibitor butyrate improves metabolism and reduces muscle atrophy during aging. Aging Cell 14(6):957–970. https://doi.org/10.1111/acel.12387

Wang K, Liu H, Hu Q, Wang L, Liu J, Zheng Z, Zhang W, Ren J, Zhu F, Liu GH (2022) Epigenetic regulation of aging: implications for interventions of aging and diseases. Signal Transduct Target Ther 7(1):374. https://doi.org/10.1038/s41392-022-01211-8

Wang X, Yu F, Ye L (2023) Epigenetic control of mesenchymal stem cells orchestrates bone regeneration. Front Endocrinol (Lausanne) 14:1126787. https://doi.org/10.3389/fendo.2023.1126787

Wapenaar H, Dekker FJ (2016) Histone acetyltransferases: challenges in targeting bi-substrate enzymes. Clin Epigenetics 8:59. https://doi.org/10.1186/s13148-016-0225-2

Williams CB, Gurd BJ (2012) Skeletal muscle SIRT1 and the genetics of metabolic health: therapeutic activation by pharmaceuticals and exercise. Appl Clin Genet 5:81–91. https://doi.org/10.2147/tacg.S31276

Wu H, Hu Y, Jiang C, Chen C (2024) Global scientific trends in research of epigenetic response to exercise: a bibliometric analysis. Heliyon 10(4):e25644

Xia C, Tao Y, Li M, Che T, Qu J (2020) Protein acetylation and deacetylation: An important regulatory modification in gene transcription (review). Exp Ther Med 20(4):2923–2940. https://doi.org/10.3892/etm.2020.9073

Xiao Y, Li W, Yang H, Pan L, Zhang L, Lu L, Chen J, Wei W, Ye J, Li J, Li G, Zhang Y, Tan M, Ding J, Wong J (2021) HBO1 is a versatile histone acyltransferase critical for promoter histone acylations. Nucleic Acids Res 49(14):8037–8059. https://doi.org/10.1093/nar/gkab607

Xiong Y, Zhang M, Li Y (2020) Recent advances in the development of CBP/p300 Bromodomain inhibitors. Curr Med Chem 27(33):5583–5598. https://doi.org/10.2174/0929867326666190731141055

Xu J, Li C, Kang X (2023) The epigenetic regulatory effect of histone acetylation and deacetylation on skeletal muscle metabolism-a review. Front Physiol 14:1267456. https://doi.org/10.3389/fphys.2023.1267456

Ying K, Liu H, Tarkhov AE, Sadler MC, Lu AT, Moqri M, Horvath S, Kutalik Z, Shen X, Gladyshev VN (2024) Causality-enriched epigenetic age uncouples damage and adaptation. Nat Aging 4(2):231–246. https://doi.org/10.1038/s43587-023-00557-0

Yoshihara T, Machida S, Tsuzuki T, Kakigi R, Chang SW, Sugiura T, Naito H (2019) Age-related changes in histone modification in rat gastrocnemius muscle. Exp Gerontol 125:110658. https://doi.org/10.1016/j.exger.2019.110658

Zheng L, Rao Z, Wu J, Ma X, Jiang Z, Xiao W (2024) Resistance exercise improves glycolipid metabolism and mitochondrial biogenesis in skeletal muscle of T2DM mice via miR-30d-5p/SIRT1/PGC-1α Axis. Int J Mol Sci 25(22). https://doi.org/10.3390/ijms252212416

Zhou VW, Goren A, Bernstein BE (2011) Charting histone modifications and the functional organization of mammalian genomes. Nat Rev Genet 12(1):7–18. https://doi.org/10.1038/nrg2905.10.1038/nrg2905

Zhou L, Pinho R, Gu Y, Radak Z (2022) The role of SIRT3 in exercise and aging. Cells 11(16). https://doi.org/10.3390/cells11162596

Zhou P, Wang W, Li J, Zheng Z, Du X, Fu L, Li X (2024) Identification of novel transcription factors regulated by H3K27 acetylation in myogenic differentiation of porcine skeletal muscle satellite cells. FASEB J 38(21):e70144

Chapter 7
Exercise-Epigenetic Axis: Metabolic Alignment to Oxidative Stress and Redox Status

Gareth W. Davison, Aya Mohammed, Rachelle Irwin, Diane Lees-Murdock, and Colum P. Walsh

Abstract A regimented exercise schedule serves to modify cell molecular components leading to muscle adaptation. These modifications are driven by transcriptional processes under the control of several epigenetic paradigms. Metabolism is an emerging factor in cell epigenetic modifications, of which alpha-ketoglutarate (α-KG), Acetyl-CoA, and reactive oxygen species (ROS) are considered key metabolites. ROS, particularly following high-intensity exercise, leads to the oxidative modification of DNA, and potentially to the oxidation of the epigenetic mark 5-methylcytosine. ROS can also interfere with several epigenetic modifying enzymes and cofactors, highlighting the important interrelationship between cell metabolism and reprogramming of gene transcription through a state of disrupted epigenetics. While our understanding of the influence of exercise-induced oxidative stress and redox biochemistry in epigenetics is in its infancy, the ensuing chapter should be viewed as a starting point to explore the multidirectional connections between exercise, oxidative stress/redox biology in the epigenetic landscape. To this end, we will initially outline the relationship between metabolism and the epigenetic nexus, prior to outlining how exercise stress can influence the epigenetics. An overview of the potential interplay between exercise-induced oxidative stress/redox biology and epigenetics will be outlined, before offering a brief review of the interactions between exercise and noncoding RNAs.

Keywords Exercise · DNA methylation · Histone modifications · Reactive oxygen species · Metabolism · Oxidative stress · Redox biology

G. W. Davison (✉) · A. Mohammed · R. Irwin · D. Lees-Murdock -
Faculty of Life and Health Sciences, Ulster University, Belfast, Northern Ireland, UK
e-mail: gw.davison@ulster.ac.uk

C. P. Walsh
Department of Biomedical and Clinical Sciences, Linköping University, Linköping, Sweden

F. Gao, X. Zhang (eds.), *Exercise, Epigenetics and Human Health*, Epigenetics and Human Health 15, https://doi.org/10.1007/978-3-032-08045-5_7

Abbreviations

ROS	Reactive oxygen species
α-KG	Alpha-ketoglutarate
SAM	S-adenosylmethionine
PTMs	Posttranslational modifications
MAT	Methionine adenosyltransferase
HATs	Histone acetyltransferases
ACLY	ATP citrate lyase
NOX	NADPH-oxidases
OGG1	8-Oxoguanine DNA glycosylase

7.1 Introduction

Regular exercise evokes several adaptive responses in muscle tissue contributing to improved muscle function, efficiency, and health. Indeed, skeletal muscle is a highly plastic tissue that readily adapts to the energetic demands placed on it through exercise stress (McGee and Walder 2017). Skeletal muscle adaptations are driven by the transcriptional regulation of genes, leading to persistent alterations in the proteome. Environmentally driven epigenetic modifications, via DNA methylation or post-translational histone changes, control the transcriptional landscape of the genome, while metabolism is an emerging factor in the control of cell epigenetic modifications. These modifications are influenced, in part, by several substrates and cofactors such as alpha-ketoglutarate (α-KG), Acetyl-CoA, and reactive oxygen species (ROS); the latter also lead to a state of disrupted redox signaling and oxidative stress, thus creating a dysfunctional epigenetic-metabolic alignment.

ROS production can increase as a function of normal metabolism and can modulate the methylome through direct hydroxylation of 5-methylcytosine (5mC) (Lewandowska and Bartoszek 2011), and via DNA oxidation and TET-mediated hydroxymethylation, potentially leading to a state of DNA hypomethylation (Chia et al. 2011; Davison et al. 2022). ROS can also indirectly influence epigenetics, since methylation and histone-modifying enzymes depend on numerous metabolites and cofactors. These include Fe^{2+} (involved in ROS production), ascorbic acid (potent antioxidant and involved in TET), and fumarate, thus indicating that epigenetic changes are tightly linked to global cell metabolism (Guillaumet-Adkins et al. 2017; Davison et al. 2021), and any fluctuations in these intermediates may lead to a reprogramming of gene transcription (Davison et al. 2022).

High-intensity exercise leads to DNA oxidation due to a metabolic overproduction of ROS in respiring mitochondria, and it is plausible that exercise-induced oxidative stress can also disrupt the epigenetic state of the cell. Although the interplay between cell metabolism and epigenetics is now well recognized, the exercise-epigenetic nexus from the perspective of oxidative stress and redox metabolism is not well established. Herein, we aim to highlight the potential interrelationship between exercise-induced metabolism/oxidative stress and epigenetics. To do this, we initially outline the relationship between metabolism and epigenetics, prior to providing a brief synopsis of exercise stress in the epigenetic landscape. An overview of the complex intricacies (albeit hypothetical) between exercise-induced oxidative stress and epigenetics will be provided, before finally reviewing how exercise can interact and influence noncoding RNAs.

7.2 Metabolism and the Epigenetic Nexus

Metabolism is a series of biochemical modifications in cells catalyzed by enzymatic reactions, that maintains homeostasis in response to environmental conditions (Etchegaray and Mostoslavsky 2016). Epigenetic modifications, which require intermediates of cell metabolism for enzymatic function, can be directly affected by changes in cell metabolism, altering specific acetyltransferases and methyltransferases (Keating and El-Osta 2015). In addition to cell metabolic changes, the epigenome may also be modified by intracellular metabolite localization, such as the tricarboxylic acid cycle (TCA) in mitochondria, which produces key regulatory molecules used as co-substrates for transcriptional and epigenetic processes (Shaughnessy et al. 2014; Keating and El-Osta 2015). Conversely, TCA enzymes can also block epigenetic processes. This interplay between metabolism and epigenetics has been termed "metaboloepigenetics" (Donohoe and Bultman 2012). However, the underlying mechanisms by which metabolic intermediates regulate gene expression remain largely unelucidated (Davison et al. 2021).

7.2.1 One-Carbon Metabolism and DNA/Histone Methylation/ Acetylation

One-carbon metabolism, which utilizes nutrients for various biological functions (Mentch and Locasale 2016), comprises two primary components: the folate and methionine cycles. Both cycles involve transferring single carbon units to acceptor molecules. S-adenosylmethionine (SAM) serves as the primary substrate for methyltransferase reactions on DNA, as well as on arginine and lysine residues of histones. These reactions result in DNA methylation and posttranslational modifications (PTMs) (Mentch and Locasale 2016; Keating and El-Osta 2015). Notably, ROS has the potential to oxidize and derail several substrates involved in the methylation pathway. For example, methionine is readily targeted by hydroxyl radicals, to yield methionine sulfoxide. This one-step oxidation process, catalyzed via a nonenzymatic reaction, can transfer the methyl group to cysteine to produce 5-methyl cysteine (Kawai et al. 2010). Moreover, the methyl donor SAM can influence redox balance alterations. Specifically, the enzyme controlling SAM production, methionine adenosyltransferase (MAT) is redox sensitive, and contains several redox-sensitive residues in its catalytic domain, and as such, any decrease in MAT substrate (due to ROS reactions) may affect global DNA methylation (García-Giménez et al. 2014).

DNA methylation, a crucial process in gene regulation, occurs in two distinct contexts, facilitated by three DNA methyltransferase (DNMT) enzymes. DNMT1, a maintenance methyltransferase, recognizes hemimethylated DNA during DNA replication, forming a symmetrically modified duplex (Bestor et al. 1988). In contrast, DNMT3A and DNMT3B are primarily de novo enzymes that deposit CH_3 moieties on previously unmodified cytosines (Okano et al. 1999; Grurnbaum et al. 1982). Methylation involves the addition of a CH_3 group at the C5 position of the nucleoside cytosine, resulting in the formation of 5-methylcytosine (5mC). Notably, 80% of cytosine residues in CpG dinucleotides (CpG) are methylated at position 5 (Bird et al. 1985). CpG is a unique dinucleotide that occurs at a low frequency but converges into CpG islands (CGIs), which span 300–3000 base pairs. These CGIs are primarily found in regulatory elements of genes, such as enhancers and promoters (Hitchler and Domann 2021). CpG islands are flanked by CpG shores and CpG shelves (Irizarry et al. 2009), which exhibit more tissue-specific methylation profiles. Interestingly, CGIs in promoter regions with active transcription are devoid of methylation. However, when promoter-associated CGIs are methylated, transcription is silenced. Recent genome-wide and functional studies have revealed that DNA methylation in the gene body also plays a role in facilitating transcription via prevention of spurious short transcript production (Wu et al. 2010; Neri et al. 2013; Irwin et al. 2014).

Methylation of histone H3 and H4 is catalyzed by histone methyltransferase (HMT) enzymes, which covalently attach methyl groups (CH_3) from SAM to the side-chain nitrogen atoms of, primarily, lysine and arginine residues (Vanzan et al. 2017; Wong et al. 2017). Lysine methyltransferases contain a conserved domain, Su(var)3–9, Enhancer of zeste, and Trithorax (SET), which is responsible for specifically adding methyl groups to H3K4, H3K9, H3K27, and H3K36 (Vanzan et al. 2017). The outcome of histone methylation depends on the specific histone residues modified, the number of methyl groups added (mono-, di-, or tri-methylation), and the location within the N-terminal regions of either H3 or H4 (Kaelin and McKnight 2013; Etchegaray and Mostoslavsky 2016). These modifications lead to a state of euchromatin (lightly packed DNA that promotes transcription) or heterochromatin (condensed DNA that suppresses transcription) (Davison et al. 2021).

Acetyl-CoA transfers an acetyl group to lysine amino acids on histone tails, yielding acetylation. HAT enzymes catalyze this process, and their sensitivity to intracellular acetyl-CoA concentrations is well known (Kaelin and McKnight 2013; Wong et al. 2017). Histone acetylation regulates gene transcription. For instance, acetylation (H3K9ac, H3K27ac) at specific gene loci activates transcription by opening chromatin. This occurs through two mechanisms: (1) neutralizing the positive charge of histones, weakening the nucleosome-DNA interaction and opening chromatin, and (2) acting as a docking station for transcriptional regulators (Vanzan et al. 2017; Wang et al. 2018; Davison et al. 2021).

7.2.2 *The TCA Metabolite-Dependent Regulation of DNA Methylation and Histone Modification*

The TCA cycle metabolites play a crucial role in regulating DNA and histone demethylation (Fig. 7.1). Cytosine residues undergo demethylation through a series of oxidation steps. 5-Methylcytosine (5mC) is oxidized to 5-hydroxymethylcytosine (5hmC) by the ten-eleven translocation (TET 1–3) dioxygenase proteins. This is followed by decarboxylation to 5-formylcytosine (5fC) and 5-carboxylcytosine (5caC). Subsequently, 5caC is excised by thymine DNA glycosylase (TDG) to

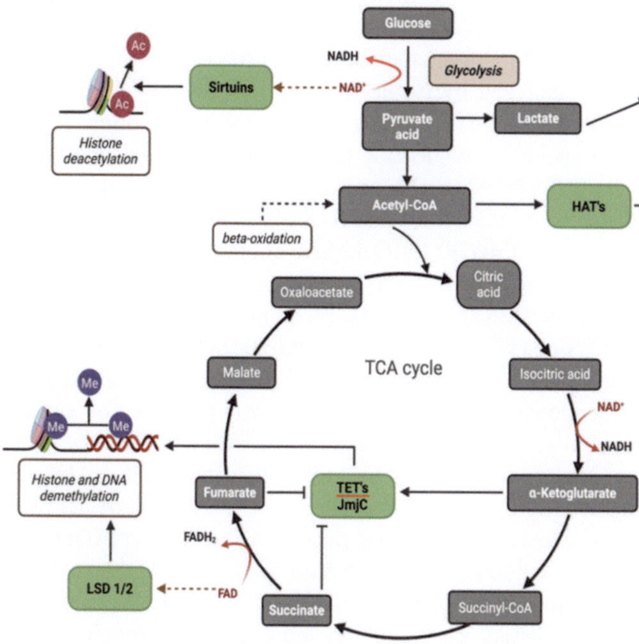

Fig. 7.1 Metabolite regulation of chromatin and DNA (de)methylation. Products of glycolysis (cytosol) and the tricarboxylic acid cycle (TCA; mitochondria) are required for chromatin and DNA modifications. Acetyl-CoA synthesized from pyruvate acid and fatty acid oxidation provides substrate for histone acetyltransferases (HATs). Lactate from pyruvate acid yields histone lactylation. Histone deacetylation requires sirtuin molecules via the cofactor nicotinamide adenine dinucleotide (NAD). Lysine-specific demethylases (LSD) utilize flavin adenine nucleotide (FAD), while α-ketoglutarate controls the ten-eleven translocation (TET) enzymes and Jumonji C domain-containing (JmjC) demethylases. Fumarate and succinate blocks TET and JmjC molecules. Molecules affected by key metabolites shown in green. (Redrawn and updated from Davison et al. 2022)

restore unmethylated cytosine (He et al. 2011; Rodriguez et al. 2017). TET proteins control these sequential oxidation steps, utilizing α-Ketoglutarate (α-KG), oxygen, and Fe^{2+} as cofactors. α-KG is a by-product of oxidative phosphorylation in the TCA cycle and is directly synthesized from isocitrate via isocitrate dehydrogenase 2 (IDH2) and 3 (IDH3). Its involvement in TET activity depends on the intracellular ratio of α-KG to succinate, fumarate, or 2-hydroxyglutarate (2-HG), and any change in substrate ratios in mitochondria can potentially block TET activity, leading to hypermethylation of DNA (Davison et al. 2021). Furthermore, α-KG-related hydroxylation reactions rely on the regeneration of Fe^{2+}, where vitamin C plays a key role in this process by controlling the reduction of oxidized Fe^{3+} to Fe^{2+}, thereby restoring TET activity (Martinez-Reynes and Chandel 2020). Mammals rely on dietary sources of vitamin C to support TET dioxygenases and sustain DNA demethylation, and in support, Blaschke et al. (2013) has shown rapid and global DNA demethylation in mouse embryonic stem cells when vitamin C was added to cell media. TET proteins exhibit physiological responses to various stimuli, including hypoxia, in response to agonists such as α-KG, Fe^{2+}, ascorbic acid, and specific antagonist molecules such as fumarate and succinate. They are also likely to be responsive to an altered and dysfunctional metabolism.

Histone demethylation, catalyzed by lysine-specific demethylases (LSDs) and JumonjiC (JmjC) domain proteins, can be either repressive or activating. LSD1, which incorporates flavin adenine dinucleotide (FAD^+), demethylates H3K4me1, H3K4me2, H3K9me1, and H3K9me2 (Etchegaray and Mostoslavsky 2016). LSD2 also utilizes FAD+ to demethylate H3K4me1 and H3K4me2 (Karytinos et al. 2009). Similar to the TET isoenzymes, *circa* 30 JmjC domain-containing histone demethylases use Fe^{2+} and α-KG to remove methyl groups from arginine and trimethylated lysine (H3K4, H3K9, H3K27, H3K36, and H4K20) in an oxidative reaction that produces hydroxymethyl lysine (Wang et al. 2018; Rodriguez et al. 2017).

Histone deacetylation leads to closed chromatin and gene silencing. These reactions are metabolically responsive, where NAD is a redox cofactor for sirtuins (HDAC enzymes). Mammalian cells have seven sirtuins, with SIRT1, 2, 6, and 7 residing in the nucleus. NAD oxidation during metabolism produces NAD^+ and NADH, affecting sirtuin activity (Miranda-Gonçalves et al. 20182018). A net positive charge (e.g., due to increased glucose flux) decreases the NAD^+/NADH ratio and inhibits sirtuin activity to regulate gene expression. Elevated NAD^+ levels activate sirtuins, and a deprivation of ATP and other substrates can lead to a rise in NAD^+ and SIRT1, leading to histone deacetylation and a potential upregulation of metabolic enzymes (Davison et al. 2021; Etchegaray and Mostoslavsky 2016).

7.3 Exercise Stress in the Epigenetic Landscape

Metabolism provides energy in the form of ATP, which is important for glycolysis and the activation of the TCA cycle to meet the demands of exercising muscle. Exercise bioenergetics may influence the epigenome by affecting metabolite availability (e.g., α-KG) and flux through the TCA cycle (Davison et al. 2022). Evidence linking exercise-induced TCA cycle intermediates and epigenetic modifications primarily involves the key mitochondrial metabolite, citrate. Acetyl-CoA is a central generator of acetyl molecules in histone acetylation induction. Mitochondrial-derived citrate is supplied to the nucleus through ATP citrate lyase (ACLY), which draws citrate from the cytoplasm into the nucleus to yield acetyl-CoA and oxaloacetate. ACYL bridges the metabolic cell state and epigenetic modifications at the histone level and may become a rate-limiting step in the transcription process (Seaborne and Sharples 2020).

In muscle tissue, ACLY silencing hinders myoblast and satellite cell differentiation, accompanied by a reduction in the expression of fast myosin heavy chain isoforms and the myogenic regulatory factor MyoD. Conversely, ACLY overexpression enhances MyoD expression by hyperacetylating H3K9, 14, and 27 (Das et al. 2017). Therefore, it's plausible that muscle stress as a function of exercise can trigger the pathway involving ACYL induction and the conversion of citrate to acetyl-CoA, and this metabolic shift may lead to hyperacetylation of key skeletal muscle remodeling transcripts (Seaborne and Sharples 2020). Sailani (2019) further demonstrated that lifelong exercise positively modifies the epigenome to safeguard skeletal muscle cells against reduced metabolic capacity and oxidative stress. Notably, DNA methylation was lower (in exercise, compared to sedentary) in 714 promoters for genes encoding crucial enzymes involved in glycogen metabolism, glycolysis, and the TCA cycle. In general, promoter regions of genes associated with the oxidative stress response were hypomethylated in the exercise group. These hypomethylation profiles suggest that lifelong exercise can regulate glycolysis and TCA cycle flux, thereby protecting cells from age-related damage and deterioration caused by oxidative stress (Davison et al. 2022). Moreover, recent work by Voisin et al. (2023) has shown that exercise training is associated with a younger epigenetic (DNA methylation) and transcriptome profile.

Lactic acid is a metabolic by-product of glycolysis, generated following high-intensity exercise, and it has been purported to act as an endogenous HDAC inhibitor, leading to histone hyperacetylation with subsequent changes in gene expression (Wagner et al. 2015). In this sense, lactate is a salient transcriptional regulator, linking the cell metabolic state to gene transcription (Latham et al. 2012). Moreover, Zhang et al. (2019) have shown lactylation of core histone lysine residues that can directly stimulate gene transcription from chromatin, implicating lactate as a precursor substrate in control of an epigenetic modification. In a further study by von Bhagat et al. (2019) using metabolite tracing, lactate produced by neoplastic cells can increase the production of α-KG in mesenchymal stem cells, leading to α-KG-mediated activation of TET and a decrease in cytosine methylation. Studies of this nature in the context of exercise are not yet available; however, it does raise the intriguing possibility that exercise-induced lactate production can regulate the epigenetic landscape of the genome and the transcriptional capacity of coding genes (Davison et al. 2022).

7.4 Exercise, Oxidative Stress, and Redox Metabolism in the Epigenetic Landscape

Oxidative stress arises from disrupted redox signaling and ROS accumulation. ROS (e.g., ·OH) cause lipid, protein, RNA and DNA damage, and can disrupt the cell's epigenetic state. For instance, ROS influence the methylome by directly hydroxylating 5-methylcytosine (5mC) (Lewandowska and Bartoszek 2011), and through DNA oxidation and TET-mediated hydroxymethylation, a state of hypomethylation can ensue (Chia et al. 2011). As stated, ROS production and substrate oxidation can influence epigenetics, as methylation and histone-modifying enzymes depend on acetyl-CoA, α-KG, Fe^{2+}, ascorbic acid, and fumarate, tightly linking, redox balance, epigenetics, and cell metabolism (Guillaumet-Adkins et al. 2017; Davison et al. 2021). Deviation from a metabolic hemostatic set point in any of these intermediates may lead to gene expression reprogramming. The intricate nexus between oxidative stress, epigenetics and gene transcription has previously been summarized by Kreuz and Fischle (2016) and Davison et al. (2022) as follows:

- *Oxidative stress and chromatin structure*: In specific cell types, oxidative stress can lead to a loss in heterochromatin and cell death. Oxidative modifications of histones may alter chromatin structure, leading to a change in transcription and DNA replication status.
- *Oxidative stress and histone modification*: ROS production inhibits Fe(II)- and α-KG-dependent histone demethylases of the JmjC family and activates histone hypermethylation.
- *Oxidative stress and DNA methylation*: Oxidative stress-induced inhibition of DNA methyltransferases and re-localization of DNA methyltransferases to CpG islands can lead to a decrease in global DNA and promoter hypomethylation. Oxidation-induced inhibition of Fe(II)- and α-KG-dependent TET demethylases, fumarate or succinate leads to altered hypermethylation in DNA.
- *Oxidative stress regulates DNA-associated proteins*: Oxidative stress sequesters the corepressor TRIM28 from chromatin leading to activation of gene expression. Oxidized HMGB1 loses its DNA-bending activity and its ability to displace histone H1, potentially inhibiting transcriptional activation. The second messenger PI(5)P is upregulated by oxidative stress, modulating the corepressor ING2, the general transcription factor TAF3, and the chromatin-modifying protein UHRF1; these changes can interfere with global transcription.

While metabolism and epigenetics interact through various factors, it's well-known that high-intensity exercise (along with other factors such as exercise volume, dietary modifications, aging and gender), leads to a state of oxidative stress, exacerbated by the production of ROS (Davison et al. 2008). Exercise-induced ROS production generates by-products of macromolecular damage, through RNA, DNA, lipid, and protein oxidation, and to date, the complex interplay between exercise, oxidative stress, and epigenetics remains largely unexplored (Davison et al. 2022). Typical examples of ROS generated as a function of exercise include superoxide anions (O_2^-), hydrogen peroxide (H_2O_2), and hydroxyl free radicals (·OH). ROS are generated in exercise and during recovery as by-products of systemic and cellular metabolism. While the mechanisms of ROS production are extensive, in brief, NADPH-oxidases (NOX) can act as a primary source of superoxide anions (O_2^-). NOX2, which has access to the cytosolic domain, and NOX4, which is localized to mitochondria, are often cited as potential suspects in the production of O_2^- and H_2O_2 (NOX4 only) (Williamson et al. 2020a). Mitochondrial generation of O_2^- and H_2O_2 involves electron leakage from donor redox centers of the electron transport chain (ETC), leading to a state of uni- or bi-valent reduction of oxygen. There are at least 11 known mitochondrial sites of O_2^- and H_2O_2 production, encompassing NADH/NAD^+ and ubiquinol/ubiquinone (UQH2/UQ). Each site has different production capabilities (Williamson et al. 2020a). Oxidative phosphorylation in mitochondria may only be a minor contributor of ROS production in exercise, as O_2^- seems to be higher in State 4 respiration (i.e., rest) compared to State 3 (i.e., exercise) (Henriquez-Olguin et al. 2020; Goncalves et al. 2015). It is possible that the cumulative production of O_2^- and H_2O_2 following exercise and in recovery leads to persistent damage over time.

Mitochondria can influence epigenetic signaling through H_2O_2 production, assuming that any epigenetic alterations in response to ROS may, in turn, lead to altered expression of genes that regulate mitochondrial metabolism (Shaughnessy et al. 2014). The lysine-specific histone demethylases 1 and 2 (LSD1, LSD2) are known to demethylate the mono- and di-methylated Lys4 and Lys9 of histone H3 via an FAD-dependent oxidative reaction involving H_2O_2 (Forneris et al. 2008). Intriguingly, the LSD1-mediated production of H_2O_2 promotes the production of 8-oxodG (Perillo et al. 2008), implying that while free iron is reduced by the nuclear ferritin pool, H_2O_2 may still lead to the production of ·OH, which likely involves iron-containing complexes associated with the DNA structure (Gorini et al. 2021). To our knowledge, the work of Barres et al. (2012) is the only study to quantify epigenetics and ROS simultaneously following exercise of up to 80% VO_2max. Exercise remodeled human skeletal muscle promoter methylation by inducing a state of global hypomethylation, resulting in dose-dependent mRNA expression in a select number of genes, including PGC-1α, PDK4, and PPAR-δ. However, promoter methylation in differentiated L6 rat myotubes remained unchanged by 1 mM H_2O_2 exposure (note that this represents an abnormal physiological dose), even though ROS increased mRNA expression. This suggests that DNA methylation alone does not control exercise-induced gene expression.

Furthermore, physiological and pathological amounts ROS production can oxidize signaling proteins and transcription factors, as well as crucial components of the epigenetic machinery, including DNA and RNA. Various modifications (e.g., to thiol groups) can cause conformational changes that affect enzymatic activity, protein-protein interactions, and protein subcellular localization (Dimauro 2020).

We have consistently demonstrated that high-intensity exercise leads to guanine nucleotide oxidation (Fogarty et al. 2013) and DNA single- and double-strand breaks; the latter quantified using a marker of histone phosphorylation, γ-H2AX, combined with 53BP1 (Williamson et al. 2020b). Other work by Williamson et al. (2020c) shows that high-intensity (95% heart rate max) exercise increases nuclear (single-cell gel electrophoresis comet assay) and mitochondrial DNA damage (Long Amplicon-Quantitative Polymerase Chain Reaction, LA-qPCR assay) in lymphocytes and muscle, respectively. DNA bases are directly modified by ROS, where the ·OH radical plays a crucial role. Indeed, the most sensitive ROS target in DNA is the guanine dinucleotide, due primarily to its low-oxidation potential. ·OH radicals are produced through Fenton reactions, which involve the reduction of H_2O_2 by either ferrous ($k \sim 76\ M^{-1}\ s^{-1}$) or copper ions ($k \sim 4.7 \times 10^3\ M^{-1}\ s^{-1}$) (Davison et al. 2021; Davison 2016). ·OH-mediated DNA changes are initiated by hydrogen abstraction or by·OH interfering with a DNA base (Davison 2016). While not explicitly demonstrated in an exercise context, ROS-induced DNA damage can potentially interfere with epigenetic processes. ·OH production indirectly affects DNA methylation through first oxidizing guanosine to 8-oxo-2′-deoxyguanosine (8-oxodG) at the C8 position of guanine. In normal cells, 8-oxodG constitutes approximately 1 in 106 guanines. However, under conditions of oxidative stress, this ratio can increase by an order of magnitude, leading to a steady-state accumulation of 8-oxodG lesions. Interestingly, 8-oxodG then inhibits the methylation of adjacent cytosines, leading to a state of hypomethylation of promoter and transcriptional activation (Dizdaroglu et al. 2008; Radak et al. 2024), while 8-oxodG, whether opposite to cytosine or in proximity (2–3 bases), can inhibit the function of DNA methyltransferases (DNMTs) (Weitzman et al. 1994; Radak et al. 2024). Moreover, the oxidation of 5mC to 5hmC can also lead to hypomethylation at CpG sites (Madugundu et al. 2014; Dimauro 2020).

Typically, intrahelical 8-oxodG is recognized and removed from an OG:C base pair in duplex DNA by 8-oxoguanine DNA glycosylase (OGG1). OGG1 cleaves the glycosidic bond between the ribose and the base, releasing 8-oxodG and generating an apurinic (AP) site in DNA. This short-patch base excision repair (BER) mechanism involves further AP site cleavage by AP-endonuclease-1 (APE1) to resynthesize undamaged DNA using the cytosine base opposite as a template for inserting an undamaged G nucleotide (Fleming and Burrows 2020a). However, on accumulation of 8-oxodG, methylation of adjacent cytosines is attenuated, leading to a state of hypomethylation and transcriptional activation (Le and Fujimori 2012). Zhou et al. (2016) proposes a model suggesting that OGG1 promotes DNA demethylation by recruiting TET1 to the oxidized lesion. This model articulates that oxidative stress recruits the OGG1/TET1 complex proteins to 8-oxodG, facilitating the conversion of 5mC to 5caC near sites of ROS-induced damage (see Fig. 7.2). Within this regulatory framework, 8-oxodG may be regarded as an epigenetic mark, with OGG1 firstly acting as an "epigenetic reader" and secondly as an "eraser" that initiates DNA BER for removing 8-oxodG (Radak et al. 2024). Of note, and as further outlined recently by Radak et al. 2024, preservation of muscle adaptation following exercise training, may be explained by the exercise-induced increase in gene expression and activity of DNA repair molecules (e.g., OGG1). In contrast, a sedentary lifestyle can downregulate mitochondrial targeting of OGG1, suggesting that the beneficial effects of a moderate production of ROS during exercise, can lead to an

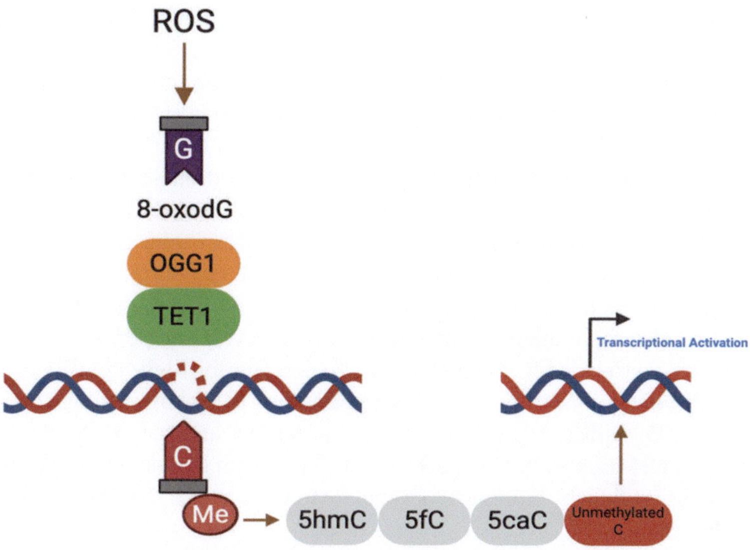

Fig. 7.2 Interface between guanine oxidation, OGG1/TET1 protein complex activation and gene expression. ROS leads to oxidative stress in the form of DNA guanine (G) damage. An accumulation of 8-oxodG recruits OGG1/TET1 complex proteins to site of damage. The adjacent methylated cytosine (C) degrades to 5caC leading to a state of unmethylated C near the site of ROS-induced damage. This facilitates an increase in transcriptional activation

enhanced state of cell adaptation and resistance against oxidative stress due to changes in methylation of promoter sequences.

In support of 8-oxodG/OGG1 acting as a key signaling complex in muscle adaptation, several studies have shown that 8-oxodG release from myofibers, activates satellite cells and promotes myoblast differentiation. This activation is mediated by the 8-oxodG/OGG1 complex, which triggers the Ras-MEK signaling pathway to enhance MyoD transcription. Unlike NEIL1 and NEIL2, OGG1 exhibits a high affinity for the free 8-oxoGua base (binding constant [Kd] = 0.56 nM) and possesses guanine nucleotide exchange factor (GEF) activity, activating small GTPases to initiate downstream cell signaling (Boldogh et al. 2012, Zheng Mattick et al. 2023, Radak et al. 2024). In exercise specifically, 8-oxodG generated from myofibers can act as a second messenger molecule to contribute to postexercise healing of injury and muscle adaptation. 8-OxodG activates satellite cells to promote myoblast differentiation via an increase in mitochondrial ROS production. In the interstitial fluid from the tibialis anterior muscle of exercising mice, 8-oxodG increases, suggesting that 8-oxodG is released from myofibers into the space surrounding satellite cells, and then subsequently imported into satellite cells to initiate muscle remodeling (Zheng et al. 2023).

Further research has revealed an intricate interplay between DNA repair, oxidatively generated 8-oxodG, and transcriptional activity in mammalian genes. For instance, Fleming et al. (2017a) demonstrated that DNA damage within

G-quadruplex-forming sequences (PQS) of a promoter can lead to a remarkable 300% increase in gene expression. This finding suggests that 8-oxodG may function as an epigenetic-like modification, while G-quadruplex-forming sequences serve as sensors of oxidative stress. G-quadruplexes are G-rich sequences that can fold into distinct DNA structures. These structures consist of four or more consecutive runs of oligo-G (specific sites that are highly reactive toward one-electron oxidation). Interspersed between these G runs are one or more loops between the G nucleotides. Within a DNA duplex, the GGG runs in one strand can fold, forming layers in G-tetrads. These exposed sites are susceptible to G oxidation through a one-electron mechanism. Central to this hypothesis is the carbonate radical anion (CO_3^-). CO_3^- is a potent one-electron oxidant formed from carbon dioxide and peroxynitrite. It can also be directly produced from bicarbonate. To note, during high-intensity exercise, muscle bicarbonate concentration can increase several fold above resting concentration. While CO_3^- (pKa < 0) exhibits greater selectivity and less oxidizing activity compared to $^{\cdot}OH$ ($E_0 = 1.78$ V vs. 2.3 V at pH 7.0), it can initiate numerous damaging reactions within a biological system (Medinas et al. 2007; Di Meo and Venditti 2020).

A G-quadruplex-forming sequence in a promoter can function as a redox sensor by transforming from a standard DNA duplex to a G-four fold upon just one G oxidation (Fleming and Burrows 2020b). Although most human gene promoters are G-rich, with approximately 40% containing potential G-4 sequences (Huppert and Balasubramanian 2007), the presence of 8-oxodG alone may not trigger a transition from a duplex to quadruplex state. However, when OGG-1 responds to an elevated generation of 8-oxodG, the AP site destabilizes the duplex structure, leading to refolding of the quadruplex if the AP is in a loop, while maintaining a stable G4. If 8-oxodG is present after G-4 folding, OGG-1 may bind for signaling purposes, and in conjunction with the binding of a redox effector factor molecule (e.g., APE1), transcription factor recruitment may occur (Fleming and Burrows 2020a). Notably, the removal of 8-oxodG from a DNA coding (sense) strand by OGG-1-mediated BER upregulates transcription, while 8-oxodG located on the template (antisense) strand of VEGF, transcription is typically downregulated by a BER-independent mechanism (Fleming et al. 2017b). In the context of cellular adaptation to oxidative stress induced by intense exercise, this data holds significant relevance. 8-OxodG in a promoter region influences the transcriptional preinitiation complex, potentially leading to the synchronization of DNA repair and the initiation of transcription. Conversely, 8-oxodG in a coding region may interfere with the progression of the transcriptional elongation complex, resulting in stalling or truncated transcription (Fleming et al. 2017b). While aspects of the proposed mechanisms as above, are speculative, particularly from an exercise perspective (also see Fig. 7.3), they warrant further scientific investigation through rigorous experimental testing in an exercise setting.

The intracellular nonenzymatic antioxidant, glutathione (GSH), is the most abundant nonprotein thiol in mammalian cells, and a major driver of redox reactions. GSH is oxidized to glutathione disulfide (GSSG) under oxidative stress conditions. Previous work has purported that any change in the GSH/GSSG redox

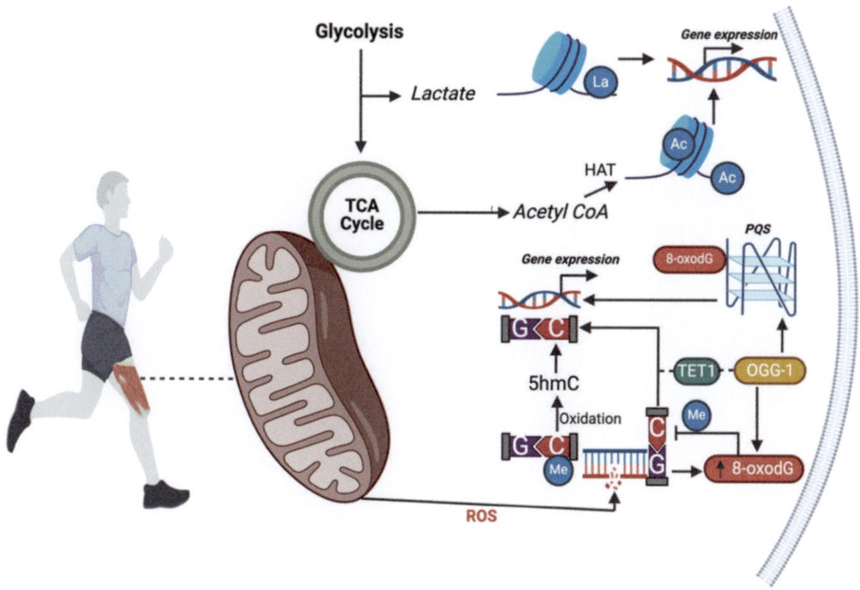

Fig. 7.3 Exercise-epigenetic nexus. Hypothetical interplay between exercise metabolism, epigenetics and gene transcription. Exercise increases ROS production leading to 5mC oxidation, DNA damage, 8-oxodG accumulation (also in DNA quadruplexes, PQS), demethylated C (cytosine) and OGG-1 activation causing transcriptional changes. OGG-1 recruits TET1 leading to hypomethylation and binds to PQS complex causing gene expression. Metabolic substrates lactate (La) and citrate control histone lactylation and acetylation respectively leading to gene expression. TCA, tricarboxylic acid cycle; Me, methyl group; Ac, acetyl group; HAT, histone acetyltransferase. (Redrawn from Davison et al. 2022)

balance, can regulate epigenetic mechanisms, through a change in (1) substrate availability, (2) enzymatic activity for DNA methylation, and (3) the expression of microRNAs (García-Giménez et al. 2014). Although strenuous exercise is known to decrease reduced GSH and increase GSSG through oxidation (Gohil et al. 1988), leading to a disrupted redox state, it is currently not known if high-intensity exercise can affect epigenetics through a decrease in GSH concentration.

7.5 Exercise and Noncoding RNAs: Mechanisms Driving Redox Metabolism and Epigenetic Modifications

MicroRNAs (miRNAs) are small noncoding RNA molecules of 19–25 nucleotides and are essential for posttranscriptional gene regulation by binding to the 3′-untranslated regions (UTRs) of target mRNAs. MiRNAs do not target a single mRNA exclusively, instead, they suppress the expression of multiple mRNAs, resulting in overlapping miRNA-regulated pathways (Torma et al. 2020). This binding allows

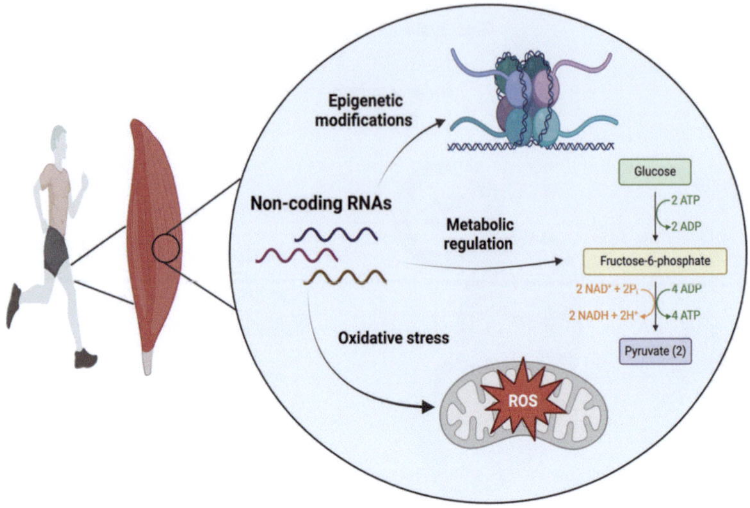

Fig. 7.4 Noncoding RNAs modulate cell processes induced by exercise. The multifaceted roles of noncoding RNAs in exercise. Noncoding RNAs modulate epigenetic modifications (e.g., chromatin remodeling), regulate oxidative stress, and influence metabolic pathways (e.g., glucose metabolism)

miRNAs to control several cell processes such as cell division, apoptosis, and the cell stress response (Bartel 2004). To the contrary, long noncoding RNAs (lncRNAs) are typically over 200 nucleotides in length and are crucial regulatory molecules involved in cell growth, cell differentiation, transcriptional regulation, RNA splicing, translation, and X chromosome inactivation (Cabili et al. 2011; Ma et al. 2016; Mattick et al. 2023).

The wide regulatory functions of noncoding RNAs (miRNAs and lncRNAs) in exercise-induced adaptations are highlighted by many tissue-specific reactions. Therefore, both these processes can play a pivotal role in skeletal muscle development (Sweta et al. 2019; Singh 2020) via redox metabolism and epigenetic alterations (Fig. 7.4), however to date, work in this area of investigation is scant.

7.5.1 miRNAs in Exercise-Induced Metabolic Regulation

The alterations in miRNA expression induced by exercise significantly impact skeletal muscle metabolism. In primary human skeletal muscle cells, overexpression of miR-29 regulates glucose uptake and insulin-stimulated glucose metabolism, as miR-29a and miR-29c overexpression can reduce expression and activity of hexokinase 2 and GLUT4 (Massart et al. 2017); two salient proteins for glucose metabolism. GLUT4 regulates cell glucose transport in response to insulin and during exercise, while hexokinase 2 catalyzes the initial step of glucose metabolism (Chadt and Al-Hasani 2020). Moreover, following endurance training, the expression of miR-29a and miR-29c is decreased in healthy young men and in the rodent model, underscoring the adaptive role of these miRNAs in modulating glucose metabolism in response to exercise.

Tissue-specific miRNA studies as a function exercise, have demonstrated that acute high-intensity resistance training alters miR-23a, miR-133a, miR-146a, miR-206, miR-378b, and miR-486 following skeletal muscle biopsy (D'Souza et al. 2017). An additional study ascertained the impact of hill running on quadriceps, gastrocnemius, and soleus muscle-miRs (miR-1, miR-133a, miR-133b, miR-206, miR-208a, miR-499) (Yin et al. 2019), showing that exercise modality is a key factor influencing muscle-miR levels. miR-1 (Wei et al. 2014) and miR-133a (Liu et al. 2008) are also important in cardiac development, while miR-206, miR-133a, and miR-133b are linked to skeletal muscle differentiation and potentially to muscle growth (Kirby and McCarthy 2013).

7.5.2 miRNAs in Exercise-Epigenetic Modifications

MiRNAs are involved in exercise-induced epigenetic changes in skeletal muscle. For example, miR-9 and miR-31 suppress the activity of histone deacetylase 4 (HDAC4) following moderate-intensity endurance cycling (Russell et al. 2013). A decrease in miR-133a expression is correlated with the upregulation of SIRT1 in muscle hypertrophy weight training (Torma et al. 2020); both HDAC4 and SIRT1 are histone deacetylases involved in chromatin condensation and gene silencing (McGee and Hargreaves 2011). SIRT1 is particularly significant in redox-sensitive histone modifications, as its deacetylation activity is linked to the cell NAD^+/NADH ratio (Davison et al. 2022), and when cells lack ATP and other metabolic substrates, NAD^+ levels tend to increase, thereby upregulating SIRT1 and other metabolic enzymes (Davison et al. 2021; Cantó et al. 2009; Etchegaray and Mostoslavsky 2016). This highlights the complex interaction among epigenetic regulation, metabolic efficiency, and redox homeostasis, with miRNAs acting as crucial mediators as a function of exercise.

7.5.3 Noncoding RNAs in Skeletal Muscle Oxidative Stress and Redox Balance

Exercise significantly impacts the redox environment in muscle tissue (Powers and Jackson 2008; Cobley and Davison 2022). The regulation of redox-sensitive ncRNAs represents a critical mechanism in exercise-mediated muscle adaptation. Oxidative stress induced by exercise can also serve as a stimulus for specific miRNA expression in skeletal muscle, while certain miRNAs actively modulate cellular redox balance. Several miRNAs enhance the cell's ability to manage oxidative stress by modulating functioning mitochondrial systems. For example, a decline in miR-31 following acute endurance and short-term exercise supports mitochondrial function via molecular targets such as nuclear respiratory factor 1 (Nrf1). Nrf1 is a key transcription factor that regulates the expression of mitochondrial-related enzymatic antioxidant genes during periods of oxidative stress (Russell et al. 2013). miR-133a is known to regulate mitochondrial biogenesis following endurance exercise training as it specifically targets the IGF-1 receptor and Akt signaling (Nie et al. 2016). This regulatory pathway can also impact oxidative stress and energy production (Piantadosi and Suliman 2012). A reduction in miRNA-7 expression coincides with an increase in SIRT3 and SOD2 expression; Sirtuin 3 (SIRT3) links directly to ATP production thus contributing to energy metabolism processes, while superoxide dismutase 2 (SOD2) reduces superoxide anions to hydrogen peroxide, thereby mitigating the effects of oxidative stress on cell damage (Dos Santos et al. 2022).

While regular exercise data compounds the significance of miRNA at controlling the oxidative stress and redox landscape, in-vivo and muscle hypertrophy models further offer controlled conditions to investigate the molecular mechanisms underpinning miRNA-mediated regulation in redox balance. For instance, a decrease in miR-133a expression is essential for the upregulation of SIRT1 activity (Torma et al. 2020); SIRT1 promotes energy production, controls the cellular antioxidant response (Juan et al. 2023), and controls select redox-sensitive proteins important for muscle remodeling (Forterre et al. 2014). Similarly, miR-214 modulates the generation of various oxidants during muscle hypertrophy through a decrease in NADPH oxidase-4 (NOX4) activity (Torma et al. 2020). In relation, NOX4 is a potential contributor to exercise-induced superoxide and basal hydrogen peroxide production (Williamson et al. 2020a). Muscle-specific deletion of myomiR miR-133a-1 and miR-133a-2 in mice leads to a reduction in PGC-1α, MtTFA (Torma et al. 2020), and citrate synthase activity (Nie et al. 2016); collectively these molecules can further impair the ROS defense mechanism.

Alternatively, lncRNAs play a significant role in controlling the skeletal muscle transcriptional response aligned to redox homeostasis. LncRNA Tug1 knockout impacts the network of genes involved in energy and redox balance. The suppression of lncRNA Tug1 disrupts the expression of the gene encoding PGC-1β (key regular of mitochondrial biogenesis) in exercise simulation in-vitro. Consequently, this disruption adversely affects oxidative stress regulation, as efficient mitochondrial biogenesis is vital for the production of antioxidant enzymes (Piantadosi and Suliman 2012). Similarly, this knockdown alters the response of several PGC-1α targets that are peripheral to core mitochondrial function, such as genes encoding the estrogen-Related Receptor family (ERRα/β/γ) (Trewin et al. 2022) and ERRα/β/γ; regulators of ROS generation and antioxidant mechanisms (Scholtes and Giguère 2021). Furthermore, Tug1 knockdown affects the expression of genes related to Ca^{2+} handling, including the downregulation of Atp2a1 which encodes sarcoplasmic/endoplasmic reticulum calcium ATPase 1 (SERCA1). Collectively these alterations cumulatively disturb mitochondrial calcium dynamics, impair ATP synthesis, and increase oxidative stress (Trewin et al. 2022), ultimately disrupting redox stability.

Similarly, several noncoding RNAs, specifically miRNAs, can indirectly enhance oxidative stress, as such, this can negatively affect redox hemostasis in muscle. To illustrate, and following endurance exercise, miR-696 and miR-23 are shown to negatively regulate PGC-1α (Aoi et al. 2010). Additionally, an increase in miR-128 expression reduces MAF BZIP Transcription Factor G (MAFG) activity in response to stress (Caggiano et al. 2017); this impairs Nrf2 activity, which may lead to a diminished muscle antioxidant response and oxidative stress.

7.6 Closing Remarks

It is abundantly clear there now exists an emerging complex relationship between metabolism, oxidative stress, and epigenetics; however, the extent to which exercise can influence and interfere with this paradigm is currently unclear. Exercise *per se* enhances the production of metabolic intermediates such as lactate, fumarate, succinate, α-KG, and ROS. ROS, as highlighted, have the potential to be regarded as a key step in shaping the epigenetic landscape, not only through their adverse (oxidation) control of DNA methylation and PTMs, but also as modulators of epigenetic mechanisms through redox signaling, and subsequent control of transcriptional regulation (Davison et al. 2022). There is clear crosstalk between exercise and miRNA function; however, the intricate relationship between exercise-induced oxidative stress and miRNA control by redox metabolism warrants in-depth scrutiny. To this end, this synopsis should be viewed as an initial point of enquiry to explore the multidirectional axis between exercise and epigenetics, and how metabolism and ROS (through oxidative stress and/or redox status) can be regarded as (emerging) key players in remodeling muscle epigenetic and transcriptional profiles. With innovation and the use of a comprehensive multi-omics approach, this area of science is ripe for scrutiny and further investigation and may include targeting specific redox states to promote beneficial epigenetic adaptations, with and without exercise training, and in clinical populations with metabolic syndrome, diabetes, or sarcopenia.

Acknowledgments Text sections of this chapter are based on previously published work by Davison G.W., and Walsh C.P. (2022) Exercise, Metabolism and Oxidative Stress in the Epigenetic Landscape. In: Cobley, JN, Davison, GW. (Eds.): Oxidative Eustress in Exercise Physiology. Boca Raton: CRC Press. pp. 209–221.

All figures created using BioRender.com

Compliance with Ethical Standards

Conflict of Interest All authors declare no conflict of interest.

Ethical Approval This chapter is a review of previously published accounts; as such, no animal or human studies were performed.

References

Aoi W, Naito Y, Mizushima K, Takanami Y, Kawai Y, Ichikawa H, Yoshikawa T (2010) The microRNA miR-696regulates PGC-1{alpha} in mouse skeletal muscle in response to physical activity. Am J Physiol Endocrinol Metab,298(4): E799-806

Barres R, Yan Y, Egan B et al (2012) Acute exercise remodels promoter methylation in human skeletal muscle. Cell Metab 15:405–411

Bartel DP (2004) MicroRNAs: genomics, biogenesis, mechanism, and function. Cell 116:281–297

Bestor TH, Laudano A, Mattaliano R, Ingram V (1988) Cloning and sequencing of a cDNA encoding DNA methyltransferase of mouse cells. The carboxyl-terminal domain of the mammalian enzyme is related to bacterial restriction methyltransferases. J Mol Biol 203:971–983

Bird AP, Taggart M, Frommer M, Miller OJ, Macleod D (1985) A fraction of the mouse genome that is derived from islands of nonmethylated, CpG-rich DNA. Cell 40:91–99

Blaschke KE, Karimi KT, MM. et al (2013) Vitamin C induces Tet-dependent DNA demethylation and a Blastocyste-like state in ES cells. Nature 500:222–226

Boldogh I, Hajas G, Aguilera-Aguirre L et al (2012) Activation of ras signaling pathway by 8-oxoguanine DNA glycosylase bound to its excision product, 8-oxoguanine. J Biol Chem 287:20769–20773

Cabili MN, Trapnell C, Goff L, Koziol M, Tazon-Vega B, Regev A, Rinn JL (2011) Integrative annotation of human large intergenic noncoding RNAs reveals global properties and specific subclasses. Genes Dev 25:1915–1927

Caggiano R, Cattaneo F, Moltedo O et al (2017) miR-128 is implicated in stress responses by targeting MAFG in skeletal muscle cells. Oxidative Med Cell Longev:9308310

Cantó C et al (2009) AMPK regulates energy expenditure by modulating NAD^+ metabolism and SIRT1 activity. Nature 458:1056–1060

Chadt A, Al-Hasani H (2020) Glucose transporters in adipose tissue, liver, and skeletal muscle in metabolic health and disease. Eur J Phys 472:1273–1298

Chia N, Wang L, Lu X et al (2011) Hypothesis: environmental regulation of 5-Hydroxymethylcytosine by oxidative stress. Epigenetics 6:853–856

Cobley JN, Davison GW (2022) Oxidative eustress in exercise physiology. CRC Press, Boca Raton

D'Souza RF et al (2017) Acute resistance exercise modulates microRNA expression profiles: combined tissue and circulatory targeted analyses. PLoS One 12(7):e0181594

Das S, Morvan F, Morozzi G et al (2017) ATP citrate Lyase regulates Myofiber differentiation and increases regeneration by altering histone acetylation. Cell Rep 21(11):3003–3011

Davison GW (2016) Exercise and oxidative damage in nucleoid DNA quantified using single cell gel electrophoresis: present and future application. Front Physiol 7:249. https://doi.org/10.3389/fphys.2016.00249

Davison GW et al (2008) Molecular detection of exercise-induced free radicals following ascorbate prophylaxis in type 1 diabetes mellitus: a randomised controlled trial. Diabetologia 51:2049–2059

Davison GW et al (2021) The metabolic-epigenetic nexus in type 2 diabetes mellitus. Free Radic Biol Med 170:194–206

Davison GW et al (2022) Exercise, metabolism and oxidative stress in the epigenetic landscape. In: Cobley JN, Davison GW (eds) Oxidative eustress in exercise physiology. CRC Press, Boca Raton, pp 209–221

Fogarty M, De Vito G, Hughes CM, Burke G, Brown JC, McEneny J, McClean C, Davison GW. (2013) Effects of α-Lipoic acid on mtDNA damage following isolated muscle contractions. Med Sci Sport Exerc 45:1469–1477

Di Meo S, Venditti P (2020) Evolution of the knowledge of free radicals and other oxidants. Oxidative Med Cell Longev

Dimauro I (2020) Exercise, redox homeostasis and the epigenetic landscape. Redox Biol 35:101477

Dizdaroglu M, Kirkali G, Jaruga P (2008) Formamidopyrimidines in DNA: mechanisms of formation, repair, and biological effects. Free Radic Biol Med 45:1610–1621

Donohoe DR, Bultman SJ (2012) Metaboloepigenetics: interrelationships between energy metabolism and epigenetics control of gene expression. J Cell Physiol 227:3169–3177

Dos Santos JAC et al (2022) Physical exercise and the functions of microRNAs. Life Sci 304:120723

Etchegaray J-P, Mostoslavsky R (2016) Interplay between metabolism and epigenetics: a nuclear adaptation to environmental changes. Mol Cell 62:695–711

Fleming AM, Burrows CJ (2020a) On the irrelevancy of hydroxyl radical to DNA damage from oxidative stress and implications for epigenetics. Chem Soc Rev 49:6524–6528

Fleming AM, Burrows CJ (2020b) Interplay of guanine oxidation and G-Quadruplex folding in gene promoters. J Am Chem Soc 142:1115–1136

Fleming AM, Ding Y, Burrows CJ (2017a) Oxidative damage is epigenetic by regulating gene transcription via base excision repair. PNAS 114:2604–2609

Fleming AM, Zhu J, Ding Y, Burrows CJ (2017b) 8-Oxo-7,8-dihydroguanine in the context of a gene promoter G-Quadruplex is an on−off switch for transcription. ACS Chem Biol 12:2417–2426

Forneris F, Binda C, Battaglioli E, Mattevi A (2008) LSD1: oxidative chemistry for multifaceted functions in chromatin regulation. Trends Biochem Sci 33:181–189

Forterre A, Jalabert A, Chikh K et al (2014) Myotube-derived exosomal miRNAs downregulate Sirtuin1 in myoblasts during muscle cell differentiation. Cell Cycle 13:78–89

García-Giménez JL et al (2014) Oxidative stress and metabolism meet epigenetic modulation in physical exercise. Free Radic Biol Med 213:123–137

Gohil K, Viguie C, Stanley WC, Brooks GA, Packer L (1988) Blood glutathione oxidation during human exercise. J Appl Physiol 1:115–119

Goncalves RLS, Quinlan CL et al (2015) Sites of superoxide and hydrogen peroxide production by muscle mitochondria assessed *ex vivo* under conditions mimicking rest and exercise. J Biol Chem 290:209–227

Gorini F, Scala G, Cooke MS et al (2021) Towards a comprehensive view of 8-oxo-7,8-dihydro-2′-deoxyguanosine: highlighting the intertwined roles of DNA damage and epigenetics in genomic instability. DNA Repair 97:103027

Grurnbaum Y, Cedar H, Razin A (1982) Substrate and sequence specificity of a eukaryotic DNA Methylase. Nature 295:620–622

Guillaumet-Adkins M, Yañez Y, Peris-Diaz MD et al (2017) Epigenetics and oxidative stress in aging. Oxidat Med Cell Aging

He Y-F, Li B-Z, Li Z et al (2011) Tet-mediated formation of 5-carboxylcytosine and its excision by TDG in mammalian DNA. Science 333:1303–1307

Henriquez-Olguin C, Meneses-Valdes R, Jensen TE (2020) Compartmentalized muscle redox signals controlling exercise metabolism – current state. Fut Challenges Redox Biol 35:101473

Hitchler MJ, Domann FE (2021) The epigenetic and morphogenetic effects of molecular oxygen and its derived reactive species in development. Free Radic Biol Med 170:70–84

Huppert JL, Balasubramanian S (2007) G-quadruplexes in promoters throughout the human genome. Nucleic Acids Res 25:406–413

Irizarry RA, Ladd-Acosta C, Wen B, Wu Z, Montano C, Onyango P, Cui H, Gabo K, Rongione M, Webster M, Ji H, Potash J, Sabunciyan S, Feinberg AP (2009) The human colon cancer methylome shows similar hypo- and hypermethylation at conserved tissue-specific CpG Island shores. Nat Genet 41:178–186

Irwin RE, Thakur A, O'Neill KM, Walsh CP (2014) 5-Hydroxymethylation marks a class of neuronal gene regulated by intragenic Methylcytosine levels. Genomics 104:383–292

Juan CG, Matchett KB, Davison GW (2023) A systematic review and meta-analysis of the SIRT1 response to exercise. Sci Rep 13(1):14752

Kaelin WG, McKnight SL (2013) Influence of metabolism on epigenetics and disease. Cell 153:56–69

Karytinos A, Forneris F, Profumo A et al (2009) A novel mammalian Flavin-dependent histone demethylase. J Biol Chem 284:17775–17782

Kawai K, Li Y-S, Song M-F, Kasai H (2010) DNA methylation by dimethyl sulfoxide and methionine sulfoxide triggered by hydroxyl radical and implications for epigenetic modifications. Bioorg Med Chem Lett 20:260–265
Keating ST, El-Osta A (2015) Epigenetics and metabolism. Circ Res 116:715–736
Kirby TJ, McCarthy JJ (2013) MicroRNAs in skeletal muscle biology and exercise adaptation. Free Radic Biol Med 64:95–105
Kreuz S, Fischle W (2016) Oxidative stress signalling to chromatin in health and disease. Epigenomics 8:843–862
Latham T, Mackay L, Sproul D et al (2012) Lactate, a product of glycolytic metabolism, inhibits histone deacetylase activity and promotes changes in gene expression. Nucleic Acids Res 40:4794 4803
Le DD, Fujimori DG (2012) Protein and nucleic acid Methylating enzymes: mechanisms and regulation. Curr Opin Chem Biol 16:507–515
Lewandowska J, Bartoszek A (2011) DNA methylation in cancer development, diagnosis and therapy – multiple opportunities for genotoxic agents to act as Methylome disruptors or Remediators. Mutagenesis 26:475–487
Liu N et al (2008) microRNA-133a regulates cardiomyocyte proliferation and suppresses smooth muscle gene expression in the heart. Genes Dev 22:3242–3254
Ma W et al (2016) miR214-regulated p53-NOX4/p66shc pathway plays a crucial role in the protective effect of Ginkgolide B against cisplatin-induced cytotoxicity in HEI-OC1 cells. Chem Biol Interact 245:72–81
Madugundu GS, Cadet J, Wagner JR (2014) Hydroxyl-radical-induced oxidation of 5-Methylcytosine in isolated and cellular DNA. Nucleic Acids Res 42:7450–7460
Martinez-Reynes I, Chandel NS (2020) Mitochondrial TCA cycle metabolites control physiology and disease. Nature Comm 11:102
Massart J et al (2017) Altered miR-29 expression in type 2 diabetes influences glucose and lipid metabolism in skeletal muscle. Diabetes 66:1807–1818
Mattick JS et al (2023) Long non-coding RNAs: definitions, functions, challenges and recommendations. Nat Rev Mol Cell Biol 24:430–447
McGee SL, Hargreaves M (2011) Histone modifications and exercise adaptations. J Appl Physiol 110:258–263
McGee SL, Walder KR (2017) Exercise and the skeletal muscle Epigenome. Cold Spring Harb Perspect Med 7(9):a029876
Medinas DB et al (2007) The carbonate radical and related oxidants derived from bicarbonate buffer. IUBMB Life 59:255–262
Mentch SJ, Locasale JW (2016) One carbon metabolism and epigenetics: understanding the specificity. Ann N Y Acad Sci 1363:91–98
Miranda-Gonçalves V, Lameirinhas A, Henrique R, Jerónimo C (2018) Metabolism and epigenetic interplay incancer: Regulation and putative therapeutic targets. Front Genet, 9:427
Neri F, Krepelova A, Incarnato D, Maldotti M, Parlato F, Galvagni F, Matarese H (2013) DNMT3L antagonizes DNA methylation at bivalent promoters and favours DNA methylation at gene bodies in ESCs. Cell 155:121–134
Nie Y et al (2016) Impaired exercise tolerance, mitochondrial biogenesis, and muscle fiber maintenance in miR-133a-deficient mice. FASEB J 30:3745–3758
Okano M et al (1999) DNA Methyltransferases Dnmt3a and Dnmt3b are essential for De novo methylation and mammalian development. Cell 99:247–257
Perillo B et al (2008) DNA oxidation as triggered by H3K9me2 demethylation drives estrogen-induced gene expression. Science 319:202–206
Piantadosi CA, Suliman HB (2012) Redox regulation of mitochondrial biogenesis. Free Radic Biol Med 53:2043–2053
Powers SK, Jackson MJ (2008) Exercise-induced oxidative stress: cellular mechanisms and impact on muscle force production. Physiol Rev 88:1243–1276

Radak Z, Pan L, Zhou L, Mozaffaritabar S, Gu YA, Pinho R, Zheng X, Ba X, Boldogh I (2024) Epigenetic and "redoxogenetic" adaptation to physical exercise. Free Radic Biol Med 210:65–74

Rodriguez H, Rafehi H, Bhave M et al (2017) Metabolism and chromatin dynamics in health and disease. Mol Asp Med 54:1–15

Russell AP et al (2013) Regulation of miRNAs in human skeletal muscle following acute endurance exercise and short-term endurance training. J Physiol 591:4637–4653

Sailani MR (2019) Lifelong physical activity is associated with promoter Hypomethylation of genes involved in metabolism, Myogenesis, contractile properties and oxidative stress resistance in aged human skeletal muscle. Sci Rep 9:3272

Scholtes C, Giguère V (2021) Transcriptional regulation of ROS homeostasis by the ERR subfamily of nuclear receptors. Antioxidants 10(3)

Seaborne RA, Sharples AP (2020) The interplay between exercise metabolism, epigenetics, and skeletal muscle Remodeling. Exerc Sport Sci Rev 48:188–200

Shaughnessy DT et al (2014) Mitochondria, energetics, epigenetics, and cellular responses to stress. Environ Health Perspect 122:1271–1278

Singh G (2020) Tiny regulators of massive tissue: MicroRNAs in skeletal muscle development, myopathies, and cancer cachexia. Front Oncol 10:598964

Sweta S et al (2019) Importance of long non-coding RNAs in the development and disease of skeletal muscle and cardiovascular lineages. Front Cell Dev Biol 7:228

Torma F, Gombos Z, Jokai M, Berkes I, Takeda M, Mimura T et al (2020) The roles of microRNA in redox metabolism and exercise-mediated adaptation. J Sport Health Sci 9:405–414

Trewin AJ et al (2022) Long non-coding RNA Tug1 modulates mitochondrial and myogenic responses to exercise in skeletal muscle. BMC Biol 20:164

Vanzan L, Sklias A, Herceg Z, Murr R (2017) Mechanisms of histone modifications. In: Tollefsbol TO (ed) Handbook of epigenetics, 2nd edn. Academic Press

Voisin S, Seale K, Jacques M, Landen S, Harvey NR, Haupt LM, Griffiths LR, Ashton KJ, Coffey VG,Thompson JL, Doering TM, Lindholm ME, Walsh C, Davison G, Irwin R, McBride C, Hansson O, AsplundO, Heikkinen AE, Eynon N (2023) Exercise is associated with younger methylome and transcriptome profilesin human skeletal muscle. Aging Cell, 00:e13859

von Bhagat TD, Ahrens D, Dawlaty M et al (2019) Lactate mediated epigenetic reprogramming regulates formation of human pancreatic cancer-associated fibroblasts. elife 8:e50663

Wagner W, Ciszewski WM, Kania KD (2015) L- and D-lactate enhance DNA repair and modulate the resistance of cervical carcinoma cells to anticancer drugs via histone deacetylase inhibition and hydroxycarboxylic acid receptor 1 activation. Cell Commun Signal 13:36

Wang Z, Long H, Chang C et al (2018) Crosstalk between metabolism and epigenetic modifications in autoimmune diseases: a comprehensive overview. Cell Mol Life Sci 75:3353–3369

Wei Y et al (2014) Multifaceted roles of miR-1s in repressing the fetal gene program in the heart. Cell Res 24:278–292

Weitzman SA et al (1994) Free radical adducts induce alterations in DNA cytosine methylation. Proc Natl Acad Sci 91:1261–1264

Williamson J et al (2020a) Targeted antioxidants in exercise-induced mitochondrial oxidative stress: emphasis on DNA damage. Antioxidants 9:1142

Williamson J et al (2020b) A combined γ-H2AX and 53BP1 approach to determine the DNA damage-repair response to exercise in hypoxia. Free Radic Biol Med 154:9–17

Williamson J et al (2020c) The mitochondrial-targeted antioxidant MitoQ, attenuates exercise-induced mitochondrial DNA damage. Redox Biol 36:101673

Wong CC, Qian Y, Yu J (2017) Interplay between epigenetics and metabolism in Oncogenesis: mechanisms and therapeutic approaches. Oncogene 36:3359–3374

Wu H, Coskun V, Tao J, Xie W, Ge W, Yoshikawa K, Li EZ, Y, Sun, YE. (2010) DNMT3a-dependent nonpromoter DNA methylation facilitates transcription of Nurosgenic genes. Science 329:444–448

Yin X et al (2019) Time-course responses of muscle-specific MicroRNAs following acute uphill or downhill exercise in Sprague-Dawley rats. Front Physiol 10:1275

Zhang D, Tang Z, Huang H et al (2019) Metabolic regulation of gene expression by histone Lactylation. Nature 574:575–583

Zheng X, Zhang W, Hu Y et al (2023) DNA repair byproduct 8-oxoguanine base promotes myoblast differentiation. Redox Biol 61:102634

Zhou X, Zhuang Z, Wang W et al (2016) OGG1 is essential in oxidative stress induced DNA demethylation. Cell Signal 28:1163–1171

Chapter 8
Exercise-Induced Epigenetics Modulation of Metabolism

Haiwang Shi and Rui Duan

Abstract Organismal metabolism commences with dietary nutrient intake, followed by digestion and absorption in the gastrointestinal tract. This process involves intricate crosstalk between host physiology and gut microbiota, with downstream effects on hepatic metabolism, pancreatic endocrine function, and skeletal muscle bioenergetics. Exercise imposes systemic metabolic stress, triggering adaptive responses in multiple tissues to meet increased energy demands of contracting muscles. Recent studies have identified exercise-derived metabolites as critical signaling molecules that modulate epigenetic machinery, including histone-modifying enzymes and DNA methyltransferases. Consequently, the complex regulatory network of metabolic epigenetic reprogramming induced by exercise has received extensive attention.

Keywords Exercise · Epigenetics · Metabolic · Metabolic reprogramming

Abbreviations

ACL	TP-citrate lyase
AMPK	AMP-activated kinase
BAT	Brown adipose tissue
CoA	Coenzyme A
C/EBP	CCAAT/enhancer-binding protein
ChIP	Chromatin immunoprecipitation
DNMTs	DNA methyltransferases
DNHDs	DNA hydroxylases
DOHaD	Developmental Origins of Health and Disease

H. Shi · R. Duan (✉)
School of Physical Education and Sports Science, South China Normal University, Guangzhou, China
e-mail: duanrui@m.scnu.edu.cn

F. Gao, X. Zhang (eds.), *Exercise, Epigenetics and Human Health*, Epigenetics and Human Health 15, https://doi.org/10.1007/978-3-032-08045-5_8

FAD	Flavin adenine dinucleotide
HATs	Histone acetyltransferases
HDACs	Histone deacetylases
HDMs	Histone demethylases
HM450K	Illumina HumanMethylation450
HMTs	Histone methyltransferases
JHDM	Jumonji-C domain-containing histone demethylases
LDH	Lactate dehydrogenase
MeCP2	Methyl-CpG binding protein
NASH	Nonalcoholic steatohepatitis
NAFLD	Nonalcoholic fatty liver disease
OGT	O-GlcNAc transferase
ROS	Reactive oxygen species
SAM	S-adenosyl methionine
SAH	S-adenosylhomocysteine
SCAN	Single Chromatin molecule Analysis in Nanochannels
SDHA	Succinate dehydrogenase complex
SOD3	Superoxide dismutase 3
SKP2	S-phase kinase-associated protein 2
TCA cycle	Tricarboxylic acid cycle
TETs	Ten-eleven translocation protein
THF	Tetrahydrofolate
T2D	Type 2 diabetes
WGBS	Whole-genome bisulfite sequencing
α-KG	α-Ketoglutarate

8.1 Introduction

Metabolism and epigenetics are fundamentally interconnected, shaping key physiological processes and mediating cellular responses to environmental stimuli. This chapter examines the modulatory role of exercise in metabolism and epigenetics, elucidating the underlying mechanisms and their implications for health and disease. To provide a structured overview, the discussion is divided into four sections: (1) mechanisms of epigenetic and metabolic regulation, (2) metabolic reprogramming across different physiological states, (3) exercise-induced epigenetic modulation, and (4) practical implications and future research directions. The integration of emerging evidence offers insights into this dynamic and evolving field.

8.2 Mechanisms of Epigenetic and Metabolic Regulation

8.2.1 Chromatin Structure and Epigenetic Modifications

As the biological carrier of eukaryotic genetic information, chromatin undergoes extensive post-translational modifications primarily targeting histone amino-terminal tails. These chemical alterations modulate DNA accessibility by influencing the recruitment of chromatin-associated effector proteins. Distinct histone modification patterns establish cooperative or competitive binding landscapes for regulatory complexes, thereby governing transitions between transcriptionally permissive and repressive chromatin states (Jenuwein and Allis 2001). This combinatorial regulatory mechanism, often referred to as the "histone code," expands the informational capacity of the genome beyond the linear nucleotide sequence.

The nucleosome, composed of DNA wrapped around an octamer of histone proteins, serves as the basic unit of chromatin organization. Chromatin exists in dynamic conformational states regulated by histone post-translational modifications (PTMs), long noncoding RNAs (lncRNAs), chromatin remodeling complexes, and molecular chaperones (Allis and Jenuwein 2016). These regulatory elements act in concert to modulate chromatin accessibility and protein-DNA interactions. Histone PTMs, such as acetylation, methylation, and phosphorylation, alter histone-DNA contacts and create binding platforms for effector proteins, thereby influencing transcription factor recruitment and chromatin remodeling enzyme activity (Yin et al. 2017). Functional genomic elements including promoters, enhancers, and exons are demarcated by distinct histone modification signatures. For instance, H3K4me3 typically marks active promoters, while H3K27 identifies enhancer regions (Neri et al. 2017). Although canonical histones provide structural stability, specialized histone variants and tail domain modifications introduce regulatory plasticity. These modifications enable spatiotemporal control of chromatin function through interactions with the underlying DNA sequence (Wolffe and Pruss 1996). The transmission of chromatin states during cell division or transgenerational inheritance-exemplified by genomic imprinting-represents a key epigenetic mechanism. Ongoing research aims to decipher how histone modifications and chromatin architecture synergistically govern epigenetic processes in health and disease (Berger et al. 2009; Allis and Jenuwein 2016).

Epigenetics refers to heritable alterations in gene expression that occur independently of DNA sequence changes, which are mediated by epigenetic machinery including DNA methylation, histone modifications, and regulatory RNAs. These molecular mechanisms govern critical biological processes such as cellular differentiation, tissue-specific gene expression, genomic imprinting, X-chromosome inactivation, and maintenance of genome stability. Meanwhile, epigenetics also serves as a significant mechanism connecting environmental factors to alterations in gene activity, thereby forging a distinct link between exercise-induced alterations and metabolic epigenetic reprogramming. In addition, considering the ongoing elevation of living standards globally and the prevalent presence of diverse fast foods and other high-energy sources, it becomes comprehensible that our bodies encounter challenges when adapting to such metabolic reprogramming.

8.2.2 Metabolic Modulation of Epigenetic Enzymes

Metabolism and epigenetics are deeply interconnected at multiple levels, with cellular metabolic flux directly influencing epigenetic regulatory networks. A central axis of this interplay involves metabolites derived from key metabolic pathways, which serve as substrates or cofactors for chromatin-modifying enzymes. For instance, S-adenosylmethionine (SAM) from one-carbon metabolism donates methyl groups for DNA and histone methylation, while α-ketoglutarate (α-KG) from the tricarboxylic acid (TCA) cycle acts as a cofactor for Jumonji domain-containing histone demethylases and ten-eleven translocation (TET) DNA hydroxylases (Lu and Thompson 2012). Numerous chromatin modifying enzymes-including DNA methyltransferases (DNMTs), histone acetyltransferases (HATs), and histone methyltransferases (HMTs)-exhibit catalytic activities dynamically regulated by metabolite availability, such as acetyl-CoA for histone acetylation, NAD^+ for sirtuin-mediated deacetylation, and ATP for chromatin remodeling complexes. Metabolic intermediates traverse nuclear pores to establish signaling conduits between cytoplasmic metabolic status and nuclear epigenetic programming: lactate produced during glycolysis activates histone deacetylases (HDACs), while kynurenine from tryptophan metabolism binds aryl hydrocarbon receptors (AhRs) to alter enhancer landscapes. This metabolite-enzyme crosstalk integrates cellular bioenergetics with chromatin accessibility and gene expression patterns, forming a regulatory framework critical for cellular function and phenotypic plasticity.

8.2.3 Acetyl-CoA and Histone Acetylation/Deacetylation

Under aerobic conditions prevailing within mammalian cells, the preponderant portion of cytosolic acetyl-CoA originates from citrate, which is translocated out of mitochondria. This translocation is mediated by a reaction catalyzed by ATP-citrate lyase (ACL). Acetyl-CoA undertakes a dual, critically important function: it powers the TCA cycle, a process indispensable for the generation of ATP, and concurrently functions as a foundational element in the synthesis of cholesterol, and other constituents essential for cell proliferation and development. Notably, in certain pathological conditions, such as cancer, where cells exhibit uncontrolled growth and proliferation, the demand for acetyl-CoA surges to support the enhanced anabolic activities. Cells may upregulate the expression of ACL and related enzymes to meet this increased demand, fueling not only the production of essential biomolecules but also potentially influencing the epigenetic regulation through histone acetylation.

Initially identified as the "activated" manifestation of acetate-owing to its energetically privileged state conducive to the donation of two-carbon moieties during anabolic biochemical reactions-acetyl-CoA serves as the substrate harnessed by HAT enzymes to effect modifications on histone tails. In mammals, HATs are segregated into five discrete families, all of which adhere to a shared enzymatic protocol: they transfer the acetyl group from acetyl-CoA onto the lysine residues of histones, with coenzyme A (CoA) being disgorged as a byproduct (Racey and Byvoet 1971). Such modifications constitute a pivotal determinant in shaping the epigenetic landscape of chromatin within eukaryotic cells (Lee and Workman 2007). Moreover, citrate arising from the metabolism of glutamine within the TCA cycle can likewise contribute to the cytoplasmic reservoir of acetyl-CoA, as corroborated by a series of studies (Takahashi et al. 2006). Notably, significant defects in histone acetylation and gene expression related to the regulation of glucose metabolism have been documented in cells where ACL expression is downregulated or knocked out (Wellen et al. 2009). In yeast, the inactivation of the enzyme responsible to produce cytoplasmic acetyl-CoA ultimately led to a widespread reduction in histone acetylation, suggesting that acetyl-CoA is a key metabolite regulating histone acetylation (Takahashi et al. 2006).

HDACs and the NAD^+-dependent Sirtuin family of deacetylases are vital enzyme families that catalyze the histone deacetylation. Among them, HDACs are recruited to their target genes through direct binding to activators and repressors of transcription and through incorporation into large multiprotein transcriptional complexes. Consequently, the specificity with which HDACs regulate distinct gene programs hinges upon multiple factors, including the identity of the cell, the range of available partner proteins within that cell, and the prevailing signaling environment of the cell (Shahbazian and Grunstein 2007). The other set of proteins belongs to the Sirtuin family of deacetylases, which display distinct functional roles, substrate specificities, and subcellular localization patterns. Despite this diversity, all isoforms harbor a conserved catalytic core that utilizes NAD^+ as a cofactor. While initially characterized as histone deacetylases, subsequent studies revealed their expanded enzymatic repertoire, including deacylase and O-ADP-ribosyltransferase activities (Blander and Guarente 2004).

The accumulation of acetyl-CoA within subcellular compartments likely demands the action of deacetylase enzymes. Their function is to counteract nonenzymatic acetylation alterations that might, either deliberately or inadvertently, disrupt the normal operation of proteins, as reported in prior research (He et al. 2012). It is postulated that the Sirtuin family of protein deacylases could potentially serve this "restorative" or "detoxifying" purpose (Shi and Tu 2015). Consistent with this hypothesis, in the absence of mitochondrial SIRT3, mitochondrial enzymes have been found to exhibit hyperacetylation, and the deacetylation of these enzymes generally results in an augmentation of their activity (Ahn et al. 2008; Hirschey et al. 2010). Moreover, the expression of SIRT3 is conspicuously upregulated under circumstances such as fasting, in response to high-fat diets, or during exercise—all of which lead to an elevation in mitochondrial acetyl-CoA levels (He et al. 2012).

Likewise, the identification of a diverse array of acylation modifications on proteins, along with the discovery of corresponding sirtuins that preferentially catalyze their removal, bolsters the notion that proteins can be modified by acyl-CoA metabolites other than acetyl-CoA (Tan et al. 2014). A prime example is the participation of the mitochondrial SIRT3 enzyme in the deacetylation of the acetyl-CoA synthase enzyme AceCS2. Under appropriate nutritional conditions, AceCS2 becomes acetylated at a specific lysine residue, which substantially impairs its capacity to convert acetate into acetyl-CoA. Furthermore, elevated levels of acetylation severely hinder the activation of PGC-1α. Fortunately, SIRT1-mediated deacetylation can reactivate it, and this process plays a coordinating role in regulation numerous genes whose products collectively govern intermediate metabolism in various tissues throughout the body (Lerin et al. 2006).

When glucose embarks on its progression through the glycolytic pathway, a fraction of it branches off into the hexosamine biosynthetic pathway. This diversion culminates in the production of N-acetylglucosamine (GlcNAc), which can be used as a substrate for O-GlcNAc transferase (OGT) to execute histone GlcNAcylation. Concurrently, the flow dynamics, or flux, within the glycolytic pathway governs the NAD^+/NADH ratio (Kierans and Taylor 2024). This ratio assumes a cardinal significance as it modulates the activity of sirtuin histone deacetylases. For instance, a diet rich in simple carbohydrates can lead to rapid and sustained increases in glucose availability, altering the glycolytic flux and subsequently influencing the activity of histone deacetylases. This, in turn, can modify the epigenetic state of chromatin, potentially affecting gene expression patterns related to metabolism and other cellular functions.

The differential activities and subcellular localizations of HDACs and sirtuins further underline the highly complex and intertwined relationship between metabolism and epigenetics, which are modulated by diverse metabolic states, offer cells a meticulously calibrated mechanism for maintaining an appropriate epigenetic landscape and metabolic balance. As cells are confronted with fluctuating nutrient availability, stressors, or developmental cues, the coordinated action of these enzymes ensurers that histone modifications, and by extension, gene expression, are adjusted in a way that sustains cellular function and enables adaptation. For instance, during exercise, the upregulation of SIRT3 in response to elevated mitochondrial acetyl-CoA levels helps to preserve mitochondrial enzyme activity and overall metabolic efficiency. At the same time, it may also exert an influence on epigenetic regulation in a manner that promotes beneficial adaptations related to energy metabolism and stress resistance. In contrast, in disease states such as cancer or neurodegenerative disorders, disruptions to the normal functioning of these enzymes can precipitate aberrant epigenetic and metabolic phenotypes, thereby spotlighting their potential as therapeutic targets. A comprehensive understanding of these processes at the molecular level paves the way for developing innovative strategies to manipulate metabolism and epigenetics, with the aim of achieving therapeutic benefits.

8.2.4 Methylation of DNA and Histones

Histone methylation rivals acetylation in epigenetic significance by modulating chromatin states (Kouzarides 2002). DNMTs and HMTs catalyze methyl group addition to cytosine residues in DNA and lysine/arginine residues in histones, respectively. Despite divergent architectures and substrate specificities, both enzyme families employ SAM as the methyl donor, transferring the group to substrates while generating S-adenosylhomocysteine (SAH) as a byproduct. Given that SAM serves as the methyl donor and SAH acts as a highly potent inhibitor of DNMTs and HMTs, and considering that the pivotal metabolic determinant of methyltransferase reactions hinges on the rate of SAH clearance (Selhub and Miller 1992), the intricate metabolism of SAM and SAH implies that multiple metabolic factors could influence the establishment of DNA and histone methylation. While methyl groups are chemically stable, recent studies have revealed dynamic methylome remodeling during embryonic development, challenging the traditional view of DNA and histone methylation as static epigenetic marks. This remodeling involves active demethylation processes mediated by enzymes such as TET hydroxylases and histone demethylases, underscoring the plasticity of epigenetic regulation during critical developmental stages.

Over the past decade, extensive research has identified multiple enzymes mediating histone and DNA methylation/demethylation, both of which rely on SAM as the principal methyl donor. While SAM serves as the universal methyl donor for methyltransferases due to its labile sulfonium-bound methyl group, tetrahydrofolate (THF) derivatives like N5-methyltetrahydrofolate (N5-MTHF) exhibit insufficient methyl transfer potential for most biosynthetic reactions. SAM biosynthesis is catalyzed by methionine adenosyltransferase (MAT), which combines methionine and ATP in the first step of the methionine cycle. The electrophilic sulfur atom in SAM activates the methyl group for efficient transfer to acceptor molecules, generating SAH as a byproduct. SAH is then hydrolyzed to homocysteine and adenosine, and the methyl cycle is reinitiated through remethylation of homocysteine using N5-MTHF, thereby regenerating methionine (Bird 2002). This metabolic network integrates one-carbon metabolism with epigenetic regulation, highlighting the dynamic interplay between cellular bioenergetics and chromatin modification.

Recent advances have identified and functionally annotated multiple histone demethylases (HDMs) and DNA demethylases, including LSD1, Jumonji-C domain-containing histone demethylases (JHDM), and TETs. LSD1 mediates FAD-dependent histone demethylation via an oxidative mechanism producing formaldehyde, with its activity linked to cellular redox status through FAD recycling (Shi et al. 2004). In contrast, JHDM enzymes utilize an α-KG/Fe(II)/O_2-dependent mechanism to remove methyl groups, generating succinate and formaldehyde (Klose and Zhang 2007). Notably, this catalytic strategy is conserved in TET enzymes, which hydroxylate 5-methylcytosine in DNA using α-KG as a cofactor (He et al. 2011; Ito et al. 2011). These findings highlight evolutionary convergence in demethylation machinery across histone and DNA substrates, underscoring the interconnectedness of epigenetic regulatory networks.

Current understanding of metabolic regulation in JHDM- and TET-dependent demethylation processes remains limited. Mechanistically, these demethylation systems are metabolically intriguing due to their reliance on α-KG and succinate-two key intermediates of the tricarboxylic acid (TCA) cycle. α-KG biosynthesis occurs during mitochondrial TCA cycle activity, where isocitrate is converted by isocitrate dehydrogenases 2 and 3 (IDH2/IDH3) to generate this critical metabolite (Chowdhury et al. 2011). α-KG biosynthesis also occurs through catabolism of spacific amino acids, including arginine, proline, histidine, and glutamine. These amino acids are first converted to glutamate, which then enters the TCA cycle. Within the mitochondrial matrix, glutamate is further oxidized by the 2-oxoglutarate dehydrogenase complex to generate succinyl-CoA. Conversely, α-KG is consumed during cysteine and lysine catabolism, which yield pyruvate and acetyl-CoA respectively. This versatile metabolite can be sourced from both glucose and glutamine, participating in multiple anabolic processes. In rapidly dividing cells, intracellular α-KG concentrations typically range from 0.5 mM to 3 mM, significantly exceeding the Michaelis constant (Km) of JHDMs (Klose and Zhang 2007; Thirstrup et al. 2011). However, under hormonal stimulation such as glucagon or vasopressin exposure, intracellular α-KG concentrations may decrease rapidly by up to 60% (Staddon and McGivan 1984). This metabolite's intercompartmental trafficking between the cytosol and mitochondria is mediated by the α-KG/malate carrier-a component of the malate-aspartate shuttle system responsible for transporting reducing equivalents (NAD^+/NADH) across the mitochondrial membrane. Recent studies indicate that the net flux through this shuttle system directly regulates α-KG exchange rates between subcellular compartments (Locasale and Cantley 2011). Collectively, these findings establish a mechanistic axis integrating cellular metabolism and epigenetic regulation through α-KG-dependent modulation of JmjC-domain histone demethylases and TET family dioxygenases. This regulatory framework not only deepens our understanding of epigenetic remodeling but also provides clues for exploring how metabolic oscillations influence transcriptional networks.

8.3 Metabolic Reprogramming Across Different Physiological States

8.3.1 Epigenetic Contributions to Obesity

The increasing prevalence of obesity is primarily driven by environmental determinants rather than solely genetic predisposition. With global improvements in living standards and the proliferation of energy-dense processed foods, modern humans face unprecedented metabolic challenges driven by chronic caloric excess. Epigenetic mechanisms provide a critical interface between environmental cues and gene expression reprogramming, linking rapid dietary shifts to obesity-related phenotypes. DNA methylation and histone acetylation-key epigenetic regulators-mediate this interplay, potentially explaining how obesity predisposes individuals to comorbidities such as insulin resistance and cardiovascular disease.

Prior investigations into obesity epigenetics employed the Infinium HumanMethylation450 BeadChip to profile DNA methylation in peripheral blood leukocytes. These studies uncovered multiple obesity-associated CpG sites in adolescent populations, with obese individuals exhibiting greater methylation variance compared to lean controls. Notably, combined analysis of differential methylation and variability achieved an accuracy of approximately 70% in predicting obesity status (Xu et al. 2013). Among these DNA methylation sites, three CpG sites were located within the HIF3A gene. The positive correlation between increased methylation and elevated BMI was further substantiated in adipose tissue. The HIF3A gene encodes hypoxia-inducible factor 3 subunit alpha (HIF-3α), which is a component of a group of heterodimeric transcription factors responsible for modulating responses to low oxygen levels, or hypoxia (Dick et al. 2014).

In another extensive research endeavor centered on DNA methylation in $CD4^+$ T cells, eight CpG sites were found to be correlated with BMI, and five were associated with waist circumference (Aslibekyan et al. 2015). The largest-scale epigenome-wide association study to date, incorporating Infinium HumanMethylation450 BeadChip data from over 10,000 whole blood samples, identified 187 CpG sites significantly associated with BMI (Wahl et al. 2017). Using Mendelian randomization analysis, investigators demonstrated that blood DNA methylation changes are generally outcomes of obesity, though they may serve as contributing factors in specific cases. Further functional validation in adipose tissue confirmed this bidirectional relationship. Notably, obesity-associated DNA methylation signatures predicted future risk of type 2 diabetes (T2D), a major obesity-related comorbidity. Among 87 BMI-correlated CpG loci, ten genes including ABCG1, SREBF1, and CPT1A exhibited altered expression, and these genes showed significant enrichment in lipid metabolism pathways (Mendelson et al. 2017).

While obesity-associated epigenetic signatures in peripheral blood are well-documented, emerging evidence highlights critical roles of tissue-specific epigenomic remodeling in obesity pathogenesis. Adipose tissue, skeletal muscle, and hypothalamic neurons exhibit distinct epigenetic landscapes that contribute to metabolic dysfunction and comorbidity development. For instance, DNA methylation patterns in adipose tissue differentiate weight loss responders from nonresponders to caloric restriction (Bouchard et al. 2010). In another study, a 5-day high-fat diet was shown to impact the DNA methylation of PGC-1α, which is involved in regulating genes related to oxidative energy metabolism (Brons et al. 2010). Additionally, early-life overnutrition also triggers epigenetic remodeling of the pro-opiomelanocortin (POMC) promoter in hypothalamic neurons, disrupting hunger-satiety signaling (Plagemann et al. 2009). While these studies have established that epigenetic modifications contribute to obesity, metabolic diseases, and their complications, it remains uncertain whether DNA methylation changes in methylation precede the onset of obesity. This lack of certainty emphasizes the need for further longitudinal studies to clarify the temporal sequence of epigenetic alterations and obesity development, which may inform new prevention and treatment strategies.

8.3.2 *Epigenetic Mechanisms in Type 2 Diabetes*

Type 2 diabetes (T2D) is characterized by chronically elevated blood glucose levels resulting from insulin resistance compounded by β-cell dysfunction. Epidemiological evidence highlights aging, physical inactivity, and adiposity as key drivers of insulin resistance in insulin-responsive tissues including skeletal muscle, liver, and adipose tissue (Hall et al. 2018). With global population aging and rising obesity rates, T2D prevalence has reached crisis proportions: the current 422 million cases are projected to escalate to 592 million by 2035. Alarmingly, diabetes accounts for 5 million deaths annually, with cardiovascular complications being the leading cause of mortality (Cho et al. 2018).

Conventional wisdom attributes aging-related pathological conditions primarily to genetic predisposition and adult lifestyle choices. However, emerging evidence underscores that the etiology of many age-associated disorders involves not only these established factors but also developmental processes occurring during early life. Mounting studies indicate that epigenetic regulatory mechanisms—responsible for dynamic gene expression control-mediate the connection between early-life adversity and subsequent risk of chronic diseases such as T2D (Bansal and Simmons 2018). The Developmental Origins of Health and Disease (DOHaD) hypothesis posits that early-life environmental challenges induce adaptive reprogramming of physiological functions and structural development, thereby increasing susceptibility to chronic disorders in adulthood (Hoffman et al. 2021). These disorders often affect key metabolic organs, including skeletal muscle, hepatic tissue, pancreatic islets, and the central nervous system (Vaiserman 2014). Organisms employ developmental plasticity to integrate environmental cues during gestation, optimizing metabolic programming based on predicted future conditions (Bateson et al. 2014). As a result of these processes, fetal exposure to malnutrition often results in pancreatic β-cell hypoplasia and systemic insulin resistance-a compensatory mechanism that may later manifest as altered energy homeostasis and dysregulated feeding behaviors in adulthood (Estampador and Franks 2018).

Exploring the long-term health impacts of prenatal exposure to the 1944–1945 Dutch Hunger Winter, investigators have identified DNA methylation alterations potentially mediating these outcomes (Fernandez-Twinn et al. 2019). Analyses of 60-year-old famine survivors identified hypomethylation at the imprinted IGF2 gene locus when compared to unexposed same-sex siblings, highlighting transgenerational epigenetic reprogramming (Heijmans et al. 2008). Conversely, the rising prevalence of maternal obesity during pregnancy drives fetal overgrowth (macrosomia/large-for-gestational-age) and adolescent-onset adiposity, processes mechanistically linked to metabolic programming and increased T2D risk in offspring (Poston et al. 2016). Epidemiological analyses of the Chinese Great Famine (1959–1961) cohort revealed that gestational famine survivors exhibit higher risks of obesity, T2D, hyperglycemia, and metabolic syndrome when compared to post-famine birth cohorts (Li and Lumey 2017). Notably, the gene-environment interaction between famine exposure and inherited genetic predisposition is thought to be a key contributor to the rapid increase in the incidence of diabetes in contemporary China (Zimmet et al. 2018).

In summary, obesity and T2D are intertwined health issues deeply influenced by multiple factors. Epigenetic mechanisms, as a bridge connecting environmental exposures and genetic regulation, offer crucial insights. Alterations in DNA methylation and histone acetylation patterns are associated with obesity and its related comorbidities, such as T2D. Studies of various tissues and cell types have revealed specific epigenetic changes associated with obesity phenotypes and predictive of T2D risk, suggesting that metabolic epigenetic reprogramming during development can induce epigenetic modifications that have lasting health effects.

8.3.3 Cancer-Related Metabolic and Epigenetic Reprogramming

Oncogenesis is fundamentally characterized by an interactive regulatory network between metabolic rewiring and epigenetic reprogramming, where metabolic intermediates act as enzymatic cofactors or substrates for chromatin modifiers-a phenomenon modulated by subcellular compartmentalization of both metabolic and epigenetic machinery. This metabolic-epigenetic axis drives oncogenic transformation by orchestrating context-dependent alterations in chromatin architecture, thereby driving tumor immune evasion (Sun et al. 2022). Concurrently, epigenetic deregulation manifests as aberrant DNA methylation profiles in human malignancies, typified by genome-wide hypomethylation and locus-specific promoter hypermethylation. Landmark studies by Vogelstein et al. first identified global DNA hypomethylation in colorectal carcinomas over four decades ago, while murine models demonstrated that systemic hypomethylation induces aggressive T-cell lymphomas (Feinberg and Vogelstein 1983). Conversely, tumor suppressor gene silencing in human cancers is mechanistically linked to promoter-specific hypermethylation (Belavy et al. 2007).

Epigenetic dysregulation modulates metabolic gene transcription, thereby coordinating metabolic reprogramming and redox homeostasis in cancer cells (Wong et al. 2017). Conversely, metabolic flux influences epigenetic regulation by modulating macromolecule biosynthesis and energy metabolism (Zheng et al. 2020). These interactions converge to drive oncogenesis. For instance, the oncoprotein c-Myc coordinates transcriptional and posttranslational regulation of glucose, glutamine, and serine metabolism while promoting SIRT3 degradation via SKP2. This leads to increased SDHA acetylation, enzyme inactivation, and succinate accumulation-a metabolite that inhibits histone demethylases, triggering H3K4me3 enrichment at oncogenic loci and subsequent tumor progression (Li et al. 2020). Notably, these regulatory events rely on metabolites including acetyl-CoA, α-KG, FAD, and succinate as enzymatic substrates or cofactors.

Developing anticancer therapies targeting histone- and DNA-associated epigenetic modifiers has garnered growing interest over recent decades. The most widely studied epigenetic inhibitors are probably HDAC inhibitors which exert their effects by blocking HDAC activity and inducing global hyperacetylation of histone and non-histone proteins, thereby modulating chromatin structure and gene transcription. Clinically, HDAC inhibitors mediate pleiotropic responses in tumor cells, eliciting direct effects such as apoptosis induction, cell cycle arrest, and differentiation while simultaneously promoting autophagy and enhancing tumor immunogenicity (Falkenberg and Johnstone 2014). Previous studies have shown that the short fatty acid n-butyrate acts as an HDAC inhibitor and induces differentiation of Friend erythroleukemia cells and robust histone hyperacetylation in HeLa cells (Riggs et al. 1977). Collectively, these findings highlight epigenetic modulation of tumor metabolism as a critical frontier in understanding cancer pathogenesis and developing novel therapeutic strategies.

8.4 Exercise-Induced Epigenetic Modulation

Contemporary research underscores the role of physical activity as a determinant of epigenetic plasticity across multiple biological systems. Consistent engagement in physical activity has been correlated with alterations in the epigenetic signatures within diverse tissues, such as skeletal muscle, adipose tissue, and the central nervous system (Fig. 8.1). In skeletal musculature, exercise-elicited epigenetic reprogramming can regulate the expression of genes implicated in metabolic pathways, mitochondrial biogenesis and function, as well as skeletal muscle fiber type specification. These epigenetic adaptations potentially underpin the enhancements in muscular performance and metabolic homeostasis. Regarding adipose tissue, exercise-induced epigenetic modifications can reconfigure the transcriptional landscape to regulate genes associated with adipocyte differentiation and lipid metabolism, thereby potentially mitigating the propensity for obesity and attendant metabolic dysregulations. Notably, certain exercise regimens have been found to improve insulin sensitivity and glucose metabolism through epigenetic regulation of key metabolic genes, which is critical for combating metabolic diseases such as T2D. In the brain, exercise-triggered epigenetic changes have been implicated in neurogenic processes, synaptic plasticity, and cognitive function. Collectively, the burgeoning body of evidence intimates that exercise functions as a potent epigenetic modulator, proffering a molecular rationale for the salutary impacts of physical activity on health promotion and disease prevention.

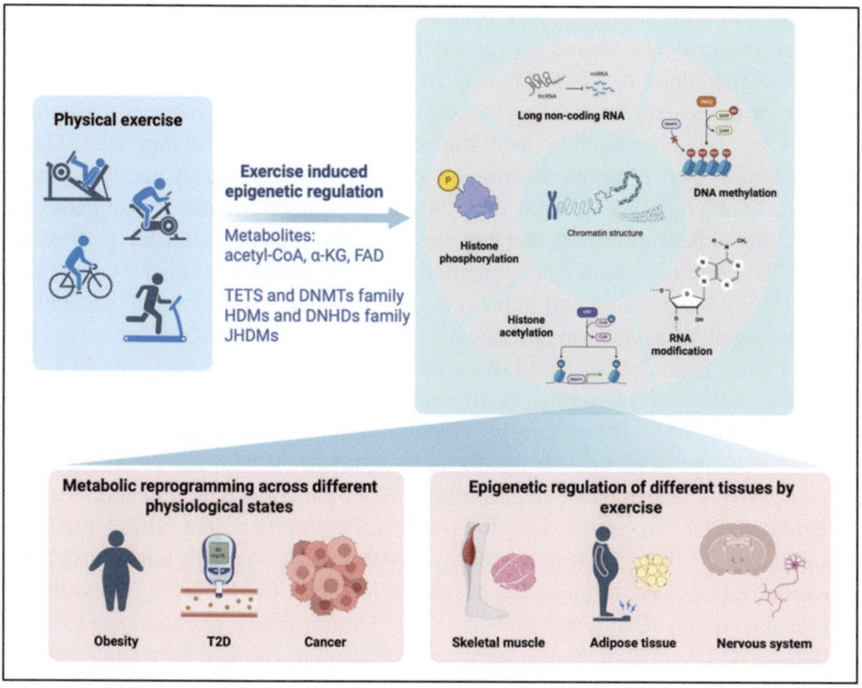

Fig. 8.1 Exercise-induced epigenetics modulation of metabolism

8.4.1 *Skeletal Muscle Epigenetics*

Skeletal muscle constitutes approximately 30–40% of adult body mass, serving as a critical metabolic organ that maintains systemic energy homeostasis. Dysfunction of this tissue is intricately linked to metabolic disorders including T2D, obesity, and sarcopenia. Exercise induces dynamic metabolic reprogramming in skeletal muscle, characterized by altered substrate utilization, cofactor availability, and redox homeostasis-processes that support both acute exercise performance and long-term adaptive remodeling. Consequently, exploring how exercise modifies the metabolic and molecular programs within cells, subsequently inducing epigenetic modifications and governing skeletal muscle remodeling post-exercise, is a potential research avenue to elucidate the role of exercise in promoting health.

Accumulating evidence demonstrates that DNA methylation-mediated epigenetic alterations underlie the pathogenesis of metabolic disorders by regulating the transcription of genes involved in systemic energy metabolism and glucose homeostasis (Nitert et al. 2012). Barres et al. systematically characterized the kinetic profiles of DNA methylation and its dose-dependent responses following acute exercise, providing novel insights into exercise-induced epigenetic dynamics (Barrès et al. 2012). Specifically, skeletal muscle biopsies were obtained pre-exercise, postexercise, and 3 hours postexercise following acute low-intensity (40% peak aerobic capacity) and high-intensity (80% peak aerobic capacity) training sessions. High-intensity exercise was shown to induce DNA hypomethylation in the promoter regions of metabolic regulators *Ppargc1α*, *Tfam*, *Mef2A*, and *Pdk4*, underscoring the role of epigenetic remodeling in exercise-mediated transcriptional reprogramming. A separate investigation involving healthy individuals with familial T2D risk demonstrated that 6-month endurance training reduced DNA methylation at *MEF2A*, *RUNX1*, *NDUFC2*, and *THADA* loci in skeletal muscle. These genes are functionally linked to retinol metabolism, calcium signaling, and the pathogenesis of T2D. Exercise-mediated hypomethylation of these loci correlated with upregulated GLUT4 expression and improved glucose uptake, suggesting a mechanistic role for epigenetic regulation in reducing T2D risk (Nitert et al. 2012).

SIRT1 has been demonstrated to establish a physical interaction with PGC-1α, thereby enabling SIRT1 to functionally modulate the acetylation status of PGC-1α (Nemoto et al. 2005). Paradoxically, a reduction in the acetylation level of PGC-1α induces a change in its nuclear localization, endowing it with enhanced transcriptional activity. Consequently, an elevation in SIRT1-mediated modification leads to upregulated PGC-1α expression. It is generally proposed that SIRT1-mediated deacetylation of PGC-1α is regulated by AMP-activated protein kinase (AMPK), whose activation depends on adenosine monophosphate (AMP) levels, as evidenced by relevant studies (Canto and Auwerx 2009). Contrastingly, in endurance exercise models, SIRT1 knockout studies have uncovered an alternative regulatory axis for PGC-1α acetylation. Specifically, it has been observed that the activity of the acetyltransferase GCN5 declines following SIRT1 KO, accompanied by a commensurate decrease in PGC-1α acetylation levels (Philp et al. 2011). Concurrently, acute exercise elicits metabolic reprogramming, generating large quantities of intermediates such as acetyl-CoA, ATP, and α-KG-key substrates for epigenetic modifiers. Moreover, during high-intensity sprinting or resistance exercise, mitochondrial acetyl-CoA supply fails to meet cellular ATP demands, triggering anaerobic glycolysis. This metabolic shift converts pyruvate to lactate via lactate dehydrogenase (LDH), providing rapid energy to working muscles (Gibb et al. 2017). Recent studies highlight lactate as a novel epigenetic metabolite capable of modulating histone lactylation-a recently discovered posttranslational modification linking metabolism to transcriptional regulation (Zhang et al. 2019). However, direct evidence linking lactate to exercise-induced epigenetic remodeling in skeletal muscle remains scarce. Given the massive lactate production during exercise, further investigation into its role in regulating muscle transcriptomic responses and functional adaptation is warranted.

In conclusion, many exercise factors secreted by skeletal muscle as an exercise organ and metabolic substrates such as lactic acid during exercise constitute a complex molecular network of exercise epigenetic regulation, which in turn regulates the exercise performance of skeletal muscle through metabolic reprogramming, which will be a potential direction of exercise-promoting health-related research.

8.4.2 Systemic Effects on Adipose Tissue

Adipose tissue functions as both an energy reservoir and an endocrine organ, secreting adipokines and cytokines that modulate systemic metabolism and physiological processes including energy expenditure, thermogenesis, appetite regulation, and glycemic control (Galic et al. 2010). Adipose tissue expansion and obesity arise through adipocyte hypertrophy, preadipocyte hyperplasia, and enhanced adipogenic differentiation. Emerging evidence highlights a bidirectional regulatory network between adipogenic transcription factors and chromatin remodeling complexes that controls adipocyte fate. For instance, during 3T3-L1 differentiation, the CCAAT/enhancer-binding protein α (C/EBPα) promoter undergoes hypermethylation, whereas peroxisome proliferator-activated receptor γ (PPARγ) promoter hypomethylation correlates with transcriptional activation (Fujiki et al. 2009). Chromatin accessibility profiling reveals histone H3 lysine 27 acetylation (H3K27ac) marks active enhancers at lipogenic gene promoters to facilitate transcription factor recruitment (Mikkelsen et al. 2010), while DNA methylation dynamics regulate adipose lineage commitment and differentiation. Demethylation of late adipogenic gene promoters including GLUT4 and LEP coincides with their expression in mature adipocytes, suggesting epigenetic alterations during adipogenesis (Yokomori et al. 1999; Pinnick and Karpe 2011). These findings establish epigenetic regulation as a key determinant of adipose tissue plasticity and associated metabolic dysfunction.

Epigenetic signals such as DNA methylation and histone modifications play an important role in regulating adipogensis. A longitudinal study demonstrated that 6-month endurance exercise training induced global DNA methylation increases in adipose tissue of sedentary males, with differential methylation observed at 4% of analyzed CpG dinucleotide sites (Barres and Zierath 2011). Functional analysis revealed altered mRNA expression in 33% of methylated gene regions, establishing a mechanistic link between epigenetic changes and transcriptional changes. Notably, exercise training modulated DNA methylation at 39 obesity- and T2D-associated loci in adipose tissue, suggesting that exercise-induced DNA methylation is involved in adipose tissue metabolism and potentially offers protection against related diseases (Ronn et al. 2013). Furthermore, epigenetic dysregulation contributes to non-alcoholic fatty liver disease (NAFLD) progression from hepatic steatosis to steatohepatitis-a precursor to hepatocellular carcinoma and cardiovascular disease (Perakakis et al. 2020). Combined exercise-diet interventions induced hypomethylation of critical hepatic genes such as GAB2 (GRB2-associated binding protein 2), a regulator of hepatic lipid accumulation (Wu et al. 2021). These epigenetic changes correlated with reduced steatosis and improved insulin sensitivity, highlighting the therapeutic potential of epigenetic reprogramming in metabolic disease management.

In conclusion, the emerging evidence strongly suggests that epigenetic modifications, especially those in adipose tissue caused by exercise and diet interventions, hold great promise as potential therapeutic targets and biomarkers for metabolic disorders and deserve further exploration and research.

8.4.3 Neurological Benefits of Exercise

Accumulating evidence highlights exercise-induced neuroprotection in both humans and preclinical models, particularly in age-related neurodegenerative conditions. Exercise-mediated improvements in cognitive function are associated with enhanced synaptic plasticity, hippocampal neurogenesis, and upregulated expression of neurotrophic factors. Preclinical studies demonstrate that exercise modulates epigenetic mechanisms, including DNA methylation and histone modifications, to promote transcriptional programs underlying neuronal survival and connectivity. These findings underscore the therapeutic potential of physical activity for brain remodeling and functional recovery in neurological disorders.

Exercise modulates histone-modifying enzymes, balancing histone acetyltransferase (HAT) and histone deacetylase (HDAC) activities to influence chromatin remodeling and gene transcription in the hippocampus (Maejima et al. 2018). Preclinical studies demonstrate that exercise upregulates HAT activity—which is mediated by CREB-binding protein (CBP) and p300—and downregulates HDAC2/3 expression, resulting in increased histone H3K9ac and H4K12ac marks at promoters of neurogenic genes (Elsner et al. 2011). These epigenetic changes correlate with reduced neuronal apoptosis and enhanced survival in stress models. Notably, in senescence-accelerated mouse prone 8 (SAMP8) models, exercise-induced HAT activation and HDAC inhibition coincide with chromatin remodeling at autophagy-related genes (Atg7, LC3) and memory-associated transcription factors (c-Fos, BDNF), leading to improved spatial learning and memory retention (Saha and Pahan 2006; Cosin-Tomas et al. 2014). In stress-exposed rats, exercise elicits histone H3S10 phosphorylation-coupled acetylation at promoters of immediate early genes (Arc, Egr1), promoting neurogenesis in the dentate gyrus and reducing stress-induced immobility behavior (Collins et al. 2009). These findings establish exercise as a regulator of epigenetic plasticity in the brain, with therapeutic implications for aging and stress-related disorders.

Beyond modulating HATs and deacetylases HDACs, physical activity enhances BDNF gene expression through epigenetic remodeling of the *Bdnf* promoter IV, particularly via exercise-induced hypomethylation (Vaynman et al. 2004). This is accompanied by increased phosphorylation of methyl-CpG binding protein 2 (MeCP2), which dissociates from methylated DNA and relieves transcriptional repression of BDNF (Gomez-Pinilla et al. 2011). Under resting conditions, MeCP2 binds methylated CpG sites to suppress BDNF transcription; however, neuronal depolarization during exercise disrupts this interaction (Chen et al. 2003). Additionally, exercise activates Ca^{2+}/calmodulin-dependent protein kinase II (CaMKII) and cyclic adenosine monophosphate (cAMP)-response element-binding protein (CREB) in the hippocampus, critical signaling nodes for activity-dependent synaptic remodeling and cognitive enhancement (Gomez-Pinilla et al. 2011). Collectively, these mechanisms demonstrate how physical activity exerts neuroprotective effects by orchestrating chromatin accessibility at the *Bdnf* locus and downstream transcriptional programs.

In summary, physical exercise not only serves to enhance overall health but also augments brain remodeling. It achieves this by regulating epigenetic information, fostering neuroplasticity, and modulating the expression of neurotrophic factors. These processes are integral to the maintenance and improvement of brain function, highlighting the significance of physical activity in promoting neurological well-being and resilience.

8.5 Practical Implications and Future Research Directions

8.5.1 Translational and Intergenerational Implications

Maternal and paternal obesity, along with T2D, have been firmly established as prominent risk factors that predispose offspring to the development of metabolic disorders. These parental health conditions can instigate lifelong metabolic dysfunctions in their children, which may endure even if the offspring lead a healthy lifestyle. Interestingly, previous studies reveal that consistent physical activity by both mothers and fathers can confer a host of beneficial effects on the well-being of their offspring. In rodent models, for instance, regular parental exercise has been demonstrated to play a preventative role against metabolic diseases, especially diabetes. Notably, it appears to exert a positive influence on offspring glucose homeostasis and prompt favorable metabolic adaptations within vital tissues like the liver and muscle.

Maternal health occupies a central position in dictating the metabolic health of offspring, significantly swaying their susceptibility to obesity, T2D, and other metabolic disorders. The dietary choices and physical activity levels of mothers during pregnancy can modify fetal development through diverse mechanisms, with epigenetic regulation being a crucial one. While suboptimal maternal nutrition, particularly high-fat diets, amplifies the risk of metabolic diseases in offspring, maternal exercise has emerged as a potent protective factor. A high-fat diet during pregnancy can trigger hypermethylation of genes such as *Ppargc1α*, a pivotal regulator of mitochondrial biogenesis and energy metabolism. This epigenetic alteration leads to persistent metabolic impairments in offspring, including compromised glucose tolerance, insulin resistance, and liver dysfunction. In contrast, engaging in exercise during pregnancy has been shown to shield against these adverse outcomes. Exercise can temper the harmful impacts of maternal obesity by modulating epigenetic marks and signaling pathways that govern gene expression in offspring (Laker et al. 2014). Specifically, it promotes the expression of superoxide dismutase 3 (SOD3) in the placenta, which in turn activates the AMPK pathway in fetal liver cells. This activation prompts DNA demethylation at critical metabolic gene promoters, consequently enhancing glucose metabolism. This process effectively thwarts the hypermethylation of genes like Ppargc1α, restoring its expression and bolstering mitochondrial function in both liver and muscle tissues (Kusuyama et al. 2022).

Additionally, maternal exercise exerts an influence on DNA methylation patterns in tissues such as brown adipose tissue (BAT) and skeletal muscle, augmenting thermogenesis, mitochondrial biogenesis, and muscle endurance in offspring. Recent studies have shown that maternal exercise upregulates the expression of the adipokine serine protease inhibitor A3C (SERPINA3C), promotes the demethylation of the promoter of the Krüppel-like factor 4 (Klf4) gene in fetal pre-adipocytes, inhibits adipose inflammation caused by a high-fat diet in the offspring, and improves the metabolic health of the offspring (Li et al. 2025). These effects vividly illustrate how exercise not only counteracts the detrimental consequences of maternal obesity but also bequeaths lasting metabolic advantages that span generations (Son et al. 2020).

Paternal diet and lifestyle, especially high-fat diets and exercise habits, also wield a significant influence on offspring metabolic health. Parental high-fat feeding has been associated with impaired glucose tolerance, insulin resistance, and increased adiposity in their offspring (McPherson et al. 2017). Conversely, paternal exercise, encompassing physical activity and resistance training, can mitigate these adverse effects. It does so by enhancing sperm physiology, modifying sperm RNA populations, and rectifying epigenetic aberrations (Claycombe-Larson et al. 2020), particularly those related to DNA methylation and microRNA (miRNA) profiles (McPherson et al. 2017). These changes potentially lower the risk of metabolic disorders in offspring. Moreover, exercise bolsters antioxidant defenses and redox signaling in fathers, which not only promotes cellular homeostasis within their own bodies but also offers protection against oxidative stress to their offspring. Collectively, these findings accentuate the intergenerational implications of parental health choices and emphasize the importance of lifestyle modifications for safeguarding the long-term metabolic well-being of future generations.

Transgenerational epigenetics in mammals remains a controversial topic. *Takahashi* et al. constructed DNA-methylated mouse embryonic stem cells, and then obtained methylated-edited mice with an aberrant metabolic phenotype by microinjection of methylated embryonic stem cells. Notably, acquired methylation and related phenotypic traits targeting CpG islands are not only maintained, but also passed on over multiple consecutive generations (Burmester et al. 2023). Our latest study shows for the first time that maternal exercise enhances the synthesis and secretion of vitamin C in the mother's liver, upregulates the expression of TET2 during fetal skeletal muscle development, reprogramming DNA methylation of skeletal muscle in offspring, and intergenerational increases the differentiation tendency of oxidative fibers in skeletal muscle and the expression of genes related to mitochondrial biogenesis. Enhance the endurance performance of the offspring (Shi et al. 2025). This finding represents preliminary evidence suggesting intergenerational inheritance of acquired DNA methylation. Parental exercise has been shown to enhance offspring health via diverse forms of epigenetic regulation. However, it is regrettable that there is a paucity of reported investigations regarding the transgenerational genetic impacts of exercise on offspring health. This area, therefore, presents a promising and worthy research direction that warrants further exploration and scrutiny to elucidate the underlying mechanisms and potential implications.

8.5.2 Emerging Technologies

With the exponential growth of epigenetics research, there is increasing interest in developing novel technologies to characterize epigenetic landscapes underlying health and disease states. The field has witnessed the development of innovative techniques capable of profiling chromatin states at multiple scales—from single-locus resolution to genome-wide mapping—enabling comprehensive analysis of histone modifications, DNA methylation, and chromatin accessibility.

DNA methylation remains the most extensively studied epigenetic modification, with a diverse array of analytical methods enabling both quantitative and qualitative assessments of methylated cytosines. Bisulfite conversion serves as the cornerstone for many methylation profiling workflows, and technological advancements over recent decades have revolutionized its application. Hybridization-based platforms like the Illumina HumanMethylation450 array and high-throughput sequencing approaches such as whole-genome bisulfite sequencing (WGBS) now allow rapid genome-wide methylation characterization. In addition, single-cell bisulfite sequencing represents a transformative development, which leverages single-cell omics platforms to profile methylation landscapes at unprecedented resolution. Integrative approaches that combine single-cell DNA methylation with transcriptomic and metabolomic analyses are uncovering functional relationships and dynamic networks across multi-omic layers, providing mechanistic insights into epigenetic regulation in complex biological systems.

Chromatin immunoprecipitation (ChIP) remains the gold-standard method for profiling histone modifications and transcription factor binding events (Bannister and Kouzarides 2011). Complementary to ChIP, chromosome conformation capture (3C) technologies characterize chromatin folding patterns that link genomic architecture to biological function (Qi and Zhang 2019). Integrative approaches combining ChIP-seq with 3C-based methods like Hi-C now map epigenetic landscapes within 3D chromatin structures, revealing how histone modification patterns and chromosomal spatial organization synergistically regulate eukaryotic gene expression (Di Pierro et al. 2017). Advanced methods combining epigenetic modifier-specific antibodies with single-molecule visualization approaches, such as Single Chromatin Analysis in Nanochannels (SCAN), enable high-throughput co-localization fluorescent profiling of combinatorial epigenomic landscapes on individual nucleosomes (Murphy et al. 2013). SCAN further captures the dynamic interplay between DNA methylation, histone modifications, and methylation-sensitive transcription factors during transcriptional regulation, providing unprecedented resolution into epigenetic crosstalk.

Noncoding RNAs (ncRNAs) represent a diverse class of RNA transcripts that do not encode proteins. Advancements in high-throughput sequencing and bioinformatics have enabled the discovery of novel ncRNAs species and helped elucidate their roles in gene regulatory networks. Given their ability to interact with coding and noncoding RNAs, DNA, and proteins, innovative methodologies integrating immunoprecipitation with next-generation sequencing (NGS) now map chromatin occupancy sites for specific ncRNAs and protein complexes. These techniques provide mechanistic insights into ncRNA-mediated epigenetic regulation (Hafner et al. 2010).

With rapid development in the epigenetics field, many powerful techniques have been discovered to access epigenetic functions and mechanisms of the epigenome and its associated proteins that have driven the remarkable advances in epigenetics. Numerous methods have been developed that help to provide better, precise, accurate and systematic understanding of biological functions of epigenetic regulations. Thus, the quick advances in a broad range of overall biological technologies will likely improve interpretation of a massive amount of information in the mapping of exercise regulates epigenomics.

8.6 Conclusion

Exercise emerges as a potent modulator of both metabolism and epigenetics, offering substantial benefits for health promotion and disease prevention. By bridging these domains, this chapter underscores the transformative potential of regular physical activity in shaping human health. Continued research into exercise-induced epigenetic changes will not only deepen our understanding of physiological adaptation but also facilitate the development of innovative therapeutic strategies for addressing chronic diseases.

Acknowledgments We thank current and former members of the Regenerative Medicine in Sports Science laboratory for their contributions.

Compliance with Ethical Standards

Conflict of Interest The authors declare no conflict of interest.

Ethical Approval This chapter is a review of previously published accounts, as such, no animal or human studies were performed.

References

Ahn BH, Kim HS, Song S, Lee IH, Liu J, Vassilopoulos A, Deng CX, Finkel T (2008) A role for the mitochondrial deacetylase Sirt3 in regulating energy homeostasis. Proc Natl Acad Sci USA 105(38):14447–14452. https://doi.org/10.1073/pnas.0803790105

Allis CD, Jenuwein T (2016) The molecular hallmarks of epigenetic control. Nat Rev Genet 17(8):487–500. https://doi.org/10.1038/nrg.2016.59

Aslibekyan S, Demerath EW, Mendelson M, Zhi D, Guan W, Liang L, Sha J, Pankow JS, Liu C, Irvin MR, Fornage M, Hidalgo B, Lin LA, Thibeault KS, Bressler J, Tsai MY, Grove ML, Hopkins PN, Boerwinkle E, Borecki IB, Ordovas JM, Levy D, Tiwari HK, Absher DM, Arnett

DK (2015) Epigenome-wide study identifies novel methylation loci associated with body mass index and waist circumference. Obesity (Silver Spring) 23(7):1493–1501. https://doi.org/10.1002/oby.21111

Bannister AJ, Kouzarides T (2011) Regulation of chromatin by histone modifications. Cell Res 21(3):381–395. https://doi.org/10.1038/cr.2011.22

Bansal A, Simmons RA (2018) Epigenetics and developmental origins of diabetes: correlation or causation? Am J Physiol Endocrinol Metab 315(1):E15–E28. https://doi.org/10.1152/ajpendo.00424.2017

Barres R, Zierath JR (2011) DNA methylation in metabolic disorders. Am J Clin Nutr 93(4):897S–900S. https://doi.org/10.3945/ajcn.110.001933

Barrès R, Yan J, Egan B, Treebak JT, Rasmussen M, Fritz T, Caidahl K, Krook A, O'Gorman DJ, Zierath JR (2012) Acute exercise remodels promoter methylation in human skeletal muscle. Cell Metab 15(3):405–411. https://doi.org/10.1016/j.cmet.2012.01.001

Bateson P, Gluckman P, Hanson M (2014) The biology of developmental plasticity and the predictive adaptive response hypothesis. J Physiol 592(11):2357–2368. https://doi.org/10.1113/jphysiol.2014.271460

Belavy DL, Richardson CA, Wilson SJ, Rittweger J, Felsenberg D (2007) Superficial lumbopelvic muscle overactivity and decreased cocontraction after 8 weeks of bed rest. Spine (Phila Pa 1976) 32(1):E23–E29. https://doi.org/10.1097/01.brs.0000250170.53746.27

Berger SL, Kouzarides T, Shiekhattar R, Shilatifard A (2009) An operational definition of epigenetics. Genes Dev 23(7):781–783. https://doi.org/10.1101/gad.1787609

Bird A (2002) DNA methylation patterns and epigenetic memory. Genes Dev 16(1):6–21. https://doi.org/10.1101/gad.947102

Blander G, Guarente L (2004) The Sir2 family of protein deacetylases. Annu Rev Biochem 73:417–435. https://doi.org/10.1146/annurev.biochem.73.011303.073651

Bouchard L, Rabasa-Lhoret R, Faraj M, Lavoie ME, Mill J, Perusse L, Vohl MC (2010) Differential epigenomic and transcriptomic responses in subcutaneous adipose tissue between low and high responders to caloric restriction. Am J Clin Nutr 91(2):309–320. https://doi.org/10.3945/ajcn.2009.28085

Brons C, Jacobsen S, Nilsson E, Ronn T, Jensen CB, Storgaard H, Poulsen P, Groop L, Ling C, Astrup A, Vaag A (2010) Deoxyribonucleic acid methylation and gene expression of PPARGC1A in human muscle is influenced by high-fat overfeeding in a birth-weight-dependent manner. J Clin Endocrinol Metab 95(6):3048–3056. https://doi.org/10.1210/jc.2009-2413

Burmester GR, Cohen SB, Winthrop KL, Nash P, Irvine AD, Deodhar A, Mysler E, Tanaka Y, Liu J, Lacerda AP, Palac H, Shaw T, Mease PJ, Guttman-Yassky E (2023) Safety profile of upadacitinib over 15 000 patient-years across rheumatoid arthritis, psoriatic arthritis, ankylosing spondylitis and atopic dermatitis. RMD Open 9(1). https://doi.org/10.1136/rmdopen-2022-002735

Canto C, Auwerx J (2009) PGC-1alpha, SIRT1 and AMPK, an energy sensing network that controls energy expenditure. Curr Opin Lipidol 20(2):98–105. https://doi.org/10.1097/MOL.0b013e328328d0a4

Chen WG, Chang Q, Lin Y, Meissner A, West AE, Griffith EC, Jaenisch R, Greenberg ME (2003) Derepression of BDNF transcription involves calcium-dependent phosphorylation of MeCP2. Science 302(5646):885–889. https://doi.org/10.1126/science.1086446

Cho NH, Shaw JE, Karuranga S, Huang Y, da Rocha Fernandes JD, Ohlrogge AW, Malanda B (2018) IDF diabetes atlas: global estimates of diabetes prevalence for 2017 and projections for 2045. Diabetes Res Clin Pract 138:271–281. https://doi.org/10.1016/j.diabres.2018.02.023

Chowdhury R, Yeoh KK, Tian YM, Hillringhaus L, Bagg EA, Rose NR, Leung IK, Li XS, Woon EC, Yang M, McDonough MA, King ON, Clifton IJ, Klose RJ, Claridge TD, Ratcliffe PJ, Schofield CJ, Kawamura A (2011) The oncometabolite 2-hydroxyglutarate inhibits histone lysine demethylases. EMBO Rep 12(5):463–469. https://doi.org/10.1038/embor.2011.43

Claycombe-Larson KG, Bundy AN, Roemmich JN (2020) Paternal high-fat diet and exercise regulate sperm miRNA and histone methylation to modify placental inflammation, nutrient transporter mRNA expression and fetal weight in a sex-dependent manner. J Nutr Biochem 81:108373. https://doi.org/10.1016/j.jnutbio.2020.108373

Collins A, Hill LE, Chandramohan Y, Whitcomb D, Droste SK, Reul JM (2009) Exercise improves cognitive responses to psychological stress through enhancement of epigenetic mechanisms and gene expression in the dentate gyrus. PLoS One 4(1):e4330. https://doi.org/10.1371/journal.pone.0004330

Cosin-Tomas M, Alvarez-Lopez MJ, Sanchez-Roige S, Lalanza JF, Bayod S, Sanfeliu C, Pallas M, Escorihuela RM, Kaliman P (2014) Epigenetic alterations in hippocampus of SAMP8 senescent mice and modulation by voluntary physical exercise. Front Aging Neurosci 6:51. https://doi.org/10.3389/fnagi.2014.00051

Di Pierro M, Cheng RR, Lieberman Aiden E, Wolynes PG, Onuchic JN (2017) De novo prediction of human chromosome structures: epigenetic marking patterns encode genome architecture. Proc Natl Acad Sci USA 114(46):12126–12131. https://doi.org/10.1073/pnas.1714980114

Dick KJ, Nelson CP, Tsaprouni L, Sandling JK, Aissi D, Wahl S, Meduri E, Morange PE, Gagnon F, Grallert H, Waldenberger M, Peters A, Erdmann J, Hengstenberg C, Cambien F, Goodall AH, Ouwehand WH, Schunkert H, Thompson JR, Spector TD, Gieger C, Tregouet DA, Deloukas P, Samani NJ (2014) DNA methylation and body-mass index: a genome-wide analysis. Lancet 383(9933):1990–1998. https://doi.org/10.1016/S0140-6736(13)62674-4

Elsner VR, Lovatel GA, Bertoldi K, Vanzella C, Santos FM, Spindler C, de Almeida EF, Nardin P, Siqueira IR (2011) Effect of different exercise protocols on histone acetyltransferases and histone deacetylases activities in rat hippocampus. Neuroscience 192:580–587. https://doi.org/10.1016/j.neuroscience.2011.06.066

Estampador AC, Franks PW (2018) Precision medicine in obesity and type 2 diabetes: the relevance of early-life exposures. Clin Chem 64(1):130–141. https://doi.org/10.1373/clinchem.2017.273540

Falkenberg KJ, Johnstone RW (2014) Histone deacetylases and their inhibitors in cancer, neurological diseases and immune disorders. Nat Rev Drug Discov 13(9):673–691. https://doi.org/10.1038/nrd4360

Feinberg AP, Vogelstein B (1983) Hypomethylation distinguishes genes of some human cancers from their normal counterparts. Nature 301(5895):89–92. https://doi.org/10.1038/301089a0

Fernandez-Twinn DS, Hjort L, Novakovic B, Ozanne SE, Saffery R (2019) Intrauterine programming of obesity and type 2 diabetes. Diabetologia 62(10):1789–1801. https://doi.org/10.1007/s00125-019-4951-9

Fujiki K, Kano F, Shiota K, Murata M (2009) Expression of the peroxisome proliferator activated receptor gamma gene is repressed by DNA methylation in visceral adipose tissue of mouse models of diabetes. BMC Biol 7:38. https://doi.org/10.1186/1741-7007-7-38

Galic S, Oakhill JS, Steinberg GR (2010) Adipose tissue as an endocrine organ. Mol Cell Endocrinol 316(2):129–139. https://doi.org/10.1016/j.mce.2009.08.018

Gibb AA, Epstein PN, Uchida S, Zheng Y, McNally LA, Obal D, Katragadda K, Trainor P, Conklin DJ, Brittian KR, Tseng MT, Wang J, Jones SP, Bhatnagar A, Hill BG (2017) Exercise-induced changes in glucose metabolism promote physiological cardiac growth. Circulation 136(22):2144–2157. https://doi.org/10.1161/CIRCULATIONAHA.117.028274

Gomez-Pinilla F, Zhuang Y, Feng J, Ying Z, Fan G (2011) Exercise impacts brain-derived neurotrophic factor plasticity by engaging mechanisms of epigenetic regulation. Eur J Neurosci 33(3):383–390. https://doi.org/10.1111/j.1460-9568.2010.07508.x

Hafner M, Landthaler M, Burger L, Khorshid M, Hausser J, Berninger P, Rothballer A, Ascano M Jr, Jungkamp AC, Munschauer M, Ulrich A, Wardle GS, Dewell S, Zavolan M, Tuschl T (2010) Transcriptome-wide identification of RNA-binding protein and microRNA target sites by PAR-CLIP. Cell 141(1):129–141. https://doi.org/10.1016/j.cell.2010.03.009

Hall E, Dekker Nitert M, Volkov P, Malmgren S, Mulder H, Bacos K, Ling C (2018) The effects of high glucose exposure on global gene expression and DNA methylation in human pancreatic islets. Mol Cell Endocrinol 472:57–67. https://doi.org/10.1016/j.mce.2017.11.019

He YF, Li BZ, Li Z, Liu P, Wang Y, Tang Q, Ding J, Jia Y, Chen Z, Li L, Sun Y, Li X, Dai Q, Song CX, Zhang K, He C, Xu GL (2011) Tet-mediated formation of 5-carboxylcytosine and its excision by TDG in mammalian DNA. Science 333(6047):1303–1307. https://doi.org/10.1126/science.1210944

He W, Newman JC, Wang MZ, Ho L, Verdin E (2012) Mitochondrial sirtuins: regulators of protein acylation and metabolism. Trends Endocrinol Metab 23(9):467–476. https://doi.org/10.1016/j.tem.2012.07.004

Heijmans BT, Tobi EW, Stein AD, Putter H, Blauw GJ, Susser ES, Slagboom PE, Lumey LH (2008) Persistent epigenetic differences associated with prenatal exposure to famine in humans. Proc Natl Acad Sci USA 105(44):17046–17049. https://doi.org/10.1073/pnas.0806560105

Hirschey MD, Shimazu T, Goetzman E, Jing E, Schwer B, Lombard DB, Grueter CA, Harris C, Biddinger S, Ilkayeva OR, Stevens RD, Li Y, Saha AK, Ruderman NB, Bain JR, Newgard CB, Farese RV Jr, Alt FW, Kahn CR, Verdin E (2010) SIRT3 regulates mitochondrial fatty-acid oxidation by reversible enzyme deacetylation. Nature 464(7285):121–125. https://doi.org/10.1038/nature08778

Hoffman DJ, Powell TL, Barrett ES, Hardy DB (2021) Developmental origins of metabolic diseases. Physiol Rev 101(3):739–795. https://doi.org/10.1152/physrev.00002.2020

Ito S, Shen L, Dai Q, Wu SC, Collins LB, Swenberg JA, He C, Zhang Y (2011) Tet proteins can convert 5-methylcytosine to 5-formylcytosine and 5-carboxylcytosine. Science 333(6047):1300–1303. https://doi.org/10.1126/science.1210597

Jenuwein T, Allis CD (2001) Translating the histone code. Science 293(5532):1074–1080. https://doi.org/10.1126/science.1063127

Kierans SJ, Taylor CT (2024) Glycolysis: a multifaceted metabolic pathway and signaling hub. J Biol Chem 300(11):107906. https://doi.org/10.1016/j.jbc.2024.107906

Klose RJ, Zhang Y (2007) Regulation of histone methylation by demethylimination and demethylation. Nat Rev Mol Cell Biol 8(4):307–318. https://doi.org/10.1038/nrm2143

Kouzarides T (2002) Histone methylation in transcriptional control. Curr Opin Genet Dev 12(2):198–209. https://doi.org/10.1016/s0959-437x(02)00287-3

Kusuyama J, Makarewicz NS, Albertson BG, Alves-Wagner AB, Conlin RH, Prince NB, Alves CRR, Ramachandran K, Kozuka C, Xiudong Y, Xia Y, Hirshman MF, Hatta T, Nagatomi R, Nozik ES, Goodyear LJ (2022) Maternal exercise-induced SOD3 reverses the deleterious effects of maternal high-fat diet on offspring metabolism through stabilization of H3K4me3 and protection against WDR82 Carbonylation. Diabetes 71(6):1170–1181. https://doi.org/10.2337/db21-0706

Laker RC, Lillard TS, Okutsu M, Zhang M, Hoehn KL, Connelly JJ, Yan Z (2014) Exercise prevents maternal high-fat diet-induced hypermethylation of the Pgc-1alpha gene and age-dependent metabolic dysfunction in the offspring. Diabetes 63(5):1605–1611. https://doi.org/10.2337/db13-1614

Lee KK, Workman JL (2007) Histone acetyltransferase complexes: one size doesn't fit all. Nat Rev Mol Cell Biol 8(4):284–295. https://doi.org/10.1038/nrm2145

Lerin C, Rodgers JT, Kalume DE, Kim SH, Pandey A, Puigserver P (2006) GCN5 acetyltransferase complex controls glucose metabolism through transcriptional repression of PGC-1alpha. Cell Metab 3(6):429–438. https://doi.org/10.1016/j.cmet.2006.04.013

Li C, Lumey LH (2017) Exposure to the Chinese famine of 1959-61 in early life and long-term health conditions: a systematic review and meta-analysis. Int J Epidemiol 46(4):1157–1170. https://doi.org/10.1093/ije/dyx013

Li ST, Huang D, Shen S, Cai Y, Xing S, Wu G, Jiang Z, Hao Y, Yuan M, Wang N, Zhu L, Yan R, Yang D, Wang L, Liu Z, Hu X, Zhou R, Qu K, Li A, Duan X, Zhang H, Gao P (2020) Myc-mediated SDHA acetylation triggers epigenetic regulation of gene expression and tumorigenesis. Nat Metab 2(3):256–269. https://doi.org/10.1038/s42255-020-0179-8

Li Y, Li RY, Zhu JY, Chen M, Mu WJ, Luo HY, Li S, Yan LJ, Yin MT, Li X, Chen HM, Guo L (2025) Maternal exercise prevents metabolic disorders in offspring mice through SERPINA3C. Nat Metab 7(2):401–420. https://doi.org/10.1038/s42255-024-01213-6

Locasale JW, Cantley LC (2011) Metabolic flux and the regulation of mammalian cell growth. Cell Metab 14(4):443–451. https://doi.org/10.1016/j.cmet.2011.07.014
Lu C, Thompson CB (2012) Metabolic regulation of epigenetics. Cell Metab 16(1):9–17. https://doi.org/10.1016/j.cmet.2012.06.001
Maejima H, Kanemura N, Kokubun T, Murata K, Takayanagi K (2018) Exercise enhances cognitive function and neurotrophin expression in the hippocampus accompanied by changes in epigenetic programming in senescence-accelerated mice. Neurosci Lett 665:67–73. https://doi.org/10.1016/j.neulet.2017.11.023
McPherson NO, Lane M, Sandeman L, Owens JA, Fullston T (2017) An exercise-only intervention in obese fathers restores glucose and insulin regulation in conjunction with the Rescue of Pancreatic Islet Cell Morphology and MicroRNA expression in male offspring. Nutrients 9(2). https://doi.org/10.3390/nu9020122
Mendelson MM, Marioni RE, Joehanes R, Liu C, Hedman AK, Aslibekyan S, Demerath EW, Guan W, Zhi D, Yao C, Huan T, Willinger C, Chen B, Courchesne P, Multhaup M, Irvin MR, Cohain A, Schadt EE, Grove ML, Bressler J, North K, Sundstrom J, Gustafsson S, Shah S, McRae AF, Harris SE, Gibson J, Redmond P, Corley J, Murphy L, Starr JM, Kleinbrink E, Lipovich L, Visscher PM, Wray NR, Krauss RM, Fallin D, Feinberg A, Absher DM, Fornage M, Pankow JS, Lind L, Fox C, Ingelsson E, Arnett DK, Boerwinkle E, Liang L, Levy D, Deary IJ (2017) Association of Body Mass Index with DNA methylation and gene expression in blood cells and relations to Cardiometabolic disease: a Mendelian randomization approach. PLoS Med 14(1):e1002215. https://doi.org/10.1371/journal.pmed.1002215
Mikkelsen TS, Xu Z, Zhang X, Wang L, Gimble JM, Lander ES, Rosen ED (2010) Comparative epigenomic analysis of murine and human adipogenesis. Cell 143(1):156–169. https://doi.org/10.1016/j.cell.2010.09.006
Murphy PJ, Cipriany BR, Wallin CB, Ju CY, Szeto K, Hagarman JA, Benitez JJ, Craighead HG, Soloway PD (2013) Single-molecule analysis of combinatorial epigenomic states in normal and tumor cells. Proc Natl Acad Sci USA 110(19):7772. https://doi.org/10.1073/pnas.1218495110
Nemoto S, Fergusson MM, Finkel T (2005) SIRT1 functionally interacts with the metabolic regulator and transcriptional coactivator PGC-1alpha. J Biol Chem 280(16):16456–16460. https://doi.org/10.1074/jbc.M501485200
Neri F, Rapelli S, Krepelova A, Incarnato D, Parlato C, Basile G, Maldotti M, Anselmi F, Oliviero S (2017) Intragenic DNA methylation prevents spurious transcription initiation. Nature 543(7643):72–77. https://doi.org/10.1038/nature21373
Nitert MD, Dayeh T, Volkov P, Elgzyri T, Hall E, Nilsson E, Yang BT, Lang S, Parikh H, Wessman Y, Weishaupt H, Attema J, Abels M, Wierup N, Almgren P, Jansson PA, Rönn T, Hansson O, Eriksson KF, Groop L, Ling C (2012) Impact of an exercise intervention on DNA methylation in skeletal muscle from first-degree relatives of patients with type 2 diabetes. Diabetes 61(12):3322–3332. https://doi.org/10.2337/db11-1653
Perakakis N, Stefanakis K, Mantzoros CS (2020) The role of omics in the pathophysiology, diagnosis and treatment of non-alcoholic fatty liver disease. Metabolism 111S:154320. https://doi.org/10.1016/j.metabol.2020.154320
Philp A, Chen A, Lan D, Meyer GA, Murphy AN, Knapp AE, Olfert IM, McCurdy CE, Marcotte GR, Hogan MC, Baar K, Schenk S (2011) Sirtuin 1 (SIRT1) deacetylase activity is not required for mitochondrial biogenesis or peroxisome proliferator-activated receptor-gamma coactivator-1alpha (PGC-1alpha) deacetylation following endurance exercise. J Biol Chem 286(35):30561–30570. https://doi.org/10.1074/jbc.M111.261685
Pinnick KE, Karpe F (2011) DNA methylation of genes in adipose tissue. Proc Nutr Soc 70(1):57–63. https://doi.org/10.1017/S0029665110004015
Plagemann A, Harder T, Brunn M, Harder A, Roepke K, Wittrock-Staar M, Ziska T, Schellong K, Rodekamp E, Melchior K, Dudenhausen JW (2009) Hypothalamic proopiomelanocortin promoter methylation becomes altered by early overfeeding: an epigenetic model of obesity and the metabolic syndrome. J Physiol 587(Pt 20):4963–4976. https://doi.org/10.1113/jphysiol.2009.176156

Poston L, Caleyachetty R, Cnattingius S, Corvalan C, Uauy R, Herring S, Gillman MW (2016) Preconceptional and maternal obesity: epidemiology and health consequences. Lancet Diabetes Endocrinol 4(12):1025–1036. https://doi.org/10.1016/S2213-8587(16)30217-0

Qi Y, Zhang B (2019) Predicting three-dimensional genome organization with chromatin states. PLoS Comput Biol 15(6):e1007024. https://doi.org/10.1371/journal.pcbi.1007024

Racey LA, Byvoet P (1971) Histone acetyltransferase in chromatin. Evidence for in vitro enzymatic transfer of acetate from acetyl-coenzyme a to histones. Exp Cell Res 64(2):366–370. https://doi.org/10.1016/0014-4827(71)90089-9

Riggs MG, Whittaker RG, Neumann JR, Ingram VM (1977) N-butyrate causes histone modification in HeLa and friend erythroleukaemia cells. Nature 268(5619):462–464. https://doi.org/10.1038/268462a0

Ronn T, Volkov P, Davegardh C, Dayeh T, Hall E, Olsson AH, Nilsson E, Tornberg A, Dekker Nitert M, Eriksson KF, Jones HA, Groop L, Ling C (2013) A six months exercise intervention influences the genome-wide DNA methylation pattern in human adipose tissue. PLoS Genet 9(6):e1003572. https://doi.org/10.1371/journal.pgen.1003572

Saha RN, Pahan K (2006) HATs and HDACs in neurodegeneration: a tale of disconcerted acetylation homeostasis. Cell Death Differ 13(4):539–550. https://doi.org/10.1038/sj.cdd.4401769

Selhub J, Miller JW (1992) The pathogenesis of homocysteinemia: interruption of the coordinate regulation by S-adenosylmethionine of the remethylation and transsulfuration of homocysteine. Am J Clin Nutr 55(1):131–138. https://doi.org/10.1093/ajcn/55.1.131

Shahbazian MD, Grunstein M (2007) Functions of site-specific histone acetylation and deacetylation. Annu Rev Biochem 76:75–100. https://doi.org/10.1146/annurev.biochem.76.052705.162114

Shi L, Tu BP (2015) Acetyl-CoA and the regulation of metabolism: mechanisms and consequences. Curr Opin Cell Biol 33:125–131. https://doi.org/10.1016/j.ceb.2015.02.003

Shi Y, Lan F, Matson C, Mulligan P, Whetstine JR, Cole PA, Casero RA, Shi Y (2004) Histone demethylation mediated by the nuclear amine oxidase homolog LSD1. Cell 119(7):941–953. https://doi.org/10.1016/j.cell.2004.12.012

Shi H, Li J, Li F, Yu H, Zhang F, Wu T, Yang L, Li Y, Hu R, Chen M, Sg N, Zhuang X, Feng S, Zhu L, Duan R (2025) Vitamin C-dependent intergenerational inheritance of enhanced endurance performance following maternal exercise. Adv Sci (Weinh):e2408912. https://doi.org/10.1002/advs.202408912

Son JS, Zhao L, Chen Y, Chen K, Chae SA, de Avila JM, Wang H, Zhu MJ, Jiang Z, Du M (2020) Maternal exercise via exerkine apelin enhances brown adipogenesis and prevents metabolic dysfunction in offspring mice. Sci Adv 6(16):eaaz0359. https://doi.org/10.1126/sciadv.aaz0359

Staddon JM, McGivan JD (1984) Distinct effects of glucagon and vasopressin on proline metabolism in isolated hepatocytes. The role of oxoglutarate dehydrogenase. Biochem J 217(2):477–483. https://doi.org/10.1042/bj2170477

Sun L, Zhang H, Gao P (2022) Metabolic reprogramming and epigenetic modifications on the path to cancer. Protein Cell 13(12):877–919. https://doi.org/10.1007/s13238-021-00846-7

Takahashi H, McCaffery JM, Irizarry RA, Boeke JD (2006) Nucleocytosolic acetyl-coenzyme a synthetase is required for histone acetylation and global transcription. Mol Cell 23(2):207–217. https://doi.org/10.1016/j.molcel.2006.05.040

Tan M, Peng C, Anderson KA, Chhoy P, Xie Z, Dai L, Park J, Chen Y, Huang H, Zhang Y, Ro J, Wagner GR, Green MF, Madsen AS, Schmiesing J, Peterson BS, Xu G, Ilkayeva OR, Muehlbauer MJ, Braulke T, Muhlhausen C, Backos DS, Olsen CA, McGuire PJ, Pletcher SD, Lombard DB, Hirschey MD, Zhao Y (2014) Lysine glutarylation is a protein posttranslational modification regulated by SIRT5. Cell Metab 19(4):605–617. https://doi.org/10.1016/j.cmet.2014.03.014

Thirstrup K, Christensen S, Moller HA, Ritzen A, Bergstrom AL, Sager TN, Jensen HS (2011) Endogenous 2-oxoglutarate levels impact potencies of competitive HIF prolyl hydroxylase inhibitors. Pharmacol Res 64(3):268–273. https://doi.org/10.1016/j.phrs.2011.03.017

Vaiserman A (2014) Early-life exposure to endocrine disrupting chemicals and later-life health outcomes: an epigenetic bridge? Aging Dis 5(6):419–429. https://doi.org/10.14336/AD.2014.0500419

Vaynman S, Ying Z, Gomez-Pinilla F (2004) Hippocampal BDNF mediates the efficacy of exercise on synaptic plasticity and cognition. Eur J Neurosci 20(10):2580–2590. https://doi.org/10.1111/j.1460-9568.2004.03720.x

Wahl S, Drong A, Lehne B, Loh M, Scott WR, Kunze S, Tsai PC, Ried JS, Zhang W, Yang Y, Tan S, Fiorito G, Franke L, Guarrera S, Kasela S, Kriebel J, Richmond RC, Adamo M, Afzal U, Ala-Korpela M, Albetti B, Ammerpohl O, Apperley JF, Beekman M, Bertazzi PA, Black SL, Blancher C, Bonder MJ, Brosch M, Carstensen-Kirberg M, de Craen AJ, de Lusignan S, Dehghan A, Elkalaawy M, Fischer K, Franco OH, Gaunt TR, Hampe J, Hashemi M, Isaacs A, Jenkinson A, Jha S, Kato N, Krogh V, Laffan M, Meisinger C, Meitinger T, Mok ZY, Motta V, Ng HK, Nikolakopoulou Z, Nteliopoulos G, Panico S, Pervjakova N, Prokisch H, Rathmann W, Roden M, Rota F, Rozario MA, Sandling JK, Schafmayer C, Schramm K, Siebert R, Slagboom PE, Soininen P, Stolk L, Strauch K, Tai ES, Tarantini L, Thorand B, Tigchelaar EF, Tumino R, Uitterlinden AG, van Duijn C, van Meurs JB, Vineis P, Wickremasinghe AR, Wijmenga C, Yang TP, Yuan W, Zhernakova A, Batterham RL, Smith GD, Deloukas P, Heijmans BT, Herder C, Hofman A, Lindgren CM, Milani L, van der Harst P, Peters A, Illig T, Relton CL, Waldenberger M, Jarvelin MR, Bollati V, Soong R, Spector TD, Scott J, McCarthy MI, Elliott P, Bell JT, Matullo G, Gieger C, Kooner JS, Grallert H, Chambers JC (2017) Epigenome-wide association study of body mass index, and the adverse outcomes of adiposity. Nature 541(7635):81–86. https://doi.org/10.1038/nature20784

Wellen KE, Hatzivassiliou G, Sachdeva UM, Bui TV, Cross JR, Thompson CB (2009) ATP-citrate lyase links cellular metabolism to histone acetylation. Science 324(5930):1076–1080. https://doi.org/10.1126/science.1164097

Wolffe AP, Pruss D (1996) Deviant nucleosomes: the functional specialization of chromatin. Trends Genet 12(2):58–62. https://doi.org/10.1016/0168-9525(96)81401-6

Wong CC, Qian Y, Yu J (2017) Interplay between epigenetics and metabolism in oncogenesis: mechanisms and therapeutic approaches. Oncogene 36(24):3359–3374. https://doi.org/10.1038/onc.2016.485

Wu G, Zhang X, Gao F (2021) The epigenetic landscape of exercise in cardiac health and disease. J Sport Health Sci 10(6):648–659. https://doi.org/10.1016/j.jshs.2020.12.003

Xu X, Su S, Barnes VA, De Miguel C, Pollock J, Ownby D, Shi H, Zhu H, Snieder H, Wang X (2013) A genome-wide methylation study on obesity: differential variability and differential methylation. Epigenetics 8(5):522–533. https://doi.org/10.4161/epi.24506

Yin Y, Morgunova E, Jolma A, Kaasinen E, Sahu B, Khund-Sayeed S, Das PK, Kivioja T, Dave K, Zhong F, Nitta KR, Taipale M, Popov A, Ginno PA, Domcke S, Yan J, Schubeler D, Vinson C, Taipale J (2017) Impact of cytosine methylation on DNA binding specificities of human transcription factors. Science 356(6337). https://doi.org/10.1126/science.aaj2239

Yokomori N, Tawata M, Onaya T (1999) DNA demethylation during the differentiation of 3T3-L1 cells affects the expression of the mouse GLUT4 gene. Diabetes 48(4):685–690. https://doi.org/10.2337/diabetes.48.4.685

Zhang D, Tang Z, Huang H, Zhou G, Cui C, Weng Y, Liu W, Kim S, Lee S, Perez-Neut M, Ding J, Czyz D, Hu R, Ye Z, He M, Zheng YG, Shuman HA, Dai L, Ren B, Roeder RG, Becker L, Zhao Y (2019) Metabolic regulation of gene expression by histone lactylation. Nature 574(7779):575–580. https://doi.org/10.1038/s41586-019-1678-1

Zheng Q, Maksimovic I, Upad A, David Y (2020) Non-enzymatic covalent modifications: a new link between metabolism and epigenetics. Protein Cell 11(6):401–416. https://doi.org/10.1007/s13238-020-00722-w

Zimmet P, Shi Z, El-Osta A, Ji L (2018) Epidemic T2DM, early development and epigenetics: implications of the Chinese famine. Nat Rev Endocrinol 14(12):738–746. https://doi.org/10.1038/s41574-018-0106-1

Chapter 9
Exercise-Induced Metabolites Influence Health via Epigenetic Modifications

Minghong Leng, Shanhu Sun, and Yangpo Cao

Abstract Physical exercise induces significant alterations in metabolic homeostasis across multiple organ systems, with emerging evidence suggesting that many of the beneficial effects of exercise are mediated through epigenetic modifications. The intricate relationship between exercise-induced metabolic changes and subsequent epigenetic modifications profoundly influences physiological adaptations and health benefits. During exercise, skeletal muscles dramatically increase energy expenditure, mobilizing various substrates including glycogen, plasma glucose, and triglycerides, which subsequently alters metabolite concentrations in blood and tissues. These exercise-associated metabolites, including TCA cycle intermediates (succinate, fumarate, α-ketoglutarate), lactate, acetyl-CoA, β-hydroxybutyrate (βOHB), S-adenosylmethionine (SAM), kynurenine, and serotonin, serve as substrates or cofactors for epigenetic modifications. These metabolites regulate chromatin structure through DNA methylation and histone modifications, contributing to improved metabolic health, reduced inflammation, enhanced stress resistance, and protection against various pathological conditions. This review highlights the emerging understanding of exercise as a powerful epigenetic modulator through metabolic reprogramming; providing mechanistic insights into exercise-mediated health benefits and potential therapeutic applications for metabolic, cardiovascular, and neurological disorders.

Keywords Exercise · Metabolites · Epigenetics · DNA methylation · Histone modification

M. Leng · S. Sun · Y. Cao (✉)
Department of Pharmacology, Joint Laboratory of Guangdong-Hong Kong Universities for Vascular Homeostasis and Diseases, School of Medicine, Southern University of Science and Technology, Shenzhen, China
e-mail: caoyp@sustech.edu.cn

F. Gao, X. Zhang (eds.), *Exercise, Epigenetics and Human Health*, Epigenetics and Human Health 15, https://doi.org/10.1007/978-3-032-08045-5_9

Abbreviations

ACLY	ATP-citrate lyase
BBB	blood–brain barrier
DNMTs	DNA methyltransferases
HATs	histone acetyltransferases
HMTs	histone methyltransferases
JMJDs	Jumonji domain (Jmjd)-containing histone demethylases
MCTs	monocarboxylate transporters
NAD^+/NADH	nicotinamide adenine dinucleotide
NAM	nicotinamide
PUFAs	polyunsaturated fatty acids
SAH	S-adenosyl-l-homocysteine
SAM	S-Adenosylmethionine
SIRT	NAD^+-dependent histone deacetylase sirtuins
TET	Ten-eleven-translocation family of enzymes
α-KG	α-ketoglutarate
βOHB	β-hydroxybutyrate

9.1 Introduction

Exercise is usually divided into three categories: aerobic exercise, resistance exercise, and complex exercise that combines the first two during the exercise process. Almost all the different kinds of exercises affect the metabolic homeostasis of body cells. In addition to providing energy for cellular life activities, the metabolic processes within cells also have a wide range of effects on intracellular molecular signals, particularly reflected in epigenetic modifications of cells. Epigenetic modifications such as chromatin structure changes, DNA fragment methylation, chemical transformation of histone chromosomal proteins (acetylation, methylation, ADP ribosylation, ubiquitination, phosphorylation, etc.), and regulation of non-coding RNA (ncRNA) are highly dependent on metabolites from complex metabolic pathways, such as S-adenosylmethionine, lactate, ATP, etc. Emerging evidence suggests that a lot of the positive effects of exercise are driven by epigenetic changes, which alter the expression levels of various genes associated with physiological and pathological conditions. In addition, there is emerging evidence showing that epigenetic modifications serve as a medium for intergenerational transmission of exercise effects. In this chapter, we will focus on discussing the metabolic changes caused by exercise, which in turn lead to epigenetic outcomes.

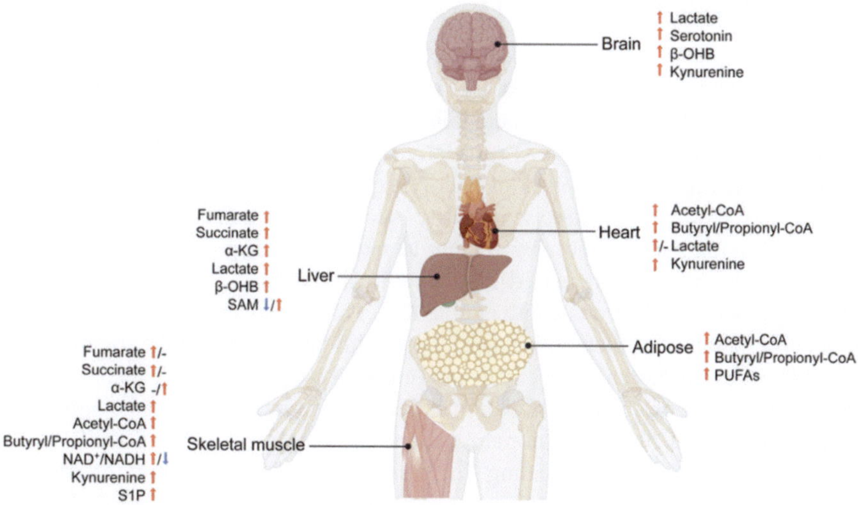

Fig. 9.1 Exercise affects metabolite contents in multiple tissues. Exercise has an impact on the metabolic processes in many organs and tissues of the human body, and then alters metabolite contents. In different tissues, the specific changes in metabolites vary. α-KG α-ketoglutarate, β-OHB β-hydroxybutyrate, NAD^+/NADH nicotinamide adenine dinucleotide, PUFAs polyunsaturated fatty acids

9.2 Exercise-Induced Alterations in Metabolites

Exercise imposes a significant challenge to the body's metabolic homeostasis (Fig. 9.1). During physical activity, the contractile function of skeletal muscles dramatically increases energy expenditure, leading to a substantial rise in energy demand. Vigorous exercise can elevate total body energy consumption by up to 25 times compared to the resting state. It also boosts the ATP turnover rate in skeletal muscles by 100 times and raises the energy demand of skeletal muscles to over 95% of the total body (Gaitanos et al. 1993). With intramuscular ATP stores at about 5 mmol per kg of wet muscle, these reserves can only sustain maximal exercise for less than 2 seconds, making them insufficient for prolonged contractile activity (Hargreaves and Spriet 2020). To meet these heightened energy needs, the body mobilizes energy reserves from various sources, including liver glycogen, adipose tissue triglycerides, plasma glucose, and the glycogen and triglycerides stored within the skeletal muscles themselves (Goodpaster and Sparks 2017). This mobilization subsequently alters the concentrations of metabolites in the blood and tissues involved in these processes. Recent advancements in metabolomics based on mass spectrometry (MS) provide a comprehensive snapshot of systemic metabolic signatures and identifies a wide range of exercise-associated metabolites (Sato et al. 2022; Starnes et al. 2017; Ohmura et al. 2021; Nayor et al. 2020).

9.2.1 *Muscle Metabolites Altered During Exercise*

Skeletal muscle serves as the primary effector organ during exercise, demonstrating a remarkable ability to adapt its metabolic state in response to varying patterns and intensities of physical activity. At rest, fatty acids are the predominant energy substrate, accounting for 80%–90% of energy expenditure in skeletal muscle (Smith et al. 2023). However, as exercise intensity escalates, ATP production derived from fatty acid oxidation becomes insufficient to meet the energy demands of skeletal muscle, resulting in an increased reliance on carbohydrate sources that facilitate more rapid ATP production (Smith et al. 2023). This shift in energy substrates leads to an accumulation of metabolites associated with glycolysis and the tricarboxylic acid (TCA) cycle. Notably, succinate and fumarate are produced in skeletal muscle during endurance exercise and function as competitive inhibitors of α-ketoglutarate (α-KG)-dependent Jumonji C-domain histone demethylases (KDMs), while α-KG does not exhibit a corresponding increase in some metabolomic profiles of endurance-exercised muscles (Ohmura et al. 2021; Huffman et al. 2014; Maurer et al. 2021). Conversely, (Yuan et al. 2020) reported an elevation in α-KG levels in resistance-trained muscles, with no corresponding increase in succinate or fumarate. These alterations are accompanied by enhanced α-KG biosynthesis activity via glutamate dehydrogenase (GDH) and isocitrate dehydrogenase (IDH), suggesting that different types of exercise are associated with distinct metabolic changes and epigenetic patterns. Another important metabolite that exemplifies this is lactate, which sharply rises during acute and vigorous exercise. Under conditions of limited oxygen availability, lactate dehydrogenase (LDH) catalyzes the conversion of pyruvate to lactate, providing energy approximately 100 times more rapidly than oxidative phosphorylation (Hargreaves and Spriet 2020; Peek et al. 2017). This by-product, lactate, emerges as a crucial substrate for histone lactylation, thereby participating in the epigenetic regulation of skeletal muscle homeostasis (Zhang et al. 2019; Maschari et al. 2022; Dai et al. 2023). As exercise duration extends, a significant amount of muscle glycogen is depleted, and blood glucose levels decrease, resulting in a relative shortage of carbohydrates available for energy metabolism. Concurrently, there is an increase in triglyceride hydrolysis from adipose tissue, leading to the release of a substantial quantity of free fatty acids into blood. These free fatty acids are subsequently taken up and utilized by skeletal muscle as an alternative energy substrate, thereby providing energy for muscle contraction and prolonging the duration of exercise until exhaustion (Hargreaves and Spriet 2020; Goodpaster and Sparks 2017). Fatty acids undergo β-oxidation in the mitochondria, where they are sequentially cleaved to remove two-carbon units and generate acetyl-CoA. Notably, acetyl-CoA generated from both fatty acid β-oxidation and pyruvate oxidation cannot be directly transported across mitochondrial membranes (Pietrocola et al. 2015). Instead, mitochondrial acetyl-CoA condenses with oxaloacetate to form citrate, which can be exported from the mitochondria. Then ATP-citrate lyase (ACLY) converts citrate back into acetyl-CoA, providing a substrate for histone acetylation in the nucleus (Choudhary et al. 2014). Additionally, fatty acyl-CoA and acylcarnitines serve as important intermediates in fatty acid catabolism within exercised muscles (Xu et al. 2016). Notably, short-chain fatty acyl-CoA molecules, such as butyryl-CoA and propionyl-CoA, emerge as active substrates for histone lysine butyrylation and propionylation, linking fatty acid catabolism to epigenetic regulation (Nshanian et al. 2025; Yang et al. 2021). In the downstream of energy metabolism, nicotinamide adenine dinucleotide (NAD^+) and its reduced form (NADH) are the key carriers for transferring electrons to oxidative phosphorylation. During moderate- to high-intensity exercise, elevated glycolytic activity promotes the reduction of NAD^+ to NADH, leading to a decreased NAD^+/NADH ratio and cytoplasmic NADH accumulation, which is attributed to the limited capacity of mitochondrial NADH import shuttles (Denis et al. 1991; Walzik et al. 2023). In contrast, low-intensity exercise predominantly relies on fatty acid oxidation, resulting in a higher NAD^+/NADH ratio, with these trends also observed during the post-exercise recovery phase (Sahlin et al. 1987). This suggests that exercise may regulate NAD^+-dependent histone deacetylase sirtuins (SIRTs) in an intensity dependent manner (Vargas-Ortiz et al. 2019). In addition, muscle protein degradation provides amino acids as an alternative energy substrate during exercise, although its contribution to ATP production is relatively low. However, several metabolites derived from amino acids play a potential role in epigenetic regulation. For instance, glutamate catabolism serves as a supplementary source of α-KG, the substrate for KDMs (Tretter and Adam-Vizi 2005; García-Giménez et al. 2024). Furthermore, exercise increases the levels of kynurenine, an intermediate in tryptophan metabolism, which has recently been reported to regulate H3K4 methylation (Sato et al. 2022; Martin et al. 2020; Hayakawa et al. 2019). Moreover, bioactive sphingolipid metabolism is altered during exercise. Acute exercise also enhances ceramide synthesis, leading to increased levels of ceramide and related metabolites, such as sphingosine and sphingosine-1-phosphate (S1P) (Tan-Chen et al. 2020). Notably, S1P has been identified as an endogenous nuclear inhibitor of histone deacetylases (HDAC) 1 and 2, playing a crucial role in regulating skeletal muscle metabolic homeostasis (Wang et al. 2018; Nguyen-Tran et al. 2014).

9.2.2 *Heart Metabolites Altered During Exercise*

Exercise presents metabolic challenges to the heart similar to those experienced by skeletal muscle. During physical activity, the heart's contractile force and oxygen consumption can increase up to tenfold compared to resting levels (Gibb and Hill 2018). This enhancement ensures efficient blood circulation, meeting the heightened oxygen and nutrient demands of active muscles. As myocardial workload intensifies, there is an increased catabolism of various substrates, particularly fatty acids and lactate. Exercise-induced lipolysis in adipose tissue elevates circulating free fatty acids, which enhances their uptake and utilization by the heart (Goodpaster and Sparks 2017). Concurrently, the rates of cardiac triglyceride catabolism rise during exercise (Lassers et al. 1971). Post-exercise adaptations further augment the expression of genes involved in fatty acid transport and oxidation, optimizing fat utilization in the heart (Strøm et al. 2005). This adaptation provides the heart with a greater supply of epigenetically modified substrate metabolites, such as acetyl-CoA and short-chain fatty acyl-CoA (Sato et al. 2022; Yang et al. 2021). However, TCA cycle intermediates like succinate, fumarate, and α-ketoglutarate do not show elevated levels in the metabolomic profiles of exercised hearts (Sato et al. 2022; Starnes et al. 2017; Fulghum et al. 2022). Glucose metabolism in the heart remains relatively stable during exercise. In response to moderate to low-intensity exercise, the increased myocardial workload slightly enhances glucose uptake and oxidation (Gibb and Hill 2018). Conversely, during high-intensity exercise, carbohydrate oxidation is inhibited by the accumulation of lactate (Fulghum and Hill 2018). Lactate can be utilized as a complementary energy substrate in exercised heart; plasma lactate can be transported into cardiomyocytes, converted to pyruvate, and oxidized in the mitochondria to generate ATP (Fulghum and Hill 2018). Plasma lactate levels rise significantly during intense activities, contributing substantially to the increased proportion of lactate as an energy substrate in the heart (Stanley 1991). The balance between lactate uptake and utilization results in dynamic changes in lactate levels within the heart. During moderate-intensity exercise, the heart takes up lactate in proportion to its consumption rate, without increasing myocardial lactate abundance beyond resting levels (Stanley 1991). In contrast, vigorous exercise leads to a significant increase in heart lactate concentration (Cunha et al. 2022). Additionally, long-term exercise training can inhibit lactate production by reducing the heart's glycolytic activity (Xu et al. 2024). This suggests that lactate may play a dynamic epigenetic regulatory role in response to varying exercise intensities and durations. Furthermore, the epigenetic regulatory amino acid intermediate, kynurenine also increases in the heart, mirroring changes observed in skeletal muscle (Sato et al. 2022). However, it remains unclear whether this increase is driven by primary alterations in heart tryptophan metabolism or by enhanced plasma uptake.

9.2.3 Liver Metabolites Altered During Exercise

The liver functions similarly to a rechargeable battery, releasing and replenishing energy according to metabolic demands. During physical activity, it plays a crucial role in providing glucose to meet the increased energy requirements of working muscles, primarily through two processes: glycogenolysis and gluconeogenesis. Glycogenolysis, the breakdown of liver glycogen, is a primary response during exercise. Activated glycogen phosphorylase converts glycogen to glucose-1-phosphate, which is then transformed into glucose-6-phosphate. Glucose-6-phosphate can enter glycolysis in muscles for immediate energy or be dephosphorylated to release free glucose into circulation, thereby ensuring a continuous energy supply for working muscles (Hargreaves and Spriet 2020; Trefts et al. 2015). In addition to glycogenolysis, gluconeogenesis—the synthesis of glucose from non-carbohydrate precursors—plays a critical role during prolonged exercise, particularly as glycogen stores begin to deplete. The liver synthesizes glucose from substrates such as lactate, glycerol, and gluconeogenic amino acids like alanine and glutamine, which are transported to the liver and undergo various metabolic pathways (Hu et al. 2020). Lactate, a byproduct of anaerobic metabolism in muscles, can be converted back to pyruvate and subsequently into glucose through gluconeogenesis, thus maintaining blood glucose levels during extended periods of exercise (Hu et al. 2020). The high flux of lactate serves as a potential source for exercise-induced histone lactylation in the liver (Huang et al. 2023a). The gluconeogenic process is energy-intensive, requiring ATP and GTP, as well as various specific enzymatic reactions to transform substrates into glucose. The energetic demands of the liver during this process are primarily met through an increased oxidation of fatty acids. This leads to higher levels of TCA cycle metabolites, including those that epigenetically regulate metabolic activity, such as succinate, fumarate, and acetyl-CoA, similar to that observed in heart and skeletal muscle (Rector et al. 2008; Huang et al. 2010). Moreover, elevated liver fatty acid oxidation results in the formation of ketone bodies. Excess acetyl-CoA is converted into acetoacetate; the primary ketone synthesized in the liver. Acetoacetate can further be reduced to β-hydroxybutyrate (βOHB) through the action of 3-hydroxybutyrate dehydrogenase. This shift toward ketone body production represents an important adaptive response, providing an alternative energy source that helps to preserve glycogen stores in muscles (Evans et al. 2017). Once produced, these ketone bodies are transported via monocarboxylate transporters (MCTs) to metabolically active tissues, such as skeletal muscle, to support energy needs during exercise (Trefts et al. 2015; Evans et al. 2017). Ketone bodies, particularly βOHB, have been shown to inhibit HDACs, leading to increased histone acetylation marks such as H3K9 and H3K14 to play a role in the epigenetic regulation of metabolic stress (Newman and Verdin 2014; Seaborne and Sharples 2020). Moreover, exercise significantly impacts methionine metabolism and the levels of S-Adenosylmethionine (SAM), the primary methyl donor for DNA and histone methylation (Riberio et al. 2018; Yan et al. 2022; Gao et al. 2018). Acute exercise upregulates the hepatic SAM-synthesizing enzyme, MAT1A, although it leads to a reduction in SAM levels, suggesting an enhanced methylation flux and the increased SAM consumption during exercise (Riberio et al. 2018). The increased methylation flux is mediated predominantly by the elevated activity of phosphatidylethanolamine N-methyltransferase (PEMT), which catalyzes the methylation of phosphatidylethanolamine (PE) to phosphatidylcholine (PC) (Riberio et al. 2018). PEMT competes for available SAM with DNA and histone methylation, resulting in a decreased substrate availability for these epigenetic processes (Haws et al. 2020). Additionally, similar metabolic alterations in liver methionine metabolism have been observed during fasting-induced energy stress, which parallels the metabolic effects of exercise and is mediated by glucagon (Capelo-Diz et al. 2023). In contrast, chronic exercise training has been shown to upregulate the expression of MAT1A and increase SAM levels in both the liver and plasma (Yan et al. 2022). This sustained increase in SAM provides a greater substrate for the methylation of DNA and histones, highlighting the complex interplay between exercise, hepatic methionine metabolism, and epigenetic regulation.

9.2.4 Adipose Metabolites Altered During Exercise

During exercise, adipose tissue metabolism undergoes significant alterations to facilitate energy supply in response to increased demands. The primary function of white adipose tissue (WAT) is to store energy in the form of triglycerides and release free fatty acids when energy is required. This mobilization is crucial during fasting and endurance activities. Lipolysis, the breakdown of triglycerides into free fatty acids and glycerol, plays a central role in fat mobilization during physical activity (Goodpaster and Sparks 2017). At rest and during low-intensity exercise, fatty acids derived from adipose tissue serve as a primary energy source. Research indicates that lipolysis increases approximately two to threefold above resting levels during low-intensity exercise, accompanied by a concurrent decrease in the re-esterification of fatty acids (Wolfe et al. 1990). The capacity for fatty acid mobilization escalates with prolonged exercise, and after 90 minutes, the rate of lipolysis continues to rise, supporting the energy requirements of skeletal muscles as endogenous carbohydrate stores diminish (Romijn et al. 1993). This shift allows a greater availability of fatty acids for energy production in working muscles. Interestingly, fatty acid mobilization tends to decrease at vigorous exercise intensities. This decline is likely attributed to reduced blood flow to adipose tissue, resulting in a greater reliance on local intramuscular fat and carbohydrate stores for immediate energy (Horowitz 2022). Long-chain fatty acids, especially monounsaturated fatty acids, are the primary species mobilized during exercise, serving as a vital energy source (Nikolaidis and Mougios 2004). In contrast, polyunsaturated fatty acids (PUFAs) become more prominent during prolonged or higher-intensity activities (Contrepois et al. 2020; Gollasch et al. 2019). This shift in fatty acid mobilization is crucial for sustaining energy demands as exercise duration and intensity increase. Importantly, PUFAs have been reported to induce demethylation of specific gene promoters, which leads to the regulation of genes involved in various physiological processes, including inflammation and metabolism (Ediriweera and Gayashani Sandamalika 2025). This regulatory effect suggests that PUFAs may play a significant role in adapting metabolic responses during physical activity and promoting overall metabolic health.

9.2.5 Brain Metabolites Altered During Exercise

Exercise has been increasingly recognized for its ability to influence brain function through the modulation of various metabolic processes and the crossing of metabolites across the blood–brain barrier (BBB). These exercise-induced metabolites not only play crucial roles in energy metabolism but also significantly impact epigenetic mechanisms, thereby altering gene expression and contributing to brain health (Wang et al. 2024a). One of the key metabolites produced during exercise is lactate, the byproduct of anaerobic metabolism. Lactate can cross the BBB via MCT2 and serve as an alternative energy source for neurons (Lee et al. 2023). Moreover, lactate functions as a signaling molecule that induces histone lactylation in the brain, which regulates gene expression, particularly in functionally important enhancer regions, thus affecting neural development and reducing inflammation (Wang et al. 2024a, b; Li et al. 2025; Han et al. 2023). Serotonin is another brain metabolite influenced by exercise, known for its role in mood regulation and cognitive functions. Aerobic exercise can stimulate brain serotonin activity, increase release and synthesis of serotonin, and trigger parallel elevations in tryptophan availability in blood (Strasser and Fuchs 2015). Recently, histones have been described as targets of serotonylation, revealing a non-canonical, neurotransmission-independent role of serotonin in brain function (Farrelly et al. 2019; Al-Kachak et al. 2024). βOHB, the ketone body produced during prolonged exercise and low carbohydrate intake, also crosses the BBB via MCTs (Wang et al. 2024a). βOHB has been shown to exert neuroprotective effects and enhance cognitive function by inhibiting HDACs and promoting histone lysine β-hydroxybutyrylation (Kbhb) (Li et al. 2024; Tanegashima et al. 2017). Additionally, kynurenine, the metabolite derived from tryptophan, and its derivatives can cross the BBB and affect histone modifications, such as H3K4 trimethylation and H2AS40 O-GlcNAcylation and upregulate gene expression at loci associated with hypothalamic functions, highlighting the role of exercise-induced kynurenine in brain health and function (Hayakawa et al. 2019; Wang et al. 2024a).

9.3 Many Biological Processes Are Accompanied by Metabolic and Epigenetic Remodeling

The connection between metabolism and epigenetic is a complex bidirectional regulation (Smith et al. 2023). The epigenetic modification targeting essential metabolic enzymes is important for metabolic remodeling. Conversely, the metabolic signals also participate in the regulation of epigenetic (Reid et al. 2017). Here, we introduce the metabolites generated from the metabolic pathways and their roles in epigenetic modification (Fig. 9.2).

9.3.1 SAM and Histone/DNA Methylation

SAM is an essential amino acid produced from one-carbon metabolism and has a significant impact on epigenetic regulation. SAM is a major methyl donor involved in DNA, RNA, and histone methylation (Yu et al. 2019). In eukaryotic cells, SAM

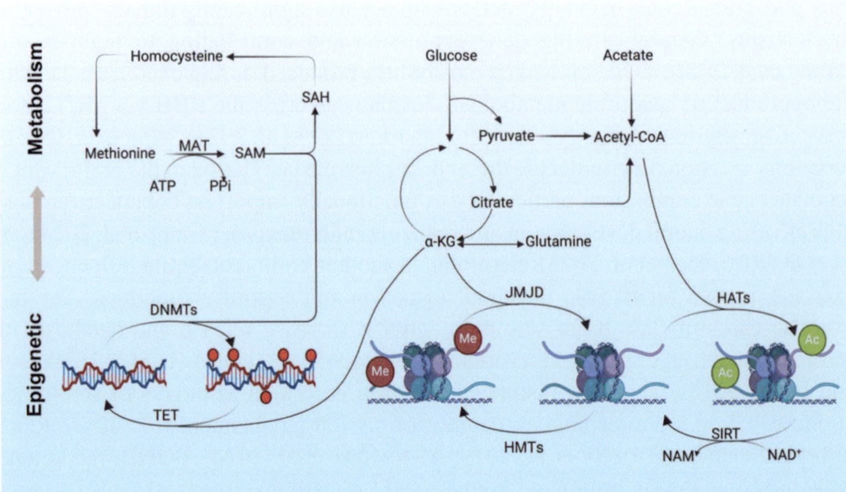

Fig. 9.2 Metabolites-mediated regulation of epigenetic modifications. Metabolites serve as critical substrates and cofactors for epigenetic regulation. SAM produced from methionine by MATs acts as the main donor for DNA and histone methylation mediated by DNMTs and HMDs. Conversely, the processes of demethylation driven by TETs and KDMs are dependent on α-KG derived from TCA cycle and glutamine. In addition, the other intermediates of glucose metabolism and TCA cycle, pyruvate and citrate, provide nuclear acetyl-CoA as a donor for HATs-dependent histone acetylation. This process is counterbalanced by NAD^+-dependent histone deacetylases SIRTs. SAM S-Adenosyl methionine, SAH S-adenosyl-l-homocysteine, MATs methionine adenosyltransferase, DNMTs DNA methyltransferases, TET ten-eleven-translocation family of enzymes, α-KG α- ketoglutarate, JMJDs Jumonji domain (Jmjd)-containing histone demethylases HMTs histone methyltransferases, HATs histone acetyltransferases, SIRT, NAD+-dependent histone deacetylase sirtuins, NAM nicotinamide

is synthesized by methionine adenosyltransferase (MAT), which consumes ATP and methionine as substrates. Methyltransferases, such as DNA methyltransferases (DNMTs) and Histone methyltransferases (HMTs), transfer the methyl from SAM to DNA or histone, accompanied by the generation of S-adnosyl-l-homocysteine (SAH). The ratio of SAM/SAH and one-carbon cycle regulates the methyltransferase activity and histone/DNA methylation in eukaryotic cells (Pham et al. 2023).

DNA methylation is one of the main regulatory mechanisms for gene transcriptional repression. DNA methylation, including 5mC, 5caC, 5hmC, and 5fC, is a common modification by which methyl molecules are transferred to DNA, especially on the cytosine phosphate guanine islands (CpG) (Mattei et al. 2022). SAM can bind the conserved SAM-binding pocket domain of DNMT directly and active the C-5 atom for electrophilic attack. The reaction mediates the transfer of methyl group to the 5-carbon and the formation of SAH (Haghani et al. 2023; Liu et al. 2024). DNA methyltransferases, such as DNMT1, DNMT3A, and DNMT3B transfer methyl group from SAM to the 5′-site of the cytosine ring of DNA, leading to transcriptional repression by inhibiting transcription factor binding or recruiting methyl-CpG-binding domain (MBD) proteins (Bogdanović and Lister 2017; Jeltsch et al. 2020).

Histone methylation is an important post-translational modification and occurs on many residues: lysine, arginine, and histidine. This histone modification regulates almost all basic biological process, such as cell cycle and stress response, by regulating heterochromatin assemble and determining the gene boundaries (Jambhekar et al. 2019; Barski et al. 2007). Compared with DNA methylation, methylation on histones can either activate or repress transcription, depending on the methylated sites and the number of methyl groups added (Barski et al. 2007; Jenuwein and Allis 2001). Although all the histone proteins, H2A, H2B, H3, and H4 are reported to be methylated, the well-studied methylation sites for histone are H3K27, H3K4, H3K9, H3K36, H3K79, and H4K20. According to the previous studies, H3K4me2, H3K4me3, and H3K79me3 are the markers for transcriptional activation, while H3K27me2, H3K27me3, and H3K9me2 are the markers for transcriptional repression (Lawrence et al. 2016). Histone methyltransferases such as DOT1-like proteins, SET-domain-containing proteins, and arginine N-methyltransferase (PRMT) family have been reported to transfer methyl group from SAM to histones by adding 1-3 methyl groups on the residues (Gui et al. 2014).

9.3.2 α-KG and Histone/DNA Demethylation

α-KG is a metabolic intermediate of the TCA cycle and serves as a cofactor for demethylases by removing methyl groups from DNA and histones. In the TCA cycle, α-KG is produced from isocitrate by isocitrate dehydrogenases (IDH1, IDH2, and IDH3) (Asadi Shahmirzadi et al. 2020; Medeiros et al. 2017). In mitochondria beyond TCA cycle, glutamate dehydrogenases (GDH) also catalyze glutamate to α-KG. α-KG-dependent demethylases include Jumonji domain (Jmjd)-containing histone demethylases, DNA demethylases of ten-eleven-translocation family of enzymes (Tet1, Tet2, and Tet3), and RNA demethylases of Alkbh family (Huang et al. 2023b).

DNA demethylation is usually associated with gene activation. For example, Tet enzyme directs the oxidation reaction by transfer of the oxygen atom from oxygen to the targeted sites of DNA with the help of α-KG (Figueroa et al. 2010; Bowman and Levine 2017). At the same time, α-KG is converted to succinate and carbon dioxide. The dynamic balance of DNA methylation and demethylation is important for cell behaviors. Disrupting this balance has been found to be a trigger for many kinds of diseases (Figueroa et al. 2010; Tefferi et al. 2009).

Different sites of histone demethylation require different histone demethylases. The Kdm5 family (Kdm5a, Kdm5b, and Kdm5c) is responsible for H3K4me2 and H3K4me3 demethylation (Deogharia et al. 2024; Pavlenko et al. 2022). In mammalian cells, H3K4me3/me2 are both activation markers but are enriched at the transcription start sites (TSSs) or 5′ end of transcribing genes. Therefore, the Kdm5 family-mediated H3K4 demethylation usually represses gene expression (Xhabija and Kidder 2019). Moreover, the Kdm6 family (UTX and JMJD3) directs H3K27me2/me3 demethylation and the Kdm2 family (Jhdm1a and Jhdm1b) directs H3K36me demethylation (Chang et al. 2019). Since the methylation at histone different sites has different functions for gene transcription, α-KG dependent histone demethylation can either increase or repress the expression of targeted genes (Pavlenko et al. 2022).

9.3.3 Acetyl-CoA and Histone Acetylation

Acetyl-CoA is a central metabolite involved in the metabolism of proteins, lipids, carbohydrates, and other substances. Acetyl-CoA has an essential impact on epigenetic modification through histone acetylation (He et al. 2023). In eukaryotic cells, acetyl-CoA can be produced in different cellular compartments through different metabolic pathways. In mitochondria, pyruvate dehydrogenase complex (PDC) converts pyruvate into acetyl-CoA. More importantly, PDC can enter the nucleus to generate acetyl-CoA for histone acetylation (Sutendra et al. 2014). Since mitochondrial generated acetyl-CoA cannot pass through mitochondrial membrane, the cytosolic acetyl-CoA is produced in the other two ways. One is the ACLY catalyzes citrate into acetyl-CoA and the other is ACSS2 catalyzes acetate into acetyl-CoA (Kaymak et al. 2024). Acetyl-CoA, especially the nuclear acetyl-CoA is believed to be the immediate and local acetyl donor for histone acetylation (He et al. 2023; Ma et al. 2025).

Histone acetylation is the biological process of adding an acetyl group to the N-terminal tail on lysine residues. Histone acetyltransferases, such as P300/CBP facilitate the transfer of acetyl group from acetyl-CoA to histones. Acetylation decreases the positive charge of histones, weakening the interaction between the histone and DNA. The acetylated histones tend to dissociate from DNA, thereby increasing the chromatin accessibility (Zhang et al. 2023). The acetyltransferase P300-mediated H3K27ac is well-studied and is recognized as a classical enhancer marker for gene expression activation (Hogg et al. 2021).

9.3.4 NAD^+-Dependent Histone Deacetylation

NAD^+ is one of the most abundant metabolites which is a cofactor involved in more than 500 enzymatic reactions (Verdin 2015). The SIRT enzyme family, directing the NAD^{+-} dependent histone deacetylation, links the metabolism and epigenetic together. NAD^+ receives the hydride ion to form its reduced version, NADH (Verdin

2015; Katsyuba et al. 2020). The balance between the NAD^+ and NADH is essential for the glucose, fatty acid, and amino acid metabolism (White and Schenk 2012).

SIRT1, one of the histone deacetylases requires the cofactor NAD^+ to regulate nucleosome positioning and gene expression (Verdin 2015). The catalytic core of SIRT1 contains two NAD^+ binding domains, and the NAD^+ binding changes the SIRT1 protein structure, thereby transferring lysine-bound acetyl from acetylated histones to ADP-ribose (Yu et al. 2024). Histone deacetylation decreases its charge and increases its affinity for DNA. The nucleosome-wrapped DNA loses its accessibility for transcription factor binding, leading to the decrease of the corresponding gene expression (Zhang et al. 2023).

9.4 Epigenetic Roles of Exercise-Induced Metabolites in Health

Physical exercise has long been recognized for its wide-range health benefits, including circulation system, respiratory system, nervous system, and so on. However, the relevant molecular mechanism has not yet been fully revealed. Recent studies have begun to highlight the epigenetics roles underlying the exercise-related benefits, with a particular focus on the role of exercise-induced metabolites in health (Fig. 9.3).

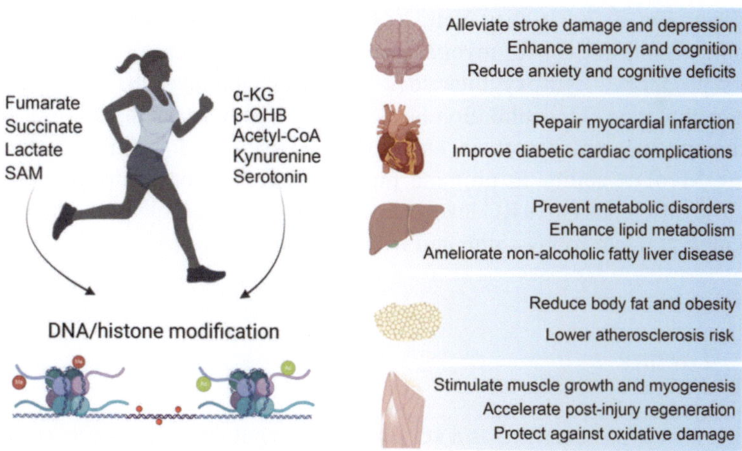

Fig. 9.3 Exercise-induced metabolites promote human health via epigenetic modifications. Exercise-induced metabolic reprogramming dynamically regulates the availability of key metabolites (e.g., SAM, α-KG, acetyl-CoA, NAD^+), which serve as substrates and cofactors for epigenetic-modifying enzymes. These metabolites regulate tissue-specific epigenetic modifications that modulate downstream gene expression. The exercise-driven epigenetic remodeling enhances organ functionality, reduces disease susceptibility, and maintains systemic physical health

9.4.1 TCA Intermediates

Exercise-induced regulation of TCA cycle intermediates is vital for muscle health, influencing muscle cell differentiation and hypertrophy through epigenetic modifications. Citrate, produced during mitochondrial metabolism, is converted to acetyl-CoA in the nucleus by ACLY (Choudhary et al. 2014). This nuclear acetyl-CoA is essential for histone acetylation, a key mechanism regulating gene expression. Research indicates that silencing ACLY impairs the differentiation of myoblasts and satellite cells, leading to decreased expression of myosin heavy chain isoforms and the myogenic regulatory factor MyoD (Das et al. 2017). Conversely, overexpressing ACLY enhances MyoD expression through hyperacetylation of specific histone lysine residues (H3K9, H3K14, and H3K27) at the MyoD promoter, thereby promoting muscle differentiation (Das et al. 2017). Recent studies suggest that different sources of acetyl-CoA, such as those derived from glucose and fatty acids, may influence histone acetylation through distinct pathways (Sebastián and Mostoslavsky 2017). This raises intriguing possibilities regarding substrate availability during exercise, and post-exercise nutrition could further modulate epigenetic changes in human skeletal muscle. Additionally, the ratio of fumarate and succinate to α-KG influences substrate availability and the histone demethylation activity of KDMs (García-Giménez et al. 2024). Different types of exercise have varying effects on the balance of these TCA cycle intermediates: endurance exercise has been reported to increase fumarate and succinate without significantly affecting α-KG levels, while resistance exercise results in a substantial shift favoring α-KG (Huffman et al. 2014; Maurer et al. 2021; Yuan et al. 2020). The α-KG-dependent histone demethylation process is critical for myogenic differentiation. Notably, the IDH2-R172K mutation produces 2-hydroxyglutarate (2-HG), which inhibits α-KG-dependent histone demethylation. This inhibition leads to increased hypermethylation of H3K9, hindering the necessary gain in chromatin accessibility for differentiation and blocking MyoD's ability to promote muscle differentiation (Schvartzman et al. 2019). The differences in α-KG levels underscore the greater role of resistance exercise in muscle hypertrophy compared with endurance exercise.

9.4.2 Lactate

Recently, lactate has emerged as a key epigenetic-regulating metabolite induced by exercise. The exercise-induced increase in lactate levels triggers a cascading effect that amplifies lactylation, enabling gene reprogramming (Zhang et al. 2019). Lactate-induced H3K9la is essential for myogenic processes; treatment with lactate enhances H3K9la levels, while blocking this modification hinders myogenesis. Neu2 has been identified as a critical target gene of H3K9la, with lactylation of the promoter region of Neu2 being essential for lactate-mediated myogenic differentiation (Dai et al. 2023). Additionally, in muscle-resident macrophages, the dynamics of histone lactylation during muscle injury are significant for muscle regeneration. Specifically, H3K18 lactylation at promoter and enhancer elements occurs at pro-restorative genes, contributing to M2 polarization of macrophages and macrophage-mediated muscle regeneration (Desgeorges et al. 2024). Similarly, histone H3K18 lactylation enhances the expression of key reparative genes such as Lrg1, VEGF-A, and IL-10 in monocyte-macrophages in the heart. This modulation creates an anti-inflammatory and pro-angiogenic environment conducive to cardiac repair following myocardial infarction (Wang et al. 2022). In the brain, lactate is taken up via MCT2 and utilized as an energy substrate during exercise. The high lactate flux leads to histone modifications such as H4K12 lactylation in the brain, which enhances the expression of genes involved in the cell cycle and neurogenesis, notably MDM2 and MDM4. This modulation influences the MDM2-p53 signaling pathway, which is essential for cell-cycle regulation, and supports neural stem cell proliferation (Li et al. 2025). Furthermore, hippocampal lactate enhances the expression of synaptic proteins crucial for spatial memory formation through induction of histone and non-histone lactylation (Wu et al. 2024). In microglia, H3K18 lactylation promotes a transition from a damaging to a reparative phenotype, thereby improving neurological functions (Han et al. 2023). Overall, exercise-induced lactate serves as a significant metabolic and signaling molecule that regulates histone lactylation, influencing gene expression across multiple organs. It supports skeletal muscle myogenic differentiation and regeneration, fosters an anti-inflammatory and angiogenic environment in the heart, and contributes to cognitive health, highlighting its potential therapeutic implications for various disorders.

9.4.3 *βOHB*

βOHB is a metabolite produced in the liver during long-duration exercise, serving as a complementary energy substrate for muscles and the brain. Exercise-induced βOHB has emerged as a significant health modulator by inhibiting class I HDACs, particularly HDAC1, HDAC2, and HDAC3 (Shimazu et al. 2013). This inhibition enhances histone acetylation, especially at H3K14 and H3K9, activating genes associated with oxidative stress resistance such as antioxidant enzymes like superoxide dismutase 2 and catalase, while downregulating pro-oxidative NADPH oxidase, thereby reducing oxidative stress and cellular damage in models of diabetes, spinal cord injury, and nonalcoholic fatty liver disease (Shimazu et al. 2013; Futatani et al. 2017; Soni et al. 2024; Kong et al. 2017). In high glucose-stimulated human cardiac microvascular endothelial cells, βOHB's inhibition of HDAC3 increases claudin-5 production, preventing diabetes-associated cardiac microvascular hyperpermeability (Li et al. 2021). Additionally, βOHB enhances H3K9 acetylation at the GLUT1 gene, boosting glucose transport to the brain during energy stress (Tanegashima et al. 2017).

βOHB also acts as a substrate for histone Kbhb, leading to specific lysine modifications that activate gene expression independently of acetylation. Elevated βOHB levels induce histone Kbhb marks, particularly H3K9bhb, enhancing the expression of genes involved in metabolic adaptation, such as amino acid catabolism, circadian rhythms, and oxidative phosphorylation (Zhou et al. 2022). βOHB's H3K9bhb modification has shown protective effects against diabetic endothelial injury by promoting VEGF generation and upregulating apolipoprotein gene expression in adipocytes, offering anti-inflammatory and anti-atherogenic benefits (Wu et al. 2020; Nishitani et al. 2018). In the brain, βOHB, taken up by MCT1, enhances local β-hydroxybutyrylation of H3K9 at the TJP1 gene promoter, upregulating tight junction protein ZO-1 expression. This modulation maintains blood–brain barrier integrity, reduces neuroinflammation, and mitigates stroke severity (Li et al. 2024). Moreover, β-hydroxybutyrylation regulates brain-derived neurotrophic factor (BDNF), alleviating symptoms in models of major depressive disorder (Chen et al. 2017). Exercise-induced βOHB modification of histones thus represents a crucial mechanism promoting metabolic and neurological health and resilience against disease.

9.4.4 SAM

SAM serves as the primary methylation donor for DNA and histones, with the liver being the principal site of its synthesis and systemic distribution (Lu and Mato 2012). Chronic exercise has been shown to enhance the expression of the SAM synthesis enzyme MAT1A, leading to increased levels of SAM in both the liver and plasma, while acute exercise results in a reduction of SAM levels (Riberio et al. 2018; Yan et al. 2022). This acute energy stress-induced decrease in hepatic SAM causes hypomethylation of CpG sites downstream of the TSS of FGF21, thereby enhancing the hepatic expression of FGF21 (Capelo-Diz et al. 2023). Liver-derived FGF21 facilitates lipid catabolism in adipose tissue and promotes loss of body fat, contributing to the exercise-induced prevention of obesity and metabolic disease risk (Capelo-Diz et al. 2023). However, the alterations in SAM levels induced by exercise have not been extensively reported in organs other than the liver and brain. Notably, SAM can cross the BBB and has been investigated in clinical trials for various psychiatric disorders, including depression, cognitive deficits, and migraines (Topham et al. 2021). The effects of SAM on DNA methylation may partially explain its therapeutic efficacy in treating brain disorders. For example, SAM can reverse DNA hypomethylation in pain-related genes such as Scn10a, Trpa1, Ntrk1, and Gfap following nerve injury, thereby alleviating the sensory and cognitive deficits associated with peripheral nerve injury in mice (Topham et al. 2021). Recent research has also shown that exercise-induced, liver-derived SAM can restore RNA N6-methyladenosine (m6A) modifications on genes related to excitatory synapses and activate medial prefrontal cortex (mPFC) neurons and their projections to the basolateral amygdala (BLA), enhancing neural activity and exhibiting anxiolytic effects (Yan et al. 2022). Collectively, these findings suggest that exercise-induced SAM establishes a liver–brain axis, providing an epigenetic mechanism for brain adaptation following exercise to counteract environmental stress.

9.4.5 Kynurenine

Kynurenine, an intermediate of tryptophan metabolism, plays a complex role in systemic health through its various derivatives. It can be metabolized by kynurenine aminotransferase (KAT) to produce anti-inflammatory and neuroprotective kynurenic acid, or by kynurenine monooxygenase (KMO) to form neurotoxic 3-hydroxy-kynurenine and quinolinic acid (Shadboorestan et al. 2023). Exercise has been reported to facilitate the production of kynurenine and its conversion favoring kynurenic acid, thereby reducing neuroinflammation and protecting against stress-induced depression (Sato et al. 2022; Agudelo et al. 2015; Strasser et al. 2016). However, research on the roles of exercise-induced kynurenine itself remains relatively limited. Kynurenine can act as an endogenous ligand for the aryl hydrocarbon receptor (Ahr), participating in immune regulation, ROS production, and aging (Shadboorestan et al. 2023). Recently, innovative research has highlighted the epigenetic role of kynurenine in the brain. Hayakawa et al. found that kynurenine and its derivatives, 3-hydroxy-kynurenine and anthranilate, increase H3K4 trimethylation and H2AS40 O-GlcNAcylation, thereby upregulating gene expression at loci encoding hypothalamic neural peptides and differentiation-related factors, thus facilitating hypothalamic function and homeostasis (Hayakawa et al. 2019). However, the mechanisms by which kynurenine regulates histone modifications remain to be elucidated in future studies.

9.4.6 Serotonin

Serotonin is a well-established exercise-induced metabolite that reduces symptoms associated with depression and enhances mood (Strasser and Fuchs 2015). These effects are traditionally understood to be mediated by interactions between serotonin and its plasma membrane receptors, which activate signaling cascades primarily through G protein-coupled receptors (GPCRs) (Anastas and Shi 2019). However, recent studies have reported that serotonin can also be covalently bound to glutamine 5 (Gln5) of histone H3 through transamidation by γ-glutamyltransferase 2 (TGM2) (Farrelly et al. 2019). This serotonylation at H3 Gln5 (H3Q5ser) occurs in close proximity to H3K4me3 and promotes the transcription of neuronal genes during neuronal cell differentiation by potentiating the binding of the H3K4me3 reader TFIID at the gene promoters (Farrelly et al. 2019). While H3Q5ser does not affect the activity of the H3K4 methyltransferase MLL1, the corresponding H3K4me3/2 erasers, KDM5B/C, and LSD1, are significantly inhibited by the presence of this modification (Zhao et al. 2021). This suggests that the adjacent H3Q5ser potentiates H3K4me3 function by stabilizing H3K4me3 and preventing its dynamic turnover. Additionally, abnormal levels of H3K4me3Q5ser have been associated with altered gene expression related to depression in chronically stressed mice (Al-Kachak et al. 2024), revealing an epigenetic mechanism by which exercise-induced serotonin can influence depression therapy. These discoveries underscore the complex interplay between exercise-induced serotonin, epigenetic modifications, and gene expression, highlighting their collective role in the regulation of mood and mental health.

The genome-wide epigenetic regulation driven by the exercise-induced metabolites remodeling fires the potential of exercise as a non-pharmacological treatment for preventing disease and promoting health. Further research is needed to fully understand the complex interplay between exercise, metabolites, and epigenetics, which could pave the way for personalized exercise regimens tailored to maximize health benefits.

Compliance with Ethical Standards

Conflict of Interest All authors declare no conflict of interest.

Ethical Approval This chapter is a review of previously published accounts; as such, no animal or human studies were performed.

Funding This research was supported by the National Natural Science Foundation of China (82470247, 32400942, K241141103), Shenzhen Medical Research Funds (A2303047, B2402048), Natural Science Foundation of Guangdong (2025A1515011660).

References

Agudelo LZ et al (2015) Skeletal muscle PGC-1α1 modulates kynurenine metabolism and mediates resilience to stress-induced depression. Cell 160:351

Al-Kachak A et al (2024) Histone serotonylation in dorsal raphe nucleus contributes to stress- and antidepressant-mediated gene expression and behavior. Nat Commun 15:5042

Anastas JN, Shi Y (2019) Histone serotonylation: can the brain have "happy" chromatin? Mol Cell 74:418–420

Asadi Shahmirzadi A et al (2020) Alpha-ketoglutarate, an endogenous metabolite, extends lifespan and compresses morbidity in aging mice. Cell Metab 32:447–456.e6

Barski A et al (2007) High-resolution profiling of histone methylations in the human genome. Cell 129:823–837

Bogdanović O, Lister R (2017) DNA methylation and the preservation of cell identity. Curr Opin Genet Dev 46:9–14

Bowman RL, Levine RL (2017) TET2 in normal and malignant hematopoiesis. Cold Spring Harb Perspect Med 7:a026518

Capelo-Diz A et al (2023) Hepatic levels of S-adenosylmethionine regulate the adaptive response to fasting. Cell Metab 35:1373–1389.e8

Chang S, Yim S, Park H (2019) The cancer driver genes IDH1/2, JARID1C/ KDM5C, and UTX/ KDM6A: crosstalk between histone demethylation and hypoxic reprogramming in cancer metabolism. Exp Mol Med 51:1–17

Chen L, Miao Z, Xu X (2017) Β-hydroxybutyrate alleviates depressive behaviors in mice possibly by increasing the histone3-lysine9-β-hydroxybutyrylation. Biochem Biophys Res Commun 490:117–122

Choudhary C, Weinert BT, Nishida Y, Verdin E, Mann M (2014) The growing landscape of lysine acetylation links metabolism and cell signalling. Nat Rev Mol Cell Biol 15:536–550

Contrepois K et al (2020) Molecular choreography of acute exercise. Cell 181:1112–1130.e16

Cunha TF et al (2022) Lactate modulates cardiac gene expression in mice during acute physical exercise. Braz J Med Biol Res 55:e11820

Dai W et al (2023) Lactate promotes myogenesis via activating H3K9 lactylation-dependent up-regulation of Neu2 expression. J Cachexia Sarcopenia Muscle 14:2851–2865

Das S et al (2017) ATP citrate lyase regulates myofiber differentiation and increases regeneration by altering histone acetylation. Cell Rep 21:3003–3011

Denis C, Dormois D, Linossier MT, Geyssant A, Lacour JR (1991) Total muscle NADH during 30-min submaximal exercise: effects of endurance training. Acta Physiol Scand 142:301–302

Deogharia M et al (2024) Histone demethylase KDM5 regulates cardiomyocyte maturation by promoting fatty acid oxidation, oxidative phosphorylation, and myofibrillar organization. Cardiovasc Res 120:630–643

Desgeorges T, Galle E, Zhang J, von Meyenn F, De Bock K (2024) Histone lactylation in macrophages is predictive for gene expression changes during ischemia induced-muscle regeneration. Mol Metab 83:101923

Ediriweera MK, Gayashani Sandamalika WM (2025) The epigenetic impact of fatty acids as DNA methylation modulators. Drug Discov Today 30:104277

Evans M, Cogan KE, Egan B (2017) Metabolism of ketone bodies during exercise and training: physiological basis for exogenous supplementation. J Physiol 595:2857–2871

Farrelly LA et al (2019) Histone serotonylation is a permissive modification that enhances TFIID binding to H3K4me3. Nature 567:535–539

Figueroa ME et al (2010) Leukemic IDH1 and IDH2 mutations result in a hypermethylation phenotype, disrupt TET2 function, and impair hematopoietic differentiation. Cancer Cell 18:553–567

Fulghum K, Hill BG (2018) Metabolic mechanisms of exercise-induced cardiac remodeling. Front Cardiovasc Med 5:127

Fulghum K, Collins HE, Jones SP, Hill BG (2022) Influence of biological sex and exercise on murine cardiac metabolism. J Sport Health Sci 11:479–494

Futatani T et al (2017) Capillary blood ketone levels as an indicator of inadequate breast milk intake in the early neonatal period. J Pediatr 191:76–81

Gaitanos GC, Williams C, Boobis LH, Brooks S (1993) Human muscle metabolism during intermittent maximal exercise. J Appl Physiol 75:712–719

Gao J et al (2018) S-adenosyl methionine and transmethylation pathways in neuropsychiatric diseases throughout life. Neurotherapeutics 15:156–175

García-Giménez JL, Cánovas-Cervera I, Pallardó FV (2024) Oxidative stress and metabolism meet epigenetic modulation in physical exercise. Free Radic Biol Med 213:123–137

Gibb AA, Hill BG (2018) Metabolic coordination of physiological and pathological cardiac remodeling. Circ Res 123:107–128

Gollasch B, Dogan I, Rothe M, Gollasch M, Luft FC (2019) Maximal exercise and plasma cytochrome P450 and lipoxygenase mediators: a lipidomics study. Physiol Rep 7:e14165

Goodpaster BH, Sparks LM (2017) Metabolic flexibility in health and disease. Cell Metab 25:1027–1036

Gui S et al (2014) A remodeled protein arginine methyltransferase 1 (PRMT1) generates symmetric dimethylarginine. J Biol Chem 289:9320–9327

Haghani A et al (2023) DNA methylation networks underlying mammalian traits. Science 381:eabq5693

Han H et al (2023) Exercise improves cognitive dysfunction and neuroinflammation in mice through histone H3 lactylation in microglia. Immun Ageing 20:63

Hargreaves M, Spriet LL (2020) Skeletal muscle energy metabolism during exercise. Nat Metab 2:817–828

Haws SA et al (2020) Methyl-metabolite depletion elicits adaptive responses to support heterochromatin stability and epigenetic persistence. Mol Cell 78:210–223.e8

Hayakawa K, Nishitani K, Tanaka S (2019) Kynurenine, 3-OH-kynurenine, and anthranilate are nutrient metabolites that alter H3K4 trimethylation and H2AS40 O-GlcNAcylation at hypothalamus-related loci. Sci Rep 9:19768

He W, Li Q, Li X (2023) Acetyl-CoA regulates lipid metabolism and histone acetylation modification in cancer. Biochim Biophys Acta Rev Cancer 1878:188837

Hogg SJ et al (2021) Targeting histone acetylation dynamics and oncogenic transcription by catalytic P300/CBP inhibition. Mol Cell 81:2183–2200.e13

Horowitz JF (2022) Adipose tissue lipid metabolism during exercise. In: Physiology in health and disease. Springer International Publishing, Cham, pp 137–159

Hu C et al (2020) Muscle-liver substrate fluxes in exercising humans and potential effects on hepatic metabolism. J Clin Endocrinol Metab 105:1196–1209

Huang C-C, Lin W-T, Hsu F-L, Tsai P-W, Hou C-C (2010) Metabolomics investigation of exercise-modulated changes in metabolism in rat liver after exhaustive and endurance exercises. Eur J Appl Physiol 108:557–566

Huang W et al (2023a) High-intensity interval training induces protein lactylation in different tissues of mice with specificity and time dependence. Metabolites 13:647

Huang F et al (2023b) Control of histone demethylation by nuclear-localized α-ketoglutarate dehydrogenase. Science 381:eadf8822

Huffman KM et al (2014) Metabolite signatures of exercise training in human skeletal muscle relate to mitochondrial remodelling and cardiometabolic fitness. Diabetologia 57:2282–2295

Jambhekar A, Dhall A, Shi Y (2019) Roles and regulation of histone methylation in animal development. Nat Rev Mol Cell Biol 20:625–641

Jeltsch A, Broche J, Lungu C, Bashtrykov P (2020) Biotechnological applications of MBD domain proteins for DNA methylation analysis. J Mol Biol 432:1816–1823

Jenuwein T, Allis CD (2001) Translating the histone code. Science 293:1074–1080

Katsyuba E, Romani M, Hofer D, Auwerx J (2020) NAD+ homeostasis in health and disease. Nat Metab 2:9–31

Kaymak I et al (2024) ACLY and ACSS2 link nutrient-dependent chromatin accessibility to CD8 T cell effector responses. J Exp Med 221:e20231820

Kong G et al (2017) The ketone metabolite β-hydroxybutyrate attenuates oxidative stress in spinal cord injury by suppression of class I histone deacetylases. J Neurotrauma 34:2645–2655

Lassers BW, Kaijser L, Wahlqvist M, Carlson LA (1971) Effect of prolonged exercise on myocardial metabolism in man. Br Heart J 33:609

Lawrence M, Daujat S, Schneider R (2016) Lateral thinking: how histone modifications regulate gene expression. Trends Genet 32:42–56

Lee S et al (2023) Physiological significance of elevated levels of lactate by exercise training in the brain and body. J Biosci Bioeng 135:167–175

Li B et al (2021) β-hydroxybutyrate inhibits histone deacetylase 3 to promote claudin-5 generation and attenuate cardiac microvascular hyperpermeability in diabetes. Diabetologia 64:226–239

Li R et al (2024) Adaptive metabolic responses facilitate blood-brain barrier repair in ischemic stroke via BHB-mediated epigenetic modification of ZO-1 expression. Adv Sci (Weinh) 11:e2400426

Li Z et al (2025) Lactate shuttling links histone lactylation to adult hippocampal neurogenesis in mice. Dev Cell. https://doi.org/10.1016/j.devcel.2024.12.021

Liu L et al (2024) Genome-wide DNA methylation regulated by AHCY through SAM/SAH axis promotes psoriasis pathogenesis. J Dermatol Sci 116:100–110

Lu SC, Mato JM (2012) S-adenosylmethionine in liver health, injury, and cancer. Physiol Rev 92:1515–1542

Ma S et al (2025) Nutrient-driven histone code determines exhausted CD8+ T cell fates. Science 387:eadj3020

Martin KS, Azzolini M, Lira Ruas J (2020) The kynurenine connection: how exercise shifts muscle tryptophan metabolism and affects energy homeostasis, the immune system, and the brain. Am J Physiol Cell Physiol 318:C818–C830

Maschari D et al (2022) Lactate-induced lactylation in skeletal muscle is associated with insulin resistance in humans. Front Physiol 13:951390

Mattei AL, Bailly N, Meissner A (2022) DNA methylation: a historical perspective. Trends Genet 38:676–707

Maurer J, Hoene M, Weigert C (2021) Signals from the circle: tricarboxylic acid cycle intermediates as myometabokines. Metabolites 11:474

Medeiros BC et al (2017) Isocitrate dehydrogenase mutations in myeloid malignancies. Leukemia 31:272–281

Nayor M et al (2020) Metabolic architecture of acute exercise response in middle-aged adults in the community. Circulation 142:1905–1924

Newman JC, Verdin E (2014) Ketone bodies as signaling metabolites. Trends Endocrinol Metab 25:42–52

Nguyen-Tran D-H et al (2014) Molecular mechanism of sphingosine-1-phosphate action in Duchenne muscular dystrophy. Dis Model Mech 7:41–54

Nikolaidis MG, Mougios V (2004) Effects of exercise on the fatty-acid composition of blood and tissue lipids. Sports Med 34:1051–1076

Nishitani S et al (2018) Metabolomic and microarray analyses of adipose tissue of dapagliflozin-treated mice, and effects of 3-hydroxybutyrate on induction of adiponectin in adipocytes. Sci Rep 8. https://doi.org/10.1038/s41598-018-27181-y

Nshanian M et al (2025) Short-chain fatty acid metabolites propionate and butyrate are unique epigenetic regulatory elements linking diet, metabolism and gene expression. Nat Metab 7:196–211

Ohmura H, Mukai K, Takahashi Y, Takahashi T (2021) Metabolomic analysis of skeletal muscle before and after strenuous exercise to fatigue. Sci Rep 11:11261

Pavlenko E, Ruengeler T, Engel P, Poepsel S (2022) Functions and interactions of mammalian KDM5 demethylases. Front Genet 13:906662

Peek CB et al (2017) Circadian clock interaction with HIF1α mediates oxygenic metabolism and anaerobic glycolysis in skeletal muscle. Cell Metab 25:86–92

Pham VN et al (2023) Formaldehyde regulates S-adenosylmethionine biosynthesis and one-carbon metabolism. Science 382:eabp9201

Pietrocola F, Galluzzi L, Bravo-San Pedro JM, Madeo F, Kroemer G (2015) Acetyl coenzyme a: a central metabolite and second messenger. Cell Metab 21:805–821

Rector RS et al (2008) Daily exercise increases hepatic fatty acid oxidation and prevents steatosis in Otsuka Long-Evans Tokushima Fatty rats. Am J Physiol Gastrointest Liver Physiol 294:G619–G626

Reid MA, Dai Z, Locasale JW (2017) The impact of cellular metabolism on chromatin dynamics and epigenetics. Nat Cell Biol 19:1298–1306

Riberio DF, Cella PS, da Silva LECM, Jordao AA, Deminice R (2018) Acute exercise alters homocysteine plasma concentration in an intensity-dependent manner due increased methyl flux in liver of rats. Life Sci 196:63–68

Romijn JA et al (1993) Regulation of endogenous fat and carbohydrate metabolism in relation to exercise intensity and duration. Am J Phys 265:E380–E391

Sahlin K, Katz A, Henriksson J (1987) Redox state and lactate accumulation in human skeletal muscle during dynamic exercise. Biochem J 245:551–556

Sato S et al (2022) Atlas of exercise metabolism reveals time-dependent signatures of metabolic homeostasis. Cell Metab 34:329–345.e8

Schvartzman J-M, Reuter VP, Koche RP, Thompson CB (2019) 2-hydroxyglutarate inhibits MyoD-mediated differentiation by preventing H3K9 demethylation. Proc Natl Acad Sci USA 116:12851–12856

Seaborne RA, Sharples AP (2020) The interplay between exercise metabolism, epigenetics, and skeletal muscle remodeling. Exerc Sport Sci Rev 48:188–200

Sebastián C, Mostoslavsky R (2017) The various metabolic sources of histone acetylation. Trends Endocrinol Metab 28:85–87

Shadboorestan A, Koual M, Dairou J, Coumoul X (2023) The role of the kynurenine/AhR pathway in diseases related to metabolism and cancer. Int J Tryptophan Res 16:11786469231185102

Shimazu T et al (2013) Suppression of oxidative stress by β-hydroxybutyrate, an endogenous histone deacetylase inhibitor. Science 339:211–214

Smith JAB, Murach KA, Dyar KA, Zierath JR (2023) Exercise metabolism and adaptation in skeletal muscle. Nat Rev Mol Cell Biol 24:607–632

Soni S, Tabatabaei Dakhili SA, Ussher JR, Dyck JRB (2024) The therapeutic potential of ketones in cardiometabolic disease: impact on heart and skeletal muscle. Am J Physiol Cell Physiol 326:C551–C566

Stanley WC (1991) Myocardial lactate metabolism during exercise. Med Sci Sports Exerc 23:920–924

Starnes JW et al (2017) Exercise-induced alterations in skeletal muscle, heart, liver, and serum metabolome identified by non-targeted metabolomics analysis. Metabolites 7:40

Strasser B, Fuchs D (2015) Role of physical activity and diet on mood, behavior, and cognition. Neurol Psychiatry Brain Res 21:118–126

Strasser B et al (2016) Effects of exhaustive aerobic exercise on tryptophan-kynurenine metabolism in trained athletes. PLoS One 11:e0153617

Strøm CC et al (2005) Expression profiling reveals differences in metabolic gene expression between exercise-induced cardiac effects and maladaptive cardiac hypertrophy. FEBS J 272:2684–2695

Sutendra G et al (2014) A nuclear pyruvate dehydrogenase complex is important for the generation of acetyl-CoA and histone acetylation. Cell 158:84–97

Tan-Chen S, Guitton J, Bourron O, Le Stunff H, Hajduch E (2020) Sphingolipid metabolism and signaling in skeletal muscle: from physiology to physiopathology. Front Endocrinol (Lausanne) 11:491

Tanegashima K et al (2017) Epigenetic regulation of the glucose transporter gene Slc2a1 by β-hydroxybutyrate underlies preferential glucose supply to the brain of fasted mice. Genes Cells 22:71–83

Tefferi A, Lim K-H, Levine R (2009) Mutation in TET2 in myeloid cancers. N Engl J Med 361:1117.; author reply 1117–8

Topham L et al (2021) The methyl donor S-adenosyl methionine reverses the DNA methylation signature of chronic neuropathic pain in mouse frontal cortex. Pain Rep 6:e944

Trefts E, Williams AS, Wasserman DH (2015) Exercise and the regulation of hepatic metabolism. Prog Mol Biol Transl Sci 135:203–225

Tretter L, Adam-Vizi V (2005) Alpha-ketoglutarate dehydrogenase: a target and generator of oxidative stress. Philos Trans R Soc Lond B Biol Sci 360:2335–2345

Vargas-Ortiz K, Pérez-Vázquez V, Macías-Cervantes MH (2019) Exercise and sirtuins: a way to mitochondrial health in skeletal muscle. Int J Mol Sci 20:2717

Verdin E (2015) NAD^+ in aging, metabolism, and neurodegeneration. Science 350:1208–1213

Walzik D et al (2023) Tissue-specific effects of exercise as NAD+ -boosting strategy: current knowledge and future perspectives. Acta Physiol (Oxf) 237:e13921

Wang W, Toran PT, Sabol R, Brown TJ, Barth BM (2018) Epigenetics and sphingolipid metabolism in health and disease. Int J Biopharm Sci 1:105

Wang N et al (2022) Histone lactylation boosts reparative gene activation post-myocardial infarction. Circ Res 131:893–908

Wang M, Nasb M, Wu T, Jia M, So K-F (2024a) Exercise-mediated circulating metabolites for brain health. Adv Exerc Health Sci. https://doi.org/10.1016/j.aehs.2024.11.001
Wang Y et al (2024b) Lactate metabolism and histone lactylation in the central nervous system disorders: impacts and molecular mechanisms. J Neuroinflammation 21:308
White AT, Schenk S (2012) NAD(+)/NADH and skeletal muscle mitochondrial adaptations to exercise. Am J Physiol Endocrinol Metab 303:E308–E321
Wolfe RR, Klein S, Carraro F, Weber JM (1990) Role of triglyceride-fatty acid cycle in controlling fat metabolism in humans during and after exercise. Am J Phys 258:E382–E389
Wu X et al (2020) β-hydroxybutyrate antagonizes aortic endothelial injury by promoting generation of VEGF in diabetic rats. Tissue Cell 64:101345
Wu Y et al (2024) Hippocampal lactate-infusion enhances spatial memory correlated with monocarboxylate transporter 2 and lactylation. Brain Sci 14:327
Xhabija B, Kidder BL (2019) KDM5B is a master regulator of the H3K4-methylome in stem cells, development and cancer. Semin Cancer Biol 57:79–85
Xu G, Hansen JS, Zhao XJ, Chen S, Hoene M, Wang XL, Clemmesen JO, Secher NH, Häring HU, Pedersen BK, Lehmann R, Weigert C, Plomgaard P (2016) Liver and muscle contribute differently to the plasma acylcarnitine pool during fasting and exercise in human. J Clin Endocrinol Metab 101(12):5044–5052
Xu G-E et al (2024) Exercise training decreases lactylation and prevents myocardial ischemia-reperfusion injury by inhibiting YTHDF2. Basic Res Cardiol 119:651–671
Yan L et al (2022) Physical exercise prevented stress-induced anxiety via improving brain RNA methylation. Adv Sci (Weinh) 9:e2105731
Yang Z, He M, Austin J, Pfleger J, Abdellatif M (2021) Histone H3K9 butyrylation is regulated by dietary fat and stress via an Acyl-CoA dehydrogenase short chain-dependent mechanism. Mol Metab 53:101249
Yu W et al (2019) One-carbon metabolism supports S-adenosylmethionine and histone methylation to drive inflammatory macrophages. Mol Cell 75:1147–1160.e5
Yu H et al (2024) α-Ketoglutarate improves cardiac insufficiency through NAD+-SIRT1 signaling-mediated mitophagy and ferroptosis in pressure overload-induced mice. Mol Med 30:15
Yuan Y et al (2020) Exercise-induced α-ketoglutaric acid stimulates muscle hypertrophy and fat loss through OXGR1-dependent adrenal activation. EMBO J 39:e103304
Zhang D et al (2019) Metabolic regulation of gene expression by histone lactylation. Nature 574:575–580
Zhang J et al (2023) PARylated PDHE1α generates acetyl-CoA for local chromatin acetylation and DNA damage repair. Nat Struct Mol Biol 30:1719–1734
Zhao S et al (2021) Histone H3Q5 serotonylation stabilizes H3K4 methylation and potentiates its readout. Proc Natl Acad Sci USA 118:e2016742118
Zhou T et al (2022) Function and mechanism of histone β-hydroxybutyrylation in health and disease. Front Immunol 13:981285

Chapter 10
Epigenetic Modulation of Mitochondria by Exercise

Ke Cao, Jie Xu, and Zhihui Feng

Abstract Mitochondria, the essential organelles within cells, play a pivotal role in determining cellular fate. Their homeostasis is regulated by genes from both the nucleus and the mitochondria. Exercise stands out as an effective non-pharmacological approach to maintaining physical health. Research has demonstrated that exercise influences mitochondrial homeostasis through various epigenetic mechanisms, such as DNA methylation, non-coding RNAs, and histone post-translational modifications. These mechanisms contribute to the overall health of organisms and can positively impact the onset and progression of numerous diseases. This chapter explores the intricate ways in which epigenetic mechanisms regulate mitochondrial homeostasis and the influence of exercise on this process. Additionally, it summarizes the diverse roles of exercise in both physiological and pathological contexts, highlighting its effects on mitochondrial homeostasis and overall health through epigenetic pathways. The goal of this chapter is to provide theoretical insights and potential intervention targets related to the epigenetic mechanisms by which exercise enhances health, with a specific focus on the role of mitochondria.

Ke Cao and Jie Xu contributed equally to this work.

K. Cao · J. Xu
Center for Mitochondrial Biology and Medicine, The Key Laboratory of Biomedical Information Engineering of Ministry of Education, School of Life Science and Technology, Xi'an Jiaotong University, Xi'an, Shaanxi, China

Z. Feng (✉)
Center for Mitochondrial Biology and Medicine, The Key Laboratory of Biomedical Information Engineering of Ministry of Education, School of Life Science and Technology, Xi'an Jiaotong University, Xi'an, Shaanxi, China

Frontier Institute of Science and Technology, Xi'an Jiaotong University, Xi'an, Shaanxi, China
e-mail: zhfeng@xjtu.edu.cn

F. Gao, X. Zhang (eds.), *Exercise, Epigenetics and Human Health*, Epigenetics and Human Health 15, https://doi.org/10.1007/978-3-032-08045-5_10

Keywords Exercise · Mitochondria · Epigenetics · DNA methylation · Non-coding RNAs · Histone post-translational modifications

Abbreviations

ALKBH1	AlkB homolog 1
AMPK	AMP-activated protein kinase
BDNF	Brain-derived neurotrophic factor
CaMK	Calmodulin-dependent protein kinase
CBP	CREB-binding protein
CREB	cAMP response element-binding protein
circRNAs	Circular RNAs
DNMTs	DNA methyltransferases
Drp1	Dynamin-related protein-1
HATs	Histone acetyltransferases
HDACs	Histone deacetylases
hMeDIP-seq	Hydroxymethylated DNA immunoprecipitation sequencing
LTM	Long-term memory
lncRNAs	Long non-coding RNAs
MEF2A	Myocyte enhancer factor 2A
METTL4	Methyltransferase like 4
MSCs	Muscle satellite cells
miRNAs	MicroRNAs
mtDNA	Mitochondrial DNA
mtTFA, TFAM	Mitochondrial transcription factor A
NRF-1	Nuclear respiratory factor 1
N6AMT1	N-6-adenine specific DNA methyltransferase 1
ncRNAs	Non-coding RNAs
SAM	S-adenosyl-methionine
PDK4	Pyruvate dehydrogenase kinase 4
PGC-1α	Peroxisome proliferators-activated receptor γ coactivator 1 alpha
PTMs	Post-translational modifications
siRNAs	Short interfering RNAs
sncRNAs	Small non-coding RNAs
TETs	Ten-eleven translocation enzymes
TFB2M	Mitochondrial transcription factor B2
T2D	Type 2 diabetes
VDAC1	Voltage-dependent anion channel 1
5hmC	5-hydroxymethylcytosine
5mC	5-methylcytosine
6 mA	N6-methyldeoxyadenosine

10.1 Introduction

Mitochondria are highly dynamic organelles found in nearly all eukaryotic cells, with the exception of mature mammalian erythrocytes. Their morphology, size, number, and distribution can vary significantly among different cell types and physiological states. As the primary sites for biological oxidation and energy conversion, mitochondria are responsible for generating more than 90% of cellular energy through oxidative phosphorylation. Decades of research have revealed that mitochondria are multifaceted organelles (Monzel et al. 2023). In addition to functioning as the hub of cellular energy metabolism, they play crucial roles in maintaining calcium homeostasis, synthesizing heme, facilitating lipid synthesis and oxidation, responding to stress, and influencing cellular fate (Chen et al. 2020; Schulz et al. 2023; Sekine et al. 2023). To carry out these diverse functions, mitochondria have established complex networks for intra- and extracellular communication. Within the cell, these pathways involve direct interactions between mitochondria and other subcellular structures, as well as the transport of ions, metabolites, and signaling molecules via vesicles. Beyond the cell, mitochondria can initiate stress responses or other cellular changes that lead to the secretion of mitochondrial cytokines, which can travel between tissues and mediate responses to external immune challenges (Shen et al. 2022). Consequently, mitochondrial homeostasis and integrity are critical factors that contribute to health and disease across various tissues. This homeostasis is regulated by the processes of mitochondrial fission and fusion, which must be carefully balanced to maintain normal mitochondrial function. Any reduction or excess in these processes can disrupt mitochondrial homeostasis, potentially leading to a range of diseases, including neurodegenerative disorders, cardiovascular diseases, and cancer (Adebayo et al. 2021).

Exercise is a highly effective non-pharmacological strategy for maintaining health. Regular physical activity has long been recognized as a crucial component of clinical interventions aimed at preventing and treating various pathological and age-related conditions. In contrast, sedentary behavior is associated with an increased risk of age-related ailments (Yan et al. 2020). Research has shown that exercise training can delay the progression of aging and mitigate the pathological effects of conditions such as type 2 diabetes, obesity, and cardiovascular disease. These health benefits are largely attributed to the positive impact of exercise on mitochondrial homeostasis, including enhancements in mitochondrial biogenesis, increased mitochondrial fusion, reduced mitochondrial fission, promotion of mitophagy, and maintenance of mitochondrial integrity (Jin et al. 2022).

Mitochondria contain their own DNA (mtDNA) in addition to the nuclear genome, which collectively regulates mitochondrial quality and quantity—referred to as mitochondrial homeostasis—through various epigenetic mechanisms, including DNA methylation, non-coding RNAs (ncRNAs), and histone modifications (Hao et al. 2020a). In contrast, metabolites produced during the TCA cycle, such as acetyl-CoA, α-ketoglutarate, and NAD^+, serve as substrates and cofactors for enzymes involved in epigenetic processes, particularly those related to histone modifications (Lian et al. 2024). For instance, acetyl-CoA, which is vital for the TCA cycle, also acts as a substrate for histone acetyltransferases (HATs). HATs relax chromatin structure and enhance gene transcription, while histone deacetylases (HDACs) condense chromatin and repress gene expression. This interplay establishes a connection between mitochondrial bioenergetics and epigenetic regulation (Uittenbogaard et al. 2018; Vizioli et al. 2020; Lian et al. 2024). Recent advancements in epigenetic research have provided consistent evidence that exercise influences mitochondrial remodeling and cellular bioenergetics through epigenetic mechanisms. These changes affect the transcription and translation of mitochondrial-related genes, ultimately impacting the physiological and pathological states of the body (Dimauro et al. 2020).

In this chapter, we will explore the complex ways in which epigenetic mechanisms regulate mitochondrial homeostasis and examine how exercise impacts this regulatory process. Additionally, we will summarize the diverse roles of exercise in modulating mitochondrial homeostasis under both physiological and pathological conditions, highlighting its potential to enhance overall health through epigenetic pathways.

10.2 Epigenetic Regulation on Mitochondrial Homeostasis

Epigenetics constitutes a mechanism capable of regulating gene expression in a heritable manner without altering the DNA sequence. This encompasses processes such as DNA methylation, ncRNAs, and post-translational modifications (PTMs) of histones (Hogg et al. 2020). Research indicates that both nuclear (Jiang et al. 2015; Wang et al. 2023) and mitochondrial epigenetic (Shock et al. 2011; Zhao et al. 2021) mechanisms play crucial roles in regulating mitochondrial homeostasis and function by modulating the expression of mitochondria-associated genes. In this part, we will provide a summary of how epigenetic mechanisms affect mitochondrial homeostasis from both the nuclear and mitochondrial levels.

10.2.1 The Regulation of DNA Methylation on Mitochondria

10.2.1.1 Nuclear DNA Methylation and Mitochondria

DNA methylation is one of the most significant epigenetic modifications which participates in the regulation of mitochondrial homeostasis via influencing both nuclear and mitochondrial genes expression (Mattei et al. 2022). In mammals, 5-methylcytosine (5mC) constitutes the predominant form of DNA methylation, a process catalyzed by DNA methyltransferases (DNMTs, mainly include DNMT1, DNMT3a, DNMT3b, and DNMT3L in mammals) and utilizing S-adenosylmethionine (SAM) as the source of methyl groups (Mattei et al. 2022).

Peroxisome proliferators-activated receptor γ coactivator 1 alpha (PGC-1α) functions as a master switch controlling mitochondrial biogenesis and is indispensable for sustaining mitochondrial homeostasis (Qian et al. 2019; Malik et al. 2023). Research has demonstrated a link between disturbed PGC-1α levels and an array of metabolic abnormalities (Patti et al. 2003; Sanchez-Ramos et al. 2017; Zhao et al. 2023). Notably, evidence indicates that the methylation status at CpG sites of the *PPARGC1A* gene, which encodes PGC-1α, is inversely related to the expression level of PGC-1α. Consequently, this methylation process contributes to the development of multiple metabolic diseases, such as obesity (Clarke-Harris et al. 2014), type 2 diabetes (Gillberg et al. 2013), and cardiovascular diseases (Jiang et al. 2015). The *DRP1* gene provides another instance where DNA methylation within its promoter region plays a role in regulating mitochondrial homeostasis (You et al. 2023). DRP1, a crucial protein involved in mitochondrial dynamics, plays a pivotal role in regulating mitochondrial fission and its activity and expression levels are closely linked to the maintenance of mitochondrial homeostasis (Zhou et al. 2024). You and colleagues (2023) reported that epigenetic hypomethylation of the *DRP1* promoter region resulted in elevated Drp1 expression, ultimately inducing vascular senescence in HUVECs. Another study showed that FENDRR may act as an upstream regulator of DRP1 by forming an RNA–DNA triplex with *DRP1* promoter, enhancing *DRP1* methylation level and subsequently inhibiting its transcriptional activity (Wang et al. 2022a). It is noteworthy that the methylation and transcriptional activity of the *DRP1* promoter region may be jointly regulated by both DNMT1 (Park et al. 2021) and DNMT3a (Yang et al. 2022), potentially depending on the cell type. Furthermore, a study in ischemia reperfusion heart revealed a correlation between increased DNMT1 expression and decreased expression of 11 mitochondrial-related proteins (PGC-1α, TFAM, POLG, MFN1 and MFN2, FIS1, PARKIN, OPTN, ND1, ND4L, Cyt B, and COX1) (Boovarahan and Kurian 2022), suggesting an epigenetic regulatory role of DNA methylation in mitochondrial homeostasis.

Apart from 5mC, 5-hydroxymethylcytosine (5hmC) is another type of epigenetic modification that is catalyzed through the Ten-Eleven-Translocation enzymes (TETs)-mediated demethylation process and also has been shown to potentially play a role in regulating mitochondrial homeostasis. Previous research conducted in our laboratory (Liu et al. 2020) showed that in the cerebral cortex of diabetic mice, the dysregulation of nuclear DNA 5hmC mediated by TET2 downregulation preceded mitochondrial dysfunction, linking reduced 5hmC to impaired mitochondrial respiration and apoptotic activation, suggesting a role for 5hmC in modulating mitochondrial homeostasis. Consistently, Li et al. (2023) discovered that vitamin C could protect cells from apoptosis that is dependent on 5hmC in the promoter regions of nuclear-encoded genes, acting through a mechanism involving the elimination of ROS and the enhancement of mitochondrial stability. Recently, by using hydroxymethylated DNA immunoprecipitation sequencing (hMeDIP-seq) to detect the 5hmC profile, researchers discovered 99 peaks of hyper-hydroxymethylation and 150 peaks of hypo-hydroxymethylation in nuclear-encoded mitochondrial genes from Obese- versus Non-obese-mesenchymal stem/stromal cells (Eirin et al.

2024). More significantly, subsequent comprehensive analysis combining hMeDIP-seq and mRNA-seq data unveiled a convergence of nuclear-encoded mitochondrial genes that play roles in mitochondrial structure and function, fatty acid metabolism within mitochondria, redox signaling, and ATP production (Eirin et al. 2024), highlighting the regulatory role of nuclear DNA 5hmC in maintaining mitochondrial homeostasis.

10.2.1.2 Mitochondrial DNA Methylation and Mitochondria

Mitochondrial DNA (mtDNA) methylation, particularly in the form of 5mC, has garnered significant attention in recent years, despite ongoing debates and uncertainties surrounding the mechanisms of DNA methylation and demethylation in mitochondria. Over five decades after its initial description (Kudriashova et al. 1976), robust evidence supporting the occurrence of mtDNA methylation has emerged, including the detection of DNMT1 (Shock et al. 2011; Saini et al. 2017), DNMT3a (Wong et al. 2013; Dou et al. 2019), and DNMT3b (Bellizzi et al. 2013; Patil et al. 2019) in mitochondria. However, controversy on mtDNA methylation persists partly due to potential issues such as nuclear DNA contamination and false positives in bisulfite-based methods (Hong et al. 2013; Mechta et al. 2017; Owa et al. 2018). Recent whole-genome sequencing studies by us (Cao et al. 2021b) and others (Dou et al. 2019; Patil et al. 2019) have confirmed mtDNA methylation at single-base resolution in human tumor cell lines (Patil et al. 2019), human brain (Dou et al. 2019), and mouse liver (Cao et al. 2021b), revealing both CpG and non-CpG (CpC, CpA, CpT) methylation patterns in mtDNA. Unlike nuclear DNA, mtDNA methylation exhibits three distinct characteristics: higher frequency in non-CpG sites, asymmetric methylation pattern with higher frequency on the L-strand, and occurrence in both D-loop and mtDNA-encoded gene regions, rather than primarily in CpG-rich promoter regions. More importantly, recent evidence has demonstrated that mtDNA 5mC methylation may represent a novel mechanism regulating the expression of mitochondrial-encoded genes and mitochondrial homeostasis, thereby participating in the progression of a range of diseases, including insulin resistance (Cao et al. 2021b), neurodegenerative disorders (Xu et al. 2019), cardiovascular diseases (Baccarelli and Byun 2015), and cancer (Ma et al. 2023b) etc. Moreover, 5hmC has also been found in mtDNA (Shock et al. 2011; Ji et al. 2018). Studies have shown substantial enrichment of 5hmC in mtDNA fragments and total mtDNA from different species and tissues (Dzitoyeva et al. 2012; Ji et al. 2018). Ghosh et al. reported the human mitochondrial 5hmC map, revealing a preference for enrichment upstream of gene start sites (Ghosh et al. 2016). Furthermore, TET1 and TET2 have been observed in mitochondria (Dzitoyeva et al. 2012; Bellizzi et al. 2013). Although the exact mechanism of their functioning in mitochondria remains unknown, alterations in TET2 expression have been linked to changes in mtDNA 5hmC levels and transcription levels of mtDNA-encoded genes (Ji et al. 2018), suggesting a potential regulatory role of mtDNA 5hmC in mitochondrial homeostasis.

Recently, an overriding type of DNA methylation in prokaryotes, known as N6-methyldeoxyadenosine (6 mA) modification, has also been discovered in eukaryotic mitochondria (Hao et al. 2020b; Ma et al. 2023a; Hahn et al. 2024). Enzymes like N-6 adenine specific DNA methyltransferase 1 (N6AMT1) and methyltransferase like 4 (METTL4) catalyze 6 mA formation, and AlkB homolog 1 (ALKBH1) and ALKBH4 are responsible for its demethylation. Of note, METTL4 (Hao et al. 2020b; Zhang et al. 2024) and ALKBH1 (Xiao et al. 2018; Hahn et al. 2024) have been observed in mitochondria, and both of them play a vital role in the modulation of mitochondrial oxidative phosphorylation. Crucially, studies have hinted at a possible connection between mtDNA 6 mA modification and mitochondrial-encoded gene expression, and METTL4-mediated mtDNA 6 mA has been demonstrated to take part in hypoxia stress (Hao et al. 2020b) and heart failure (Zhang et al. 2024), while ALKBH1-mediated mtDNA 6 mA demethylation is involved in regulating life span (Hahn et al. 2024) and pancreatic tumorigenesis (Cai et al. 2023). Generally, research on 6 mA modification in mitochondrial genomes is limited, with unclear mechanisms regulating gene expression and biological significance needing further investigation.

10.2.2 The Regulation of Non-coding RNAs on Mitochondria

ncRNAs, which lack protein-coding ability, have emerged as pivotal regulators in various biological processes by interacting with nucleic acids and proteins. Based on length, ncRNAs are categorized into long ncRNAs (lncRNAs, >200 nt, including circRNAs) and small ncRNAs (sncRNAs, <200 nt, including miRNAs, siRNAs, and piRNAs) (Soutschek and Schratt 2023). Depending on their encoding location, ncRNAs can be divided into nuclear-encoded ncRNAs and mitochondrial-encoded ncRNAs. Both lncRNAs and sncRNAs exist in mitochondria, and both are capable of participating in the regulation of mitochondrial homeostasis and function.

Specific lncRNAs within mitochondria function as regulators, influencing mitochondrial-related gene expression both locally and in the nucleus, thereby enabling retrograde signaling by traversing from mitochondria to the nucleus (Dong et al. 2017). Apart from some proteins encoded by the nucleus that translocate to mitochondria to perform their functions, a portion of nuclear-encoded lncRNAs also have the ability to translocate to mitochondria, participating in the regulation of mitochondrial functions and homeostasis (Noh et al. 2016). Additionally, some lncRNAs encoded by mitochondrial genome may similarly translocate to the nucleus to exert regulatory effects (Rackham et al. 2011). However, the mechanisms underlying the transport of lncRNAs between the nucleus and mitochondria remain to be elucidated. For example, lncRMRP is a nuclear-encoded lncRNA that can translocate to mitochondria with the assistance of two nuclear-encoded proteins human antigen R and G-rich RNA sequence-binding factor 1, ultimately regulating mitochondrial respiration and mitochondrial DNA replication (Noh et al. 2016).

In a study by Rackham and colleagues (2011), three lncRNAs (lncND5, lncND6, and lncCyt b) were revealed to be transcribed from the mitochondrial genome and modulated by five key proteins encoded in the nucleus. These lncRNAs exhibited tissue-specific abundance levels comparable to their corresponding mRNAs, pointing to a regulatory function in mtDNA-encoded gene expression (Rackham et al. 2011). Even more intriguing, lncND5 was also detected within the nucleus, hinting at a potential regulatory function of lncND5 in nucleus as well (Rackham et al. 2011).

Similar to lncRNAs, some ncRNAs could participate in the regulation of mitochondrial homeostasis as well. Ro et al. (2013) uncovered numerous mtDNA-encoded sncRNAs derived from sense transcripts, with alterations in some of the sncRNAs linked to dysregulated host mitochondrial gene expression, suggesting their regulatory role in mitochondrial homeostasis. Mitochondrial miRNAs, belonging to a distinct category of sncRNAs, hold pivotal functions in modulating mitochondrial-associated gene expression within mitochondria and are implicated in a spectrum of diseases, such as cardiovascular diseases (Bhatti et al. 2023) and neurodegenerative diseases (Gowda et al. 2022). For instance, mitochondrial miR-7 could target on voltage-dependent anion channel 1 (VDAC1) mRNA to regulate mitochondrial permeability transition pore and overexpression of miR-7 in human neuroblastoma SH-SY5Y cells suppressed 1-methyl-4-phenylpyridinium-induced mitochondrial dysfunction (Chaudhuri et al. 2016). Another example is miR-30, which can bind to p53 and suppress its downstream effector, dynamin-related protein-1 (Drp1), thereby modulating mitochondrial fission and apoptosis processes (Li et al. 2010).

Generally, the ncRNAs that can act on mitochondria, regardless of whether they originate from the nucleus or mitochondria, remain largely unknown and their mechanisms of action and biological functions still need further elucidation. Some of the recently reported mitochondrial ncRNAs are reviewed by us (Cao et al. 2021a) and by Patange et.al (2025).

10.2.3 The Regulation of Post-translational Modifications of Histones on Mitochondria

The PTMs undergone by histones hold a pivotal role in regulating gene transcription and determining cellular fate (Wang et al. 2022b). There exists an interregulatory relationship between the histones PTMs and mitochondria. The PTMs of histones can regulate the expression of mitochondria-related genes, thereby potentially influencing mitochondrial homeostasis and function. Conversely, mitochondria-derived metabolites, including acetyl-CoA, α-ketoglutarate, and NAD^+, function as essential substrates for various histone modifications, thereby participating in the regulation of histone PTMs. Wang et al. (2023) reported that the deficiency of histone acetyltransferase, MOF, impaired histone H4 lysine 16 acetylation (H4K16ac), thereby leading to disturbing expression of genes responsible for maintaining essential mitochondrial function and eventually causing abnormal skin development. Of note, acetylcarnitine shuttling may be involved in the crosstalk between histone acetylation and mitochondrial metabolism (Izzo et al. 2023). In addition to histone acetyltransferase, histone deacetylases have also been demonstrated to play a part in regulating mitochondrial homeostasis. Marinho and colleagues (2023) assessed the impact of knocking out or inhibiting different class I histone deacetylases (HDACs) on histone acetylation and the communication between the endoplasmic reticulum and mitochondria. They observed that the application of a targeted inhibitor for class I HDACs, tacedinaline, significantly reduced the enhanced Ca^{2+} sequestration in endoplasmic reticulum, elevated Ca^{2+} level in mitochondria, promoted mitochondrial depolarization, and disturbed the interaction between endoplasmic reticulum and mitochondria (Marinho et al. 2023), suggesting a role of histone acetylation in modulating mitochondrial homeostasis. Another study showed that suppression of histone demethylase KDM5 not only resulted in an augmentation of H3K4me3 deposits but also elicited a substantial alteration in gene expression patterns, characterized by the upregulation of genes related to fatty acid oxidation, oxidative phosphorylation, and myogenesis, therefore participating in myofibril organization and cardiomyocyte maturation (Deogharia et al. 2024). Additionally, research has highlighted the crucial involvement of mitochondria in the regulation of various histone modifications, such as acetylation (Si et al. 2024), methylation (Salminen et al. 2014), succinylation (Du et al. 2011), and lactylation (He et al. 2023). Hence, a network of mutual regulation emerges from the interplay between histone PTMs and mitochondria, collectively influencing cellular destiny.

Nuclear DNA employs histones for packaging, where their PTMs are crucial for gene expression regulation. In contrast, mtDNA, devoid of histones, is organized into nucleoids primarily by TFAM (Liu et al. 2024). Interestingly, PTMs of TFAM, such as acetylation (Deng et al. 2024) and phosphorylation (Zhao et al. 2021), potentially modulating mtDNA-encoded gene expression via impacting mtDNA binding, in a manner similar to histone modifications. Although not definitively proven, TFAM PTMs may affect mtDNA transcription and replication as well. In addition to TFAM, two other key factors involved in mtDNA transcription are mitochondrial RNA polymerase (POLRMT) and mitochondrial transcription factor B2 (TFB2M). It is noteworthy that the TFAM-to-TFB2M/POLRMT ratio has been demonstrated to determine the initiation of mtDNA transcription (Lodeiro et al. 2012; Zollo et al. 2012). Furthermore, it could be speculated that PTMs of other mtDNA-associated proteins, such as Twinkle and mtSSB, may also modulate mtDNA transcription or replication as TFAM (Cao et al. 2021a). Of note, there may be interactions among different epigenetic regulatory mechanisms, which collectively participate in regulating the expression of mitochondrial genes and mitochondrial homeostasis. For example, TFAM's PTMs may work in coordination with mtDNA methylation to collectively regulated mtDNA transcription. Although the exact mechanism remains unclear, it is speculated that the PTMs of TFAM may hinder the binding of DNA methyltransferases, and in turn, mtDNA methylation may affect TFAM's binding (Cao et al. 2021a).

10.3 Exercise-Associated Epigenetic Modulation on Mitochondria

Exercise can exert beneficial effects on various systems throughout the body, not only enhancing physical endurance but also aiding in the prevention of various diseases, such as cardiovascular diseases (Paluch et al. 2024), neurodegenerative diseases (Feng et al. 2020), and metabolic-related disorders (Ostman et al. 2017). More importantly, these beneficial effects of exercise may be achieved by regulating mitochondrial homeostasis, a crucial aspect of which involves epigenetic mechanisms such as DNA methylation (Bajpeyi et al. 2017), ncRNAs (Trewin et al. 2022), and histone modifications (Joseph et al. 2017) (as illustrated in Fig. 10.1).

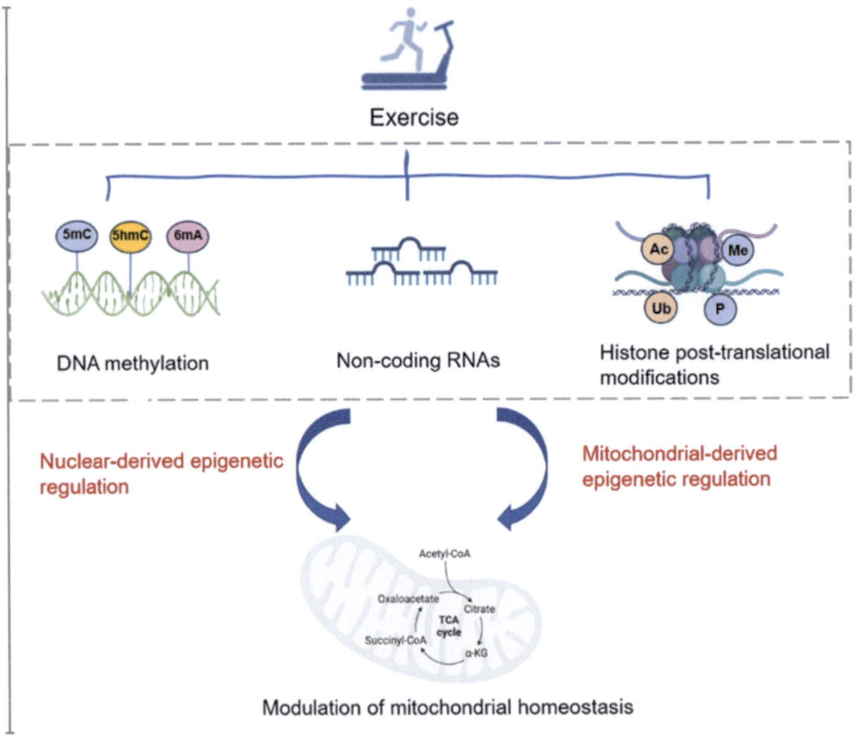

Fig. 10.1 Epigenetic regulation on mitochondria. Both nuclear and mitochondrial epigenetic mechanisms including DNA methylation, non-coding RNAs, and post-translational modifications of histones jointly contribute to the modulation of mitochondrial homeostasis exerted by exercise

10.3.1 Exercise-Associated DNA Methylation

Exercise may regulate mitochondrial homeostasis by modulating the DNA methylation of mitochondria-related genes, contributing to its health benefits. Key findings have revealed that *PPARGC1A* gene may undergo demethylation in skeletal muscle after exercise, which upregulates PGC-1α level and boosts mitochondrial biogenesis (Bajpeyi et al. 2017). This process may involve downregulation of DNMT3a and DNMT3b, leading to genome-wide demethylation (Hunter et al. 2019). More importantly, sex differences in DNA methylation responses are observed, with females showing more pronounced changes after exercise (Landen et al. 2019). Inflammation is tightly linked to disruptions in mitochondrial homeostasis. Robson-Ansley et al. (2014) explored the relationship between IL-6 levels and DNA methylation status in peripheral blood mononuclear cells of runners after intense exercise. Although they found no immediate global DNA methylation changes after exercise in blood samples, they observed that plasma IL-6 levels were closely linked to DNA methylation of 11 genes. This study suggests that DNA methylation modifications of inflammation-related genes induced by exercise may also be involved in the regulation of mitochondrial homeostasis and overall body health. Furthermore, additional genes involved in regulating mitochondrial homeostasis may be subject to DNA methylation modulation triggered by exercise. AMP-activated protein kinase (AMPK) is central to mitochondrial-regulated energy metabolism and becomes activated following exercise due to an elevated AMP/ATP ratio. King-Himmelreich et al. focused on AMPK promoter methylation in both human blood and mouse skeletal muscle by comparing pre- and post-exercise samples. Their findings revealed no alterations in global DNA methylation but did show specific methylation changes in CpG site within the *AMPKα2* gene. More importantly, these methylation changes were associated with reduced AMPKα2 expression, suggesting a regulatory role of DNA methylation in AMPK activity. Overall, there is still limited research on how exercise regulates mitochondrial homeostasis through DNA methylation, and more exploration is needed in this area.

10.3.2 Exercise-Associated Non-coding RNAs

Exercise, beyond influencing DNA methylation, may also regulate mitochondrial homeostasis by affecting ncRNA expression. A study (Trewin et al. 2022) demonstrated a significant increase in lncRNA TUG1 level in human skeletal muscle after exercise, which positively correlated with PGC-1α expression. Notably, *Tug1* knockdown in mouse myotubes resulted in reduced PGC-1α expression, disrupted mitochondrial function and morphology, and hindered muscle differentiation. RNA sequencing analysis further showed that lncRNA TUG1 primarily regulated mitochondrial Ca^{2+} handling and myogenesis-related genes, underscoring its crucial role in maintaining mitochondrial homeostasis in response to exercise. Another study by (Wang et al. 2024) reported a mitochondrial-related ncRNA networks in diabetes cardiomyopathy with or without exercise. By using RNA sequencing, they discovered 144 lncRNAs and 5 miRNAs that potentially interacted with these lncRNAs in the heart of excised rats. Further bioinformatics analysis revealed *Pdk4*, a mitochondrial-localized gene, as the target of these lncRNAs-miRNAs axis. This study indicates that exercise may modulate the expression of mitochondrial-associated genes via the regulation of specific lncRNAs and miRNAs networks. Mitochondria occupy a pivotal position in the modulation of glucose and lipid metabolic processes. A study conducted by Massart et al. (2021) revealed the miRNA expression profiles in human skeletal muscle before or after a 14-day aerobic exercise. Notably, they pinpointed miR-19b-3p as a potentially vital player in the regulation of glucose uptake and insulin sensitivity induced by aerobic exercise. In general, the regulatory mechanisms of exercise-related ncRNAs and their roles in modulating mitochondrial homeostasis still require further elucidation.

10.3.3 Exercise-Associated Post-translational Modifications of Histones

Exercise also has the potential to modulate the expression of mitochondria-related genes by means of PTMs of histones, such as acetylation and methylation. Joseph et al. (2017) reported that the activation of CaMKII triggered by exercise augmented the acetylation status of histones within the nuclear respiratory factor 1 (NRF-1) binding site near the *Mef2a* gene, exhibiting a positive correlation with the elevated binding affinity of NRF-1 to the *Mef2a* genomic sequence and promoting both *NRF-1* and *MEF2A* expressions. Given the role of NRF-1 in regulating mitochondrial biogenesis, these findings suggest that exercise may participate in the regulation of mitochondrial homeostasis through histone acetylation modifications. Another study revealed that exercise can upregulate the levels of histone H3K4me3 in the promoter region of PGC-1α by several fold (Lochmann et al. 2015), suggesting a potential regulatory role of histone methylation in the expression of PGC-1α. Taken together, these studies highlighted above underscore the significance of histone PTMs in the regulation of mitochondrial-related gene expression and maintaining mitochondrial homeostasis in response to exercise. Yet, the intricate interplay among diverse histone modifications and their collective contribution to the exercise-induced effects still await deeper exploration and clarification.

10.4 Benefits of Exercise via Epigenetic Modulation of Mitochondria Under Physiological and Pathological Conditions

10.4.1 Exercise and Muscle Regeneration

Exercise triggers a remodeling program in skeletal muscles that progressively improves muscle mass and strength, and enhances performance in both young and elderly individuals. Physical exercise is the most effective intervention for sarcopenia in elderly individuals. Many studies have shown that exercise shift the muscle mitochondria morphology profile toward that of a more oxidative status, activates a large number of transcriptional regulators in skeletal muscles that contribute to metabolic re-programming (Hood et al. 2019).

Upon stimulation or injury, quiescent muscle satellite cells (MSCs) undergo activation, a process essential for supporting muscle growth and regeneration. According to Chen et al. (2020), exercise acts as a mechanical stimulus that promotes MSC activation and proliferation, thereby preventing their exhaustion via the Igfbp7-Akt-mTOR signaling pathway. These findings highlight the interconnected of mechanical signaling, mitochondrial metabolism, epigenetic modification, and stem cell fate decisions.

10.4.2 Exercise and Skeletal Muscle Adaption

Epigenetic modifications of histones and DNA play a critical role in regulating chromatin states and gene expression in response to exercise. Endurance and resistance exercise induce histone acetylation and phosphorylation, such as H3K36 acetylation and H3S10 phosphorylation, which promote chromatin accessibility and transcription. Exercise also dynamically alters DNA methylation at CpG sites, with transient changes observed after acute sessions and some sustained modifications supporting muscle "memory" for future adaptations. These epigenetic changes are influenced by exercise-induced metabolic fluxes, which modulate acetyl and methyl group availability, redox states, and enzyme activities. Emerging evidence, such as lactate-induced histone lactylation, highlights additional layers of regulation, although their roles in exercise adaptation requires further study (Smith et al. 2023). High-intensity exercise remarkably led to DNA hypomethylation in the promoter regions of PPAR- PGC-1α, TFAM, myocyte enhancer factor 2A (MEF2A), and pyruvate dehydrogenase kinase 4 (PDK4) immediately after exercise, meanwhile PPAR-δ methylation was decreased 3 hours after exercise. These effects suggest that DNA methylation is essential to the exercise-induced effect on many metabolic gene expressions (Barrès et al. 2012). And the epigenetic status guides the susceptibility of the genome for exercise (Finke et al. 2022). These findings highlight the interplay between exercise, mitochondrial function, and epigenetic modifications, contributing to enhanced skeletal muscles transcriptional and metabolic adaptability.

10.4.3 Exercise and Cardiovascular Disease

It is well-documented that exercise is recognized as the essential preventative strategy against heart disease. Several studies have demonstrated the association between exercise and epigenetic modulation of mitochondria in cardiac tissue. Exercise induces changes in miRNA expression. Wardle et al. examined the differences in circulating miRNAs between endurance and strength athletes, identifying the upregulation of 14 miRNAs, including miR-222, miR-21, miR-146a, and miR-221, in endurance athletes compared to strength athletes (Wardle et al. 2015). Liu et al. demonstrated that miRNA-222 upregulation in mice following exercise was critical for cardiomyocyte growth, with its inhibition completely blocking this process (Liu et al. 2015). Additionally, a study found that exercise enhanced the interaction between O-linked β-N-acetylglucosamine transferase and the mSin3A/HDAC1/HDAC2 complex, reversing diabetic heart damage (Cox and Marsh 2013). Another study showed that an exercise-induced N-terminal fragment of HDAC4 (HDAC4-NT) played a protective role in preventing heart failure (Lehmann et al. 2018). In summary, exercise regulates mitochondrial bioenergetics through epigenetic mechanisms, such as miRNA expression and histone modifications, supporting the energy demands of cardiac tissue and enhancing its resilience against disease.

10.4.4 Exercise and Type 2 Diabetes

Type 2 diabetes (T2D) is increasingly prevalent, with insulin resistance as a core feature, often linked to mitochondrial dysfunction. Exercise improves insulin sensitivity and glucose uptake in skeletal muscle through pathways partially independent of insulin signaling, involving regulators like AMPK, calmodulin-dependent protein kinase (CaMK), and nitric oxide. Aerobic exercise induces adaptations in gene expression, possibly mediated by epigenetic changes such as DNA methylation. In T2D, epigenetic modifications like increased DNA methylation of PPARGC1A and histone changes impair mitochondrial function and gene expression. Exercise reduces DNA methylation at genes like PGC-1α and TFAM, enhancing their expression and mitochondrial biogenesis. HDAC inhibition and miRNA dynamics also influence PGC1 pathways, suggesting potential therapeutic targets (Santos et al. 2014).

Exercise induces epigenetic modifications in skeletal muscle, influencing chromatin structure, gene expression, and mitochondrial biogenesis. Studies show that acute exercise increases histone 3 acetylation at lysine 36, reduces HDAC5 abundance via ubiquitination, and enhances AMPK- and CaMK-mediated HDAC5 phosphorylation. These processes release MEF2, enabling transcription of PGC1 and its downstream genes. Regular exercise decreases DNA methylation at promoters of PGC1, MEF2, and NRF1, boosting mitochondrial density and function, while down-regulating miRNAs like miR-29a that suppress PGC1. These changes improve insulin sensitivity and lipid metabolism in skeletal muscle, with potential intergenerational benefits. Enhance GLUT4 expression in skeletal muscle to improve glucose uptake and insulin sensitivity is also a way of exercise-induced epigenetic modifications particularly in T2D. Regular exercise increases GLUT4 content by reducing DNA methylation at its promoter and through histone acetylation mediated by AMPK. These pathways inhibit HDAC5, enabling MEF2 to bind to the GLUT4 promoter and open chromatin structure for transcription. CaMK also facilitates GLUT4 translocation and its expression, with exercise-induced effects diminished by CaMK inhibition (dos Santos et al. 2015).

10.4.5 Exercise and Brain Plasticity and Cognition

Extensive research has suggested that epigenetic modification including DNA methylation and hydroxymethylation, histone posttranslational modifications, and non-coding RNA regulation in the course and development of neurodegeneration (Nikolac Perkovic et al. 2021). Recent studies also indicated the positive actions of exercise on learning and cognitive performance (Ni et al. 2024). Among all populations, older adults showed the greatest exercise benefits on cognition and memory (Zhang et al. 2023). In older adults, the aerobic exercise training increased the size of the anterior hippocampus which counteract the aging-related mental decline (Erickson et al. 2011; Wan et al. 2022). Some studies link the exercise benefits on cognitive to epigenetic mechanisms that regulate gene expression associated with synaptic plasticity and long-term memory (LTM). Exercise increases global HAT activity in the cortex and hippocampus, enhanced expression of CREB-binding protein (CBP), which possesses HAT activity, facilitates chromatin remodeling necessary for activity-dependent gene expression, synaptic plasticity, and LTM. CBP-mediated histone acetylation at cAMP response element-binding protein (CREB)-target promoters activates transcription critical for cognition (Zhong et al. 2016). Mutations in CBP result in deficits in synaptic plasticity and memory, highlighting CBP's essential role in neurocognitive processes (Barrett et al. 2011). Exercise also increased histone acetylation at specific sites (Zhong et al. 2016). Acetylation of H4K12 is particularly relevant, as its deregulation is associated with age-related memory impairments. Restoring H4K12 acetylation, either through exercise or HDAC inhibitors, alleviates cognitive deficits. Treadmill exercise enhances memory and hippocampal H4K12ac levels in young and aged rodents, also promoting histone H4 acetylation in the prefrontal cortex (Peleg et al. 2010).

Brain-derived neurotrophic factor (BDNF) plays a key role in hippocampus-dependent spatial task. Exercise enhanced BDNF expression through epigenetic modifications, including histone acetylation and methylation. BDNF promotes synaptic plasticity and long-term potentiation by increased H4K8 acetylation at bdnf H4K8 acetylation at bdnf I and IV promoters, activating CREB and reducing HDAC expression (Intlekofer et al. 2013). Exercises activates signaling pathways such as CaMKII, ERK1/2, and PI3K/AKT, leading to CREB phosphorylation and transcription of genes related to cognition. Blocking these pathways impairs exercise-induced cognitive enhancement, emphasizing their importance in BDNF-mediated synaptic plasticity (Reddy et al. 2023).

Aging impairs brain function by disrupted DNA methylation patterns, histone acetylation and methylation, and miRNAs level, leading to cognitive decline. Exercise restores DNA methylation and histone modifications (Li et al. 2022). Exercise enhances histone acetylation, inhibits HDAC2 mRNA and protein levels, and increases Sirtuins activity, improving synaptic plasticity and mitochondrial function. Sirtuins is a family of NAD^+ dependent deacetylase, participate in regulating mitochondrial homeostasis (Huang et al. 2020). Exercise improves mitochondrial homeostasis via NAD^+ restoration, modulates miRNAs. In addition to the benefits of physical activity, the byproduct of exercise lactate is also considered as a modulator of epigenetic landscape. Exercise-induced lactate promotes neuronal health by activating TET demethylases, increasing SIRT1, and stimulating BDNF expression (Gomez-Pinilla and Thapak 2024).

Overall, exercise-induced epigenetic modifications, particularly histone acetylation and BDNF-mediated signaling, play critical roles in enhancing synaptic plasticity, memory, and cognitive resilience. These findings underscore the therapeutic potential of physical activity in mitigating cognitive decline and promoting brain health.

10.4.6 Benefits of Exercise Transmitted Across Generation

The transgenerational transmission of epigenetic changes is well-documented in invertebrates and is increasingly being demonstrated in vertebrates. These changes can be triggered by various environmental factors, including stress, infection, drugs, diet, and exercise. Environmental stimuli, such as viral mimics and maternal separation, have been shown to induce epigenetic modifications like DNA methylation and chromatin remodeling in offspring, with some of these changes persisting across generations. For example, exercise in obese male mice not only restored insulin sensitivity but also improved the metabolic health of their offspring, while maternal exercise offset the negative epigenetic effects of a high-fat diet on offspring muscle tissue and enhanced their metabolic and cognitive functions (Laker et al. 2014; McPherson et al. 2015). A distinction is made between transgenerational and multigenerational effects: transgenerational effects involve epigenetic signatures observed in unexposed generations, whereas multigenerational effects result from direct exposure to environmental factors in F1 and F2 generations. Both maternal and paternal exercise have been shown to confer cognitive and health benefits to offspring. Clinical studies report that maternal exercise during pregnancy enhances offspring's cognitive function and academic performance (Esteban-Cornejo et al. 2016). Similarly, animal studies reveal that parental exercise improves offspring's memory, learning abilities, and resilience to neurodegenerative diseases by promoting the expression of BDNF, supporting neurogenesis, and boosting antioxidant capacity (Yin et al. 2013).

10.5 Conclusions and Perspectives

In summary, mitochondria serve as the energy hubs of cells and are central to regulating numerous metabolic pathways, playing a vital role in the beneficial effects of exercise. The maintenance of mitochondrial homeostasis is coordinated at both nuclear and mitochondrial epigenetic levels. Exercise influences this homeostasis through various epigenetic mechanisms, ultimately enhancing overall health under both physiological and pathological conditions.

Despite progress in understanding these processes, many questions remain regarding the epigenetic mechanisms by which exercise regulates mitochondrial function. For instance, the interplay between different epigenetic modifications, such as DNA methylation and histone PTMs, needs further exploration. Additionally, the interactions between nuclear-derived and mitochondrial-derived epigenetic mechanisms in regulating exercise-related mitochondrial homeostasis and physical health are active areas of research. Another important avenue for investigation is whether exercise-mediated epigenetic modifications exhibit tissue specificity. In this chapter, we primarily focus on the beneficial effects of moderate exercise on mitochondrial homeostasis and overall health. However, the potential negative impacts of excessive exercise on mitochondrial function and health through epigenetic mechanisms represent a critical research direction that requires further examination.

Compliance with Ethical Standards

Competing Interests All authors declare no conflict of interest.

Ethical Approval This chapter is a review of previously published accounts; as such, no animal or human studies were performed.

Funding This work was supported by the National Natural Science Foundations of China (No. 82271727 to K.C., No. 82300902 to J.X.), the Youth Talent Support Program of Shaanxi Association for Science and Technology (No. 20230214 to K.C.), the Natural Science Foundation of Sichuan Province (No. 2025ZNSFSC1704 to K.C.), and the Fundamental Research Funds for the Central Universities (No. xzy012023012 to J.X.).

References

Adebayo M, Singh S, Singh AP, Dasgupta S (2021) Mitochondrial fusion and fission: the fine-tune balance for cellular homeostasis. FASEB J 35(6):e21620. https://doi.org/10.1096/fj.202100067R

Baccarelli AA, Byun HM (2015) Platelet mitochondrial DNA methylation: a potential new marker of cardiovascular disease. Clin Epigenetics 7:44. https://doi.org/10.1186/s13148-015-0078-0. Available at: http://www.ncbi.nlm.nih.gov/pubmed/25901189

Bajpeyi S, Covington JD, Taylor EM, Stewart LK, Galgani JE, Henagan TM (2017) Skeletal muscle PGC1alpha -1 nucleosome position and -260 nt DNA methylation determine exercise response and prevent ectopic lipid accumulation in men. Endocrinology 158(7):2190–2199. https://doi.org/10.1210/en.2017-00051. Available at: https://www.ncbi.nlm.nih.gov/pubmed/28398573

Barrès R, Yan J, Egan B, Treebak JT, Rasmussen M, Fritz T, Caidahl K, Krook A, O'Gorman DJ, Zierath JR (2012) Acute exercise remodels promoter methylation in human skeletal muscle. Cell Metab 15(3):405–411. https://doi.org/10.1016/j.cmet.2012.01.001

Barrett RM, Malvaez M, Kramar E, Matheos DP, Arrizon A, Cabrera SM, Lynch G, Greene RW, Wood MA (2011) Hippocampal focal knockout of CBP affects specific histone modifications, long-term potentiation, and long-term memory. Neuropsychopharmacology 36(8):1545–1556. https://doi.org/10.1038/npp.2011.61

Bellizzi D, D'Aquila P, Scafone T, Giordano M, Riso V, Riccio A, Passarino G (2013) The control region of mitochondrial DNA shows an unusual CpG and non-CpG methylation pattern. DNA Res 20(6):537–547. https://doi.org/10.1093/dnares/dst029. Available at: https://www.ncbi.nlm.nih.gov/pubmed/23804556

Bhatti JS, Khullar N, Vijayvergiya R, Navik U, Bhatti GK, Reddy PH (2023) Mitochondrial miRNA as epigenomic signatures: visualizing aging-associated heart diseases through a new lens. Ageing Res Rev 86:101882. https://doi.org/10.1016/j.arr.2023.101882. Available at: https://www.ncbi.nlm.nih.gov/pubmed/36780957

Boovarahan SR, Kurian GA (2022) Investigating the role of DNMT1 gene expression on myocardial ischemia reperfusion injury in rat and associated changes in mitochondria. Biochim Biophys Acta Bioenerg 1863(6):148566. https://doi.org/10.1016/j.bbabio.2022.148566. Available at: https://www.ncbi.nlm.nih.gov/pubmed/35489443

Cai J, Shen W, Zhang G, Li X, Shen H, Li W, Tan C, Zhang T, Shi M, Yang Z, Li Y, Liu H, Zhao X (2023) Xiao Chai Hu Tang alleviates the pancreatic tumorigenesis via improving the mtDNA N6-Methyladenine modification mediated mitochondrial dysfunction in Syrian hamster model. Phytomedicine 116:154840. https://doi.org/10.1016/j.phymed.2023.154840. Available at: https://www.ncbi.nlm.nih.gov/pubmed/37172477

Cao K, Feng Z, Gao F, Zang W, Liu J (2021a) Mitoepigenetics: an intriguing regulatory layer in aging and metabolic-related diseases. Free Radic Biol Med 177:337–346. https://doi.org/10.1016/j.freeradbiomed.2021.10.031. Available at: https://www.ncbi.nlm.nih.gov/pubmed/34715295

Cao K, Lv W, Wang X, Dong S, Liu X, Yang T, Xu J, Zeng M, Zou X, Zhao D, Ma Q, Lin M, Long J, Zang W, Gao F, Feng Z, Liu J (2021b) Hypermethylation of hepatic mitochondrial ND6 provokes systemic insulin resistance. Adv Sci (Weinh) 8(11):2004507. https://doi.org/10.1002/advs.202004507. Available at: https://www.ncbi.nlm.nih.gov/pubmed/34141522

Chaudhuri AD, Choi DC, Kabaria S, Tran A, Junn E (2016) MicroRNA-7 regulates the function of mitochondrial permeability transition pore by targeting VDAC1 expression. J Biol Chem 291(12):6483–6493. https://doi.org/10.1074/jbc.M115.691352. Available at: https://www.ncbi.nlm.nih.gov/pubmed/26801612

Chen Z, Li L, Wu W, Liu Z, Huang Y, Yang L, Luo Q, Chen J, Hou Y, Song G (2020) Exercise protects proliferative muscle satellite cells against exhaustion via the Igfbp7-Akt-mTOR axis. Theranostics 10(14):6448–6466. https://doi.org/10.7150/thno.43577

Clarke-Harris R, Wilkin TJ, Hosking J, Pinkney J, Jeffery AN, Metcalf BS, Godfrey KM, Voss LD, Lillycrop KA, Burdge GC (2014) PGC1alpha promoter methylation in blood at 5-7 years predicts adiposity from 9 to 14 years (EarlyBird 50). Diabetes 63(7):2528–2537. https://doi.org/10.2337/db13-0671. Available at: https://www.ncbi.nlm.nih.gov/pubmed/24622795

Cox EJ, Marsh SA (2013) Exercise and diabetes have opposite effects on the assembly and O-GlcNAc modification of the mSin3A/HDAC1/2 complex in the heart. Cardiovasc Diabetol 12:101. https://doi.org/10.1186/1475-2840-12-101

Deng Z, He M, Hu H, Zhang W, Zhang Y, Ge Y, Ma T, Wu J, Li L, Sun M, An S, Li J, Huang Q, Gong S, Zhang J, Chen Z, Zeng Z (2024) Melatonin attenuates sepsis-induced acute kidney injury by promoting mitophagy through SIRT3-mediated TFAM deacetylation. Autophagy 20(1):151–165. https://doi.org/10.1080/15548627.2023.2252265

Deogharia M, Venegas-Zamora L, Agrawal A, Shi M, Jain AK, McHugh KJ, Altamirano F, Marian AJ, Gurha P (2024) Histone demethylase KDM5 regulates cardiomyocyte maturation by promoting fatty acid oxidation, oxidative phosphorylation, and myofibrillar organization. Cardiovasc Res 120(6):630–643. https://doi.org/10.1093/cvr/cvae014

Dimauro I, Paronetto MP, Caporossi D (2020) Exercise, redox homeostasis and the epigenetic landscape. Redox Biol 35:101477. https://doi.org/10.1016/j.redox.2020.101477

Dong Y, Yoshitomi T, Hu JF, Cui J (2017) Long noncoding RNAs coordinate functions between mitochondria and the nucleus. Epigenetics Chromatin 10(1):41. https://doi.org/10.1186/s13072-017-0149-x. Available at: https://www.ncbi.nlm.nih.gov/pubmed/28835257

dos Santos JM, Moreli ML, Tewari S, Benite-Ribeiro SA (2015) The effect of exercise on skeletal muscle glucose uptake in type 2 diabetes: an epigenetic perspective. Metabolism 64(12):1619–1628. https://doi.org/10.1016/j.metabol.2015.09.013. Available at: https://www.sciencedirect.com/science/article/pii/S002604951500267X

Dou X, Boyd-Kirkup JD, McDermott J, Zhang X, Li F, Rong B, Zhang R, Miao B, Chen P, Cheng H, Xue J, Bennett D, Wong J, Lan F, Han JJ (2019) The strand-biased mitochondrial DNA methylome and its regulation by DNMT3A. Genome Res 29(10):1622–1634. https://doi.org/10.1101/gr.234021.117. Available at: https://www.ncbi.nlm.nih.gov/pubmed/31537639

Du J, Zhou Y, Su X, Yu JJ, Khan S, Jiang H, Kim J, Woo J, Kim JH, Choi BH, He B, Chen W, Zhang S, Cerione RA, Auwerx J, Hao Q, Lin H (2011) Sirt5 is a NAD-dependent protein lysine demalonylase and desuccinylase. Science 334(6057):806–809. https://doi.org/10.1126/science.1207861. Available at: https://www.ncbi.nlm.nih.gov/pubmed/22076378

Dzitoyeva S, Chen H, Manev H (2012) Effect of aging on 5-hydroxymethylcytosine in brain mitochondria. Neurobiol Aging 33(12):2881–2891. https://doi.org/10.1016/j.neurobiolaging.2012.02.006. Available at: https://www.ncbi.nlm.nih.gov/pubmed/22445327

Eirin A, Thaler R, Glasstetter LM, Xing L, Zhu XY, Osborne AC, Mondesir R, Bhagwate AV, Lerman A, van Wijnen AJ, Lerman LO (2024) Obesity-driven mitochondrial dysfunction in human adipose tissue-derived mesenchymal stem/stromal cells involves epigenetic changes. Cell Death Dis 15(6):387. https://doi.org/10.1038/s41419-024-06774-8. Available at: https://www.ncbi.nlm.nih.gov/pubmed/38824145

Erickson KI, Voss MW, Prakash RS, Basak C, Szabo A, Chaddock L, Kim JS, Heo S, Alves H, White SM, Wojcicki TR, Mailey E, Vieira VJ, Martin SA, Pence BD, Woods JA, McAuley E, Kramer AF (2011) Exercise training increases size of hippocampus and improves memory. Proc Natl Acad Sci USA 108(7):3017–3022. https://doi.org/10.1073/pnas.1015950108

Esteban-Cornejo I, Martinez-Gomez D, Tejero-González CM, Izquierdo-Gomez R, Carbonell-Baeza A, Castro-Piñero J, Sallis JF, Veiga OL (2016) Maternal physical activity before and during the prenatal period and the offspring's academic performance in youth. The UP&DOWN study. J Matern Fetal Neonatal Med 29(9):1414–1420. https://doi.org/10.3109/14767058.2015.1049525

Feng YS, Yang SD, Tan ZX, Wang MM, Xing Y, Dong F, Zhang F (2020) The benefits and mechanisms of exercise training for Parkinson's disease. Life Sci 245:117345. https://doi.org/10.1016/j.lfs.2020.117345. Available at: https://www.ncbi.nlm.nih.gov/pubmed/31981631

Finke D, Schanze LM, Schreiter F, Kreußer MM, Katus HA, Backs J, Lehmann LH (2022) Histone deacetylase 4 deletion broadly affects cardiac epigenetic repression and regulates transcriptional susceptibility via H3K9 methylation. J Mol Cell Cardiol 162:119–129. https://doi.org/10.1016/j.yjmcc.2021.09.001. Accessed 2024/12/16

Ghosh S, Sengupta S, Scaria V (2016) Hydroxymethyl cytosine marks in the human mitochondrial genome are dynamic in nature. Mitochondrion 27:25–31. https://doi.org/10.1016/j.mito.2016.01.003. Available at: https://www.ncbi.nlm.nih.gov/pubmed/26826294

Gillberg L, Jacobsen SC, Ribel-Madsen R, Gjesing AP, Boesgaard TW, Ling C, Pedersen O, Hansen T, Vaag A (2013) Does DNA methylation of PPARGC1A influence insulin action in first degree relatives of patients with type 2 diabetes? PLoS One 8(3):e58384. https://doi.org/10.1371/journal.pone.0058384. Available at: https://www.ncbi.nlm.nih.gov/pubmed/23505498

Gomez-Pinilla F, Thapak P (2024) Exercise epigenetics is fueled by cell bioenergetics: supporting role on brain plasticity and cognition. Free Radic Biol Med 220:43–55. https://doi.org/10.1016/j.freeradbiomed.2024.04.237. Available at: https://www.sciencedirect.com/science/article/pii/S0891584924004246

Gowda P, Reddy PH, Kumar S (2022) Deregulated mitochondrial microRNAs in Alzheimer's disease: focus on synapse and mitochondria. Ageing Res Rev 73:101529. https://doi.org/10.1016/j.arr.2021.101529. Available at: https://www.ncbi.nlm.nih.gov/pubmed/34813976

Hahn A, Hung GCC, Ahier A, Dai CY, Kirmes I, Forde BM, Campbell D, Lee RSY, Sucic J, Onraet T, Zuryn S (2024) Misregulation of mitochondrial 6mA promotes the propagation of mutant mtDNA and causes aging in C. elegans. Cell Metab 36(12):2528–2541 e11. https://doi.org/10.1016/j.cmet.2024.07.020. Available at: https://www.ncbi.nlm.nih.gov/pubmed/39173633

Hao Z, Wu T, Cui X, Zhu P, Tan C, Dou X, Hsu KW, Lin YT, Peng PH, Zhang LS, Gao Y, Hu L, Sun HL, Zhu A, Liu J, Wu KJ, He C (2020a) N(6)-deoxyadenosine methylation in mammalian mitochondrial DNA. Mol Cell 78(3):382–395.e8. https://doi.org/10.1016/j.molcel.2020.02.018

Hao Z, Wu T, Cui X, Zhu P, Tan C, Dou X, Hsu KW, Lin YT, Peng PH, Zhang LS, Gao Y, Hu L, Sun HL, Zhu A, Liu J, Wu KJ, He C (2020b) N(6)-deoxyadenosine methylation in mammalian mitochondrial DNA. Mol Cell. https://doi.org/10.1016/j.molcel.2020.02.018. Available at: http://www.ncbi.nlm.nih.gov/pubmed/32183942

He Y, Ji Z, Gong Y, Fan L, Xu P, Chen X, Miao J, Zhang K, Zhang W, Ma P, Zhao H, Cheng C, Wang D, Wang J, Jing N, Liu K, Zhang P, Dong B, Zhuang G, Fu Y, Xue W, Gao WQ, Zhu HH (2023) Numb/Parkin-directed mitochondrial fitness governs cancer cell fate via metabolic regulation of histone lactylation. Cell Rep 42(2):112033. https://doi.org/10.1016/j.celrep.2023.112033. Available at: https://www.ncbi.nlm.nih.gov/pubmed/36724072

Hogg SJ, Beavis PA, Dawson MA, Johnstone RW (2020) Targeting the epigenetic regulation of antitumour immunity. Nat Rev Drug Discov 19(11):776–800. https://doi.org/10.1038/s41573-020-0077-5. Available at: https://www.ncbi.nlm.nih.gov/pubmed/32929243

Hong EE, Okitsu CY, Smith AD, Hsieh CL (2013) Regionally specific and genome-wide analyses conclusively demonstrate the absence of CpG methylation in human mitochondrial DNA. Mol Cell Biol 33(14):2683–2690. https://doi.org/10.1128/MCB.00220-13. Available at: https://www.ncbi.nlm.nih.gov/pubmed/23671186

Hood DA, Memme JM, Oliveira AN, Triolo M (2019) Maintenance of skeletal muscle mitochondria in health, exercise, and aging. Annu Rev Physiol 81:19–41. https://doi.org/10.1146/annurev-physiol-020518-114310

Huang B, Zhong D, Zhu J, An Y, Gao M, Zhu S, Dang W, Wang X, Yang B, Xie Z (2020) Inhibition of histone acetyltransferase GCN5 extends lifespan in both yeast and human cell lines. Aging Cell 19(4):e13129. https://doi.org/10.1111/acel.13129. Available at: https://onlinelibrary.wiley.com/doi/abs/10.1111/acel.13129

Hunter DJ, James L, Hussey B, Wadley AJ, Lindley MR, Mastana SS (2019) Impact of aerobic exercise and fatty acid supplementation on global and gene-specific DNA methylation. Epigenetics 14(3):294–309. https://doi.org/10.1080/15592294.2019.1582276. Available at: https://www.ncbi.nlm.nih.gov/pubmed/30764736

Intlekofer KA, Berchtold NC, Malvaez M, Carlos AJ, McQuown SC, Cunningham MJ, Wood MA, Cotman CW (2013) Exercise and sodium butyrate transform a subthreshold learning event into long-term memory via a brain-derived neurotrophic factor-dependent mechanism. Neuropsychopharmacology 38(10):2027–2034. https://doi.org/10.1038/npp.2013.104

Izzo LT, Trefely S, Demetriadou C, Drummond JM, Mizukami T, Kuprasertkul N, Farria AT, Nguyen PTT, Murali N, Reich L, Kantner DS, Shaffer J, Affronti H, Carrer A, Andrews A, Capell BC, Snyder NW, Wellen KE (2023) Acetylcarnitine shuttling links mitochondrial metabolism to histone acetylation and lipogenesis. Sci Adv 9(18):eadf0115. https://doi.org/10.1126/sciadv.adf0115

Ji F, Zhao C, Wang B, Tang Y, Miao Z, Wang Y (2018) The role of 5-hydroxymethylcytosine in mitochondria after ischemic stroke. J Neurosci Res 96(10):1717–1726. https://doi.org/10.1002/jnr.24274. Available at: https://www.ncbi.nlm.nih.gov/pubmed/30043506

Jiang Y, Xia W, Yang J, Zhu Y, Chang H, Liu J, Huo W, Xu B, Chen X, Li Y, Xu S (2015) BPA-induced DNA hypermethylation of the master mitochondrial gene PGC-1alpha contributes to cardiomyopathy in male rats. Toxicology 329:21–31. https://doi.org/10.1016/j.tox.2015.01.001. Available at: https://www.ncbi.nlm.nih.gov/pubmed/25572651

Jin L, Geng L, Ying L, Shu L, Ye K, Yang R, Liu Y, Wang Y, Cai Y, Jiang X, Wang Q, Yan X, Liao B, Liu J, Duan F, Sweeney G, Woo CWH, Wang Y, Xia Z, Lian Q, Xu A (2022) FGF21–Sirtuin 3 axis confers the protective effects of exercise against diabetic cardiomyopathy by governing mitochondrial integrity. Circulation 146(20):1537–1557. https://doi.org/10.1161/CIRCULATIONAHA.122.059631. Accessed 2024/12/25

Joseph JS, Ayeleso AO, Mukwevho E (2017) Exercise increases hyper-acetylation of histones on the Cis-element of NRF-1 binding to the Mef2a promoter: implications on type 2 diabetes. Biochem Biophys Res Commun 486(1):83–87. https://doi.org/10.1016/j.bbrc.2017.03.002

Kudriashova IB, Kirnos MD, Vaniushin BF (1976) DNA-methylase activities from animal mitochondria and nuclei: different specificity of DNA methylation. Biokhimiia 41(11):1968–1977. Available at: https://www.ncbi.nlm.nih.gov/pubmed/1022267

Laker RC, Lillard TS, Okutsu M, Zhang M, Hoehn KL, Connelly JJ, Yan Z (2014) Exercise prevents maternal high-fat diet–induced hypermethylation of the Pgc-1α gene and age-dependent metabolic dysfunction in the offspring. Diabetes 63(5):1605–1611. https://doi.org/10.2337/db13-1614. Accessed 12/26/2024

Landen S, Voisin S, Craig JM, McGee SL, Lamon S, Eynon N (2019) Genetic and epigenetic sex-specific adaptations to endurance exercise. Epigenetics 14(6):523–535. https://doi.org/10.1080/15592294.2019.1603961. Available at: https://www.ncbi.nlm.nih.gov/pubmed/30957644

Lehmann LH, Jebessa ZH, Kreusser MM, Horsch A, He T, Kronlage M, Dewenter M, Sramek V, Oehl U, Krebs-Haupenthal J, von der Lieth AH, Schmidt A, Sun Q, Ritterhoff J, Finke D, Völkers M, Jungmann A, Sauer SW, Thiel C, Nickel A, Kohlhaas M, Schäfer M, Sticht C, Maack C, Gretz N, Wagner M, El-Armouche A, Maier LS, Londoño JEC, Meder B, Freichel M, Gröne HJ, Most P, Müller OJ, Herzig S, Furlong EEM, Katus HA, Backs J (2018) A proteolytic fragment of histone deacetylase 4 protects the heart from failure by regulating the hexosamine biosynthetic pathway. Nat Med 24(1):62–72. https://doi.org/10.1038/nm.4452

Li C, Pang M, Li Y, Han L, Fan Y, Xin X, Zhang X, Zhang N, Qin Y (2023) Protective effect of vitamin C against tetrachlorobenzoquinone-induced 5-hydroxymethylation-dependent apoptosis in HepG2 cells mainly via the mitochondrial apoptosis pathway. Ecotoxicol Environ Saf 260:115097. https://doi.org/10.1016/j.ecoenv.2023.115097. Available at: https://www.ncbi.nlm.nih.gov/pubmed/37271103

Li J, Donath S, Li Y, Qin D, Prabhakar BS, Li P (2010) miR-30 regulates mitochondrial fission through targeting p53 and the dynamin-related protein-1 pathway. PLoS Genet 6(1):e1000795. https://doi.org/10.1371/journal.pgen.1000795. Available at: https://www.ncbi.nlm.nih.gov/pubmed/20062521

Li J, Wang Z, Li C, Song Y, Wang Y, Bo H, Zhang Y (2022) Impact of exercise and aging on mitochondrial homeostasis in skeletal muscle: roles of ROS and epigenetics. Cells 11(13). https://doi.org/10.3390/cells11132086

Lian WS, Wu RW, Lin YH, Chen YS, Jahr H, Wang FS (2024) Tricarboxylic acid cycle regulation of metabolic program, redox system, and epigenetic remodeling for bone health and disease. Antioxidants (Basel) 13(4). https://doi.org/10.3390/antiox13040470

Liu H, Zhen C, Xie J, Luo Z, Zeng L, Zhao G, Lu S, Zhuang H, Fan H, Li X, Liu Z, Lin S, Jiang H, Chen Y, Cheng J, Cao Z, Dai K, Shi J, Wang Z, Hu Y, Meng T, Zhou C, Han Z, Huang H, Zhou Q, He P, Feng D (2024) TFAM is an autophagy receptor that limits inflammation by binding to cytoplasmic mitochondrial DNA. Nat Cell Biol 26(6):878–891. https://doi.org/10.1038/s41556-024-01419-6

Liu R, Chen L, Wang Z, Zheng X, Wang Y, Li H, Noda M, Liu J, Long J (2020) Downregulation of the DNA 5-hydroxymethylcytosine is involved in mitochondrial dysfunction and neuronal impairment in high fat diet-induced diabetic mice. Free Radic Biol Med 148:42–51. https://doi.org/10.1016/j.freeradbiomed.2019.12.042. Available at: https://www.ncbi.nlm.nih.gov/pubmed/31899342

Liu X, Xiao J, Zhu H, Wei X, Platt C, Damilano F, Xiao C, Bezzerides V, Boström P, Che L, Zhang C, Spiegelman BM, Rosenzweig A (2015) miR-222 is necessary for exercise-induced cardiac growth and protects against pathological cardiac remodeling. Cell Metab 21(4):584–595. https://doi.org/10.1016/j.cmet.2015.02.014

Lochmann TL, Thomas RR, Bennett JP Jr, Taylor SM (2015) Epigenetic modifications of the PGC-1alpha promoter during exercise induced expression in mice. PLoS One 10(6):e0129647. https://doi.org/10.1371/journal.pone.0129647. Available at: https://www.ncbi.nlm.nih.gov/pubmed/26053857

Lodeiro MF, Uchida A, Bestwick M, Moustafa IM, Arnold JJ, Shadel GS, Cameron CE (2012) Transcription from the second heavy-strand promoter of human mtDNA is repressed by transcription factor A in vitro. Proc Natl Acad Sci USA 109(17):6513–6518. https://doi.org/10.1073/pnas.1118710109. Available at: https://www.ncbi.nlm.nih.gov/pubmed/22493245

Ma CJ, Li G, Shao WX, Min YH, Wang P, Ding JH, Xie NB, Wang M, Tang F, Feng YQ, Ci W, Wang Y, Yuan BF (2023a) Single-Nucleotide resolution mapping of N(6)-Methyladenine in genomic DNA. ACS Cent Sci 9(9):1799–1809. https://doi.org/10.1021/acscentsci.3c00481. Available at: https://www.ncbi.nlm.nih.gov/pubmed/37780356

Ma Y, Du J, Chen M, Gao N, Wang S, Mi Z, Wei X, Zhao J (2023b) Mitochondrial DNA methylation is a predictor of immunotherapy response and prognosis in breast cancer: scRNA-seq and bulk-seq data insights. Front Immunol 14:1219652. https://doi.org/10.3389/fimmu.2023.1219652. Available at: https://www.ncbi.nlm.nih.gov/pubmed/37457713

Malik N, Ferreira BI, Hollstein PE, Curtis SD, Trefts E, Weiser Novak S, Yu J, Gilson R, Hellberg K, Fang L, Sheridan A, Hah N, Shadel GS, Manor U, Shaw RJ (2023) Induction of lysosomal and mitochondrial biogenesis by AMPK phosphorylation of FNIP1. Science 380(6642):eabj5559. https://doi.org/10.1126/science.abj5559. Available at: https://www.ncbi.nlm.nih.gov/pubmed/37079666

Marinho D, Ferreira IL, Lorenzoni R, Cardoso SM, Santana I, Rego AC (2023) Reduction of class I histone deacetylases ameliorates ER-mitochondria cross-talk in Alzheimer's disease. Aging Cell 22(8):e13895. https://doi.org/10.1111/acel.13895

Massart J, Sjogren RJO, Egan B, Garde C, Lindgren M, Gu W, Ferreira DMS, Katayama M, Ruas JL, Barres R, O'Gorman DJ, Zierath JR, Krook A (2021) Endurance exercise training-responsive miR-19b-3p improves skeletal muscle glucose metabolism. Nat Commun 12(1):5948. https://doi.org/10.1038/s41467-021-26095-0. Available at: https://www.ncbi.nlm.nih.gov/pubmed/34642330

Mattei AL, Bailly N, Meissner A (2022) DNA methylation: a historical perspective. Trends Genet 38(7):676–707. https://doi.org/10.1016/j.tig.2022.03.010. Available at: https://www.ncbi.nlm.nih.gov/pubmed/35504755

McPherson NO, Owens JA, Fullston T, Lane M (2015) Preconception diet or exercise intervention in obese fathers normalizes sperm microRNA profile and metabolic syndrome in female offspring. Am J Physiol Endocrinol Metab 308(9):E805–E821. https://doi.org/10.1152/ajpendo.00013.2015. Available at: https://journals.physiology.org/doi/abs/10.1152/ajpendo.00013.2015

Mechta M, Ingerslev LR, Fabre O, Picard M, Barres R (2017) Evidence suggesting absence of mitochondrial DNA methylation. Front Genet 8:166. https://doi.org/10.3389/fgene.2017.00166. Available at: https://www.ncbi.nlm.nih.gov/pubmed/29163634

Monzel AS, Enriquez JA, Picard M (2023) Multifaceted mitochondria: moving mitochondrial science beyond function and dysfunction. Nat Metab 5(4):546–562. https://doi.org/10.1038/s42255-023-00783-1. Available at: https://www.ncbi.nlm.nih.gov/pubmed/37100996

Ni J, Zhang R, Cao X, Li X (2024) A meta-analytic study of complex exercise interventions for cognitive performance in older adults. Arch Gerontol Geriatr 130:105725. https://doi.org/10.1016/j.archger.2024.105725

Nikolac Perkovic M, Videtic Paska A, Konjevod M, Kouter K, Svob Strac D, Nedic Erjavec G, Pivac N (2021) Epigenetics of Alzheimer's disease. Biomolecules 11(2). https://doi.org/10.3390/biom11020195

Noh JH, Kim KM, Abdelmohsen K, Yoon JH, Panda AC, Munk R, Kim J, Curtis J, Moad CA, Wohler CM, Indig FE, de Paula W, Dudekula DB, De S, Piao Y, Yang X, Martindale JL, de Cabo R, Gorospe M (2016) HuR and GRSF1 modulate the nuclear export and mitochondrial localization of the lncRNA RMRP. Genes Dev 30(10):1224–1239. https://doi.org/10.1101/gad.276022.115. Available at: https://www.ncbi.nlm.nih.gov/pubmed/27198227

Ostman C, Smart NA, Morcos D, Duller A, Ridley W, Jewiss D (2017) The effect of exercise training on clinical outcomes in patients with the metabolic syndrome: a systematic review and meta-analysis. Cardiovasc Diabetol 16(1):110. https://doi.org/10.1186/s12933-017-0590-y. Available at: https://www.ncbi.nlm.nih.gov/pubmed/28854979

Owa C, Poulin M, Yan L, Shioda T (2018) Technical adequacy of bisulfite sequencing and pyrosequencing for detection of mitochondrial DNA methylation: sources and avoidance of false-positive detection. PLoS One 13(2):e0192722. https://doi.org/10.1371/journal.pone.0192722. Available at: https://www.ncbi.nlm.nih.gov/pubmed/29420656

Paluch AE, Boyer WR, Franklin BA, Laddu D, Lobelo F, Lee DC, McDermott MM, Swift DL, Webel AR, Lane A (2024) Resistance exercise training in individuals with and without cardiovascular disease: 2023 update: a scientific statement from the American heart association. Circulation 149(3):e217–e231. https://doi.org/10.1161/cir.0000000000001189

Park YJ, Lee S, Lim S, Nahmgoong H, Ji Y, Huh JY, Alfadda AA, Kim S, Kim JB (2021) DNMT1 maintains metabolic fitness of adipocytes through acting as an epigenetic safeguard of mitochondrial dynamics. Proc Natl Acad Sci USA 118(11). https://doi.org/10.1073/pnas.2021073118. Available at: https://www.ncbi.nlm.nih.gov/pubmed/33836591

Patange V, Ahirwar K, Tripathi T, Tripathi P, Shukla R (2025) Scientific investigation of non-coding RNAs in mitochondrial epigenetic and aging disorders: current nanoengineered approaches for their therapeutic improvement. Mitochondrion 80:101979. https://doi.org/10.1016/j.mito.2024.101979. Available at: https://www.ncbi.nlm.nih.gov/pubmed/39505245

Patil V, Cuenin C, Chung F, Aguilera JRR, Fernandez-Jimenez N, Romero-Garmendia I, Bilbao JR, Cahais V, Rothwell J, Herceg Z (2019) Human mitochondrial DNA is extensively methylated in a non-CpG context. Nucleic Acids Res 47(19):10072–10085. https://doi.org/10.1093/nar/gkz762. Available at: https://www.ncbi.nlm.nih.gov/pubmed/31665742

Patti ME, Butte AJ, Crunkhorn S, Cusi K, Berria R, Kashyap S, Miyazaki Y, Kohane I, Costello M, Saccone R, Landaker EJ, Goldfine AB, Mun E, DeFronzo R, Finlayson J, Kahn CR, Mandarino LJ (2003) Coordinated reduction of genes of oxidative metabolism in humans with insulin resistance and diabetes: potential role of PGC1 and NRF1. Proc Natl Acad Sci USA 100(14):8466–8471. https://doi.org/10.1073/pnas.1032913100. Available at: https://www.ncbi.nlm.nih.gov/pubmed/12832613

Peleg S, Sananbenesi F, Zovoilis A, Burkhardt S, Bahari-Javan S, Agis-Balboa RC, Cota P, Wittnam JL, Gogol-Doering A, Opitz L, Salinas-Riester G, Dettenhofer M, Kang H, Farinelli L, Chen W, Fischer A (2010) Altered histone acetylation is associated with age-dependent memory impairment in mice. Science 328(5979):753–756. https://doi.org/10.1126/science.1186088

Qian X, Li X, Shi Z, Bai X, Xia Y, Zheng Y, Xu D, Chen F, You Y, Fang J, Hu Z, Zhou Q, Lu Z (2019) KDM3A senses oxygen availability to regulate PGC-1alpha-mediated mitochondrial biogenesis. Mol Cell 76(6):885–895 e7. https://doi.org/10.1016/j.molcel.2019.09.019. Available at: https://www.ncbi.nlm.nih.gov/pubmed/31629659

Rackham O, Shearwood AM, Mercer TR, Davies SM, Mattick JS, Filipovska A (2011) Long noncoding RNAs are generated from the mitochondrial genome and regulated by nuclear-encoded proteins. RNA 17(12):2085–2093. https://doi.org/10.1261/rna.029405.111. Available at: https://www.ncbi.nlm.nih.gov/pubmed/22028365

Reddy I, Yadav Y, Dey CS (2023) Cellular and molecular regulation of exercise-A neuronal perspective. Cell Mol Neurobiol 43(4):1551–1571. https://doi.org/10.1007/s10571-022-01272-x

Ro S, Ma HY, Park C, Ortogero N, Song R, Hennig GW, Zheng H, Lin YM, Moro L, Hsieh JT, Yan W (2013) The mitochondrial genome encodes abundant small noncoding RNAs. Cell Res 23(6):759–774. https://doi.org/10.1038/cr.2013.37. Available at: https://www.ncbi.nlm.nih.gov/pubmed/23478297

Robson-Ansley PJ, Saini A, Toms C, Ansley L, Walshe IH, Nimmo MA, Curtin JA (2014) Dynamic changes in dna methylation status in peripheral blood Mononuclear cells following an acute bout of exercise: potential impact of exercise-induced elevations in interleukin-6 concentration. J Biol Regul Homeost Agents 28(3):407–417. Available at: https://www.ncbi.nlm.nih.gov/pubmed/25316129

Saini SK, Mangalhara KC, Prakasam G, Bamezai RNK (2017) DNA Methyltransferase1 (DNMT1) Isoform3 methylates mitochondrial genome and modulates its biology. Sci Rep 7(1):1525. https://doi.org/10.1038/s41598-017-01743-y. Available at: http://www.ncbi.nlm.nih.gov/pubmed/28484249

Salminen A, Kauppinen A, Hiltunen M, Kaarniranta K (2014) Krebs cycle intermediates regulate DNA and histone methylation: epigenetic impact on the aging process. Ageing Res Rev 16:45–65. https://doi.org/10.1016/j.arr.2014.05.004. Available at: https://www.ncbi.nlm.nih.gov/pubmed/24910305

Sanchez-Ramos C, Prieto I, Tierrez A, Laso J, Valdecantos MP, Bartrons R, Rosello-Catafau J, Monsalve M (2017) PGC-1alpha downregulation in steatotic liver enhances ischemia-reperfusion injury and impairs ischemic preconditioning. Antioxid Redox Signal 27(16):1332–1346. https://doi.org/10.1089/ars.2016.6836. Available at: https://www.ncbi.nlm.nih.gov/pubmed/28269997

Santos JM, Tewari S, Benite-Ribeiro SA (2014) The effect of exercise on epigenetic modifications of PGC1: the impact on type 2 diabetes. Med Hypotheses 82(6):748–753. https://doi.org/10.1016/j.mehy.2014.03.018. Available at: https://www.sciencedirect.com/science/article/pii/S0306987714001212

Schulz V, Basu S, Freibert SA, Webert H, Boss L, Mühlenhoff U, Pierrel F, Essen LO, Warui DM, Booker SJ, Stehling O, Lill R (2023) Functional spectrum and specificity of mitochondrial ferredoxins FDX1 and FDX2. Nat Chem Biol 19(2):206–217. https://doi.org/10.1038/s41589-022-01159-4

Sekine Y, Houston R, Eckl EM, Fessler E, Narendra DP, Jae LT, Sekine S (2023) A mitochondrial iron-responsive pathway regulated by DELE1. Mol Cell 83(12):2059–2076.e6. https://doi.org/10.1016/j.molcel.2023.05.031

Shen K, Pender CL, Bar-Ziv R, Zhang H, Wickham K, Willey E, Durieux J, Ahmad Q, Dillin A (2022) Mitochondria as cellular and organismal signaling hubs. Annu Rev Cell Dev Biol 38:179–218. https://doi.org/10.1146/annurev-cellbio-120420-015303

Shock LS, Thakkar PV, Peterson EJ, Moran RG, Taylor SM (2011) DNA methyltransferase 1, cytosine methylation, and cytosine hydroxymethylation in mammalian mitochondria. Proc Natl Acad Sci USA 108(9):3630–3635. https://doi.org/10.1073/pnas.1012311108. Available at: https://www.ncbi.nlm.nih.gov/pubmed/21321201

Si X, Shao M, Teng X, Huang Y, Meng Y, Wu L, Wei J, Liu L, Gu T, Song J, Jing R, Zhai X, Guo X, Kong D, Wang X, Cai B, Shen Y, Zhang Z, Wang D, Hu Y, Qian P, Xiao G, Huang H (2024) Mitochondrial isocitrate dehydrogenase impedes CAR T cell function by restraining antioxidant metabolism and histone acetylation. Cell Metab 36(1):176–192 e10. https://doi.org/10.1016/j.cmet.2023.12.010. Available at: https://www.ncbi.nlm.nih.gov/pubmed/38171332

Smith JAB, Murach KA, Dyar KA, Zierath JR (2023) Exercise metabolism and adaptation in skeletal muscle. Nat Rev Mol Cell Biol 24(9):607–632. https://doi.org/10.1038/s41580-023-00606-x

Soutschek M, Schratt G (2023) Non-coding RNA in the wiring and remodeling of neural circuits. Neuron 111(14):2140–2154. https://doi.org/10.1016/j.neuron.2023.04.031. Available at: https://www.ncbi.nlm.nih.gov/pubmed/37230080

Trewin AJ, Silver J, Dillon HT, Della Gatta PA, Parker L, Hiam DS, Lee YP, Richardson M, Wadley GD, Lamon S (2022) Long non-coding RNA Tug1 modulates mitochondrial and myogenic responses to exercise in skeletal muscle. BMC Biol 20(1):164. https://doi.org/10.1186/s12915-022-01366-4. Available at: https://www.ncbi.nlm.nih.gov/pubmed/35850762

Uittenbogaard M, Brantner CA, Chiaramello A (2018) Epigenetic modifiers promote mitochondrial biogenesis and oxidative metabolism leading to enhanced differentiation of neuroprogenitor cells. Cell Death Dis 9(3):360. https://doi.org/10.1038/s41419-018-0396-1

Vizioli MG, Liu T, Miller KN, Robertson NA, Gilroy K, Lagnado AB, Perez-Garcia A, Kiourtis C, Dasgupta N, Lei X, Kruger PJ, Nixon C, Clark W, Jurk D, Bird TG, Passos JF, Berger SL, Dou Z, Adams PD (2020) Mitochondria-to-nucleus retrograde signaling drives formation of cytoplasmic chromatin and inflammation in senescence. Genes Dev 34(5–6):428–445. https://doi.org/10.1101/gad.331272.119

Wan M, Xia R, Lin H, Ye Y, Qiu P, Zheng G (2022) Baduanjin exercise modulates the hippocampal subregion structure in community-dwelling older adults with cognitive frailty. Front Aging Neurosci 14:956273. https://doi.org/10.3389/fnagi.2022.956273

Wang D, Li H, Chandel NS, Dou Y, Yi R (2023) MOF-mediated histone H4 Lysine 16 acetylation governs mitochondrial and ciliary functions by controlling gene promoters. Nat Commun 14(1):4404. https://doi.org/10.1038/s41467-023-40108-0

Wang S, Li J, Zhao Y (2024) Construction and analysis of a network of exercise-induced mitochondria-related non-coding RNA in the regulation of diabetic cardiomyopathy. PLoS One 19(3):e0297848. https://doi.org/10.1371/journal.pone.0297848. Available at: https://www.ncbi.nlm.nih.gov/pubmed/38547044

Wang X, Li Q, He S, Bai J, Ma C, Zhang L, Guan X, Yuan H, Li Y, Zhu X, Mei J, Gao F, Zhu D (2022a) LncRNA FENDRR with m6A RNA methylation regulates hypoxia-induced pulmonary artery endothelial cell pyroptosis by mediating DRP1 DNA methylation. Mol Med 28(1):126. https://doi.org/10.1186/s10020-022-00551-z. Available at: https://www.ncbi.nlm.nih.gov/pubmed/36284300

Wang Z, Chivu AG, Choate LA, Rice EJ, Miller DC, Chu T, Chou SP, Kingsley NB, Petersen JL, Finno CJ, Bellone RR, Antczak DF, Lis JT, Danko CG (2022b) Prediction of histone post-translational modification patterns based on nascent transcription data. Nat Genet 54(3):295–305. https://doi.org/10.1038/s41588-022-01026-x. Available at: https://www.ncbi.nlm.nih.gov/pubmed/35273399

Wardle SL, Bailey ME, Kilikevicius A, Malkova D, Wilson RH, Venckunas T, Moran CN (2015) Plasma microRNA levels differ between endurance and strength athletes. PLoS One 10(4):e0122107. https://doi.org/10.1371/journal.pone.0122107

Wong M, Gertz B, Chestnut BA, Martin LJ (2013) Mitochondrial DNMT3A and DNA methylation in skeletal muscle and CNS of transgenic mouse models of ALS. Front Cell Neurosci 7:279. https://doi.org/10.3389/fncel.2013.00279. Available at: https://www.ncbi.nlm.nih.gov/pubmed/24399935

Xiao CL, Zhu S, He M, Chen D, Zhang Q, Chen Y, Yu G, Liu J, Xie SQ, Luo F, Liang Z, Wang DP, Bo XC, Gu XF, Wang K, Yan GR (2018) N(6)-methyladenine DNA modification in the human genome. Mol Cell 71(2):306–318 e7. https://doi.org/10.1016/j.molcel.2018.06.015. Available at: https://www.ncbi.nlm.nih.gov/pubmed/30017583

Xu Y, Xu L, Han M, Liu X, Li F, Zhou X, Wang Y, Bi J (2019) Altered mitochondrial DNA methylation and mitochondrial DNA copy number in an APP/PS1 transgenic mouse model of Alzheimer disease. Biochem Biophys Res Commun 520(1):41–46. https://doi.org/10.1016/j.bbrc.2019.09.094. Available at: https://www.ncbi.nlm.nih.gov/pubmed/31564416

Yan S, Fu W, Wang C, Mao J, Liu B, Zou L, Lv C (2020) Association between sedentary behavior and the risk of dementia: a systematic review and meta-analysis. Transl Psychiatry 10(1):112. https://doi.org/10.1038/s41398-020-0799-5

Yang JJ, Wang J, Yang Y, Sun F, Yang Y, Li J, Tao H, Lu C (2022) Epigenetic control of mitochondrial fission enables hepatic stellate cells activation in liver fibrosis via PGC-1alpha-Drp1 pathway. Mitochondrion 66:38–50. https://doi.org/10.1016/j.mito.2022.07.005. Available at: https://www.ncbi.nlm.nih.gov/pubmed/35905890

Yin MM, Wang W, Sun J, Liu S, Liu XL, Niu YM, Yuan HR, Yang FY, Fu L (2013) Paternal treadmill exercise enhances spatial learning and memory related to hippocampus among male offspring. Behav Brain Res 253:297–304. https://doi.org/10.1016/j.bbr.2013.07.040. Available at: https://www.sciencedirect.com/science/article/pii/S0166432813004397

You Y, Chen X, Chen Y, Pang J, Chen Q, Liu Q, Xue H, Zeng Y, Xiao J, Mi J, Tang Y, Ling W (2023) Epigenetic modulation of Drp1-mediated mitochondrial fission by inhibition of

S-adenosylhomocysteine hydrolase promotes vascular senescence and atherosclerosis. Redox Biol 65:102828. https://doi.org/10.1016/j.redox.2023.102828. Available at: https://www.ncbi.nlm.nih.gov/pubmed/37517319

Zhang F, Zhang L, Hu G, Chen X, Liu H, Li C, Guo X, Huang C, Sun F, Li T, Cui Z, Guo Y, Yan W, Xia Y, Liu Z, Lin Z, Duan W, Lu L, Wang X, Wang Z, Wang S, Tao L (2024) Rectifying METTL4-Mediated N(6)-methyladenine excess in mitochondrial DNA alleviates heart failure. Circulation 150(18):1441–1458. https://doi.org/10.1161/CIRCULATIONAHA.123.068358. Available at: https://www.ncbi.nlm.nih.gov/pubmed/38686562

Zhang M, Jia J, Yang Y, Zhang L, Wang X (2023) Effects of exercise interventions on cognitive functions in healthy populations: a systematic review and meta-analysis. Ageing Res Rev 92:102116. https://doi.org/10.1016/j.arr.2023.102116

Zhao M, Li Y, Lu C, Ding F, Xu M, Ge X, Li M, Wang Z, Yin J, Zhang J, Wang X, Ge Z, Xiao H, Xiao Y, Liu H, Liu W, Cao Y, Wang Q, You Y, Wang X, Yang K, Shi Z, Qian X (2023) PGC1alpha degradation suppresses mitochondrial biogenesis to confer radiation resistance in Glioma. Cancer Res 83(7):1094–1110. https://doi.org/10.1158/0008-5472.CAN-22-3083. Available at: https://www.ncbi.nlm.nih.gov/pubmed/36696363

Zhao Y, Wang Y, Zhao J, Zhang Z, Jin M, Zhou F, Jin C, Zhang J, Xing J, Wang N, He X, Ren T (2021) PDE2 inhibits PKA-Mediated phosphorylation of TFAM to promote mitochondrial Ca(2+)-Induced colorectal cancer growth. Front Oncol 11:663778. https://doi.org/10.3389/fonc.2021.663778. Available at: https://www.ncbi.nlm.nih.gov/pubmed/34235078

Zhong T, Ren F, Huang CS, Zou WY, Yang Y, Pan YD, Sun B, Wang E, Guo QL (2016) Swimming exercise ameliorates neurocognitive impairment induced by neonatal exposure to isoflurane and enhances hippocampal histone acetylation in mice. Neuroscience 316:378–388. https://doi.org/10.1016/j.neuroscience.2015.12.049

Zhou Z, Ma A, Moore TM, Wolf DM, Yang N, Tran P, Segawa M, Strumwasser AR, Ren W, Fu K, Wanagat J, van der Bliek AM, Crosbie-Watson R, Liesa M, Stiles L, Acin-Perez R, Mahata S, Shirihai O, Goodarzi MO, Handzlik M, Metallo CM, Walker DW, Hevener AL (2024) Drp1 controls complex II assembly and skeletal muscle metabolism by Sdhaf2 action on mitochondria. Sci Adv 10(14):eadl0389. https://doi.org/10.1126/sciadv.adl0389. Available at: https://www.ncbi.nlm.nih.gov/pubmed/38569044

Zollo O, Tiranti V, Sondheimer N (2012) Transcriptional requirements of the distal heavy-strand promoter of mtDNA. Proc Natl Acad Sci USA 109(17):6508–6512. https://doi.org/10.1073/pnas.1118594109. Available at: https://www.ncbi.nlm.nih.gov/pubmed/22454497

Chapter 11
Gut Microbiota and Exercise: Implications for Health Promotion and Athletic Performance

Lin Shi, Rikard Landberg, Yizhe Qu, Jiayao Lv, Lanqi Zhou, Tianqi Liu, Yousen Cao, and Di Wang

Abstract The gut microbiota has emerged as a pivotal mediator of human health and physiological performance, wherein exercise acts as a potent modulator of microbial composition and function. This chapter focuses on the latest research examining the relationship between physical activity, exercise training, and the gut microbiota, exploring its implications for health enhancement and athletic performance. This chapter begins with a detailed description of gut microbiome in human, outlining its complex ecosystem and metabolic capacity to influence host physiology. Emerging evidence from population-based studies is summarized to illustrate how physical activity influences the lifespan trajectory of the gut microbiome. We then provide mechanistic insights from human trials and animal studies, showcasing the exercise-induced enrichment of anti-inflammatory, butyrate-producing, and lactate-utilizing bacteria (e.g., *Lactobacillus, Faecalibacterium prausnitzii, Roseburia* spp., and *Akkermansia muciniphila*), which enhance intestinal barrier integrity, metabolic efficiency, and recovery processes. These findings position the gut microbiota as a new frontier in precision sports medicine. Moreover, we provide a compelling review of how the interplay between exercise and individual factors, such as baseline fitness, age, and dietary habits, can influence the pre-exercise microbiota profile and subsequently alter the microbial response to exercise training. Furthermore, the interplay between gut microbiota along with their metabolites that have shown to be modulated by exercise, and host epigenetics is dissected,

L. Shi (✉) · J. Lv · L. Zhou · T. Liu
College of Food Engineering and Nutritional Science, Shaanxi Normal University, Xi'an, China
e-mail: linshi198808@snnu.edu.cn

R. Landberg
Food and Nutrition Science, Department of Life Sciences, Chalmers University of Technology, Gothenburg, Sweden

Y. Qu · D. Wang
School of Physical Education, Shaanxi Normal University, Xi'an, China

Y. Cao
School of Aerospace Medicine, Fourth Military Medical University, Xi'an, China

F. Gao, X. Zhang (eds.), *Exercise, Epigenetics and Human Health*, Epigenetics and Human Health 15, https://doi.org/10.1007/978-3-032-08045-5_11

attempting to reveal the mysteries of health maintenance and disease prevention from this new perspective. By integrating multidisciplinary evidence, this chapter provides a roadmap for harnessing microbiome-exercise interactions to enhance healthspan and athletic potential through personalized microbiota-targeted interventions.

Keywords Gut microbiota · Physical activity · Exercise training · Precision sports medicine · Epigenetic regulations

Abbreviations

CRC	Colorectal Cancer
HIIT	High-intensity Interval Training
HPA	Hypothalamic–pituitary–adrenal
LPS	Lipopolysaccharide
PA	Physical Activity
SCFAs	Short-chain Fatty Acids
TMAO	Trimethylamine-N-oxide
IEC	Intestinal Epithelial Cells

11.1 The World of Human Microbiome

The human microbiome, a complex ecosystem comprising more than 100 trillion microorganisms, plays a pivotal role in maintaining human health and contributing to disease processes. These microorganisms, predominately including bacteria, archaea, fungi and viruses, inhabit various body sites such as the skin, gut, respiratory tract, and urogenital tract, forming a symbiotic relationship with the host. The microbiome's influence extends to numerous physiological functions, including digestion, immune system regulation, and protection against pathogens.

The world of microorganism was first unveiled in 1676 when Antonie van Leeuwenhoek peered through his handmade lens at a drop of rainwater, discovering a hidden universe of tiny, squirming "animalcules," invisible to the naked eye. Seven years later, he made an even more profound observation on these creatures thriving inside the human body. In his September 17, 1683, letter to the Royal Society, he documented "many small living animals moving very gracefully" in his own dental plaque, revealing an intricate ecosystem within us. This microbial community, weighing approximately 200 grams (comparable to the human brain), functions as a continuous biochemical factory. The unseen workforce of this community, later classified as bacteria, archaea, and fungi, marked humanity's first glimpse into the symbiotic microbial life that silently influences our existence.

The description of gastrointestinal microorganism was firstly provided in 1842 by an Edinburgh surgeon John Goodsir, who discovered a microorganism and named it *Sarcina ventriculi* during an examination of the ejected fluid from a 19-year-old patient's stomach. Goodsir hypothesized that *Sarcina ventriculi* was the cause of the patient's condition, and attempted to eliminate it using creosote drops, a commonly used gastrointestinal antiseptic (Goodsir and Wilson 1842). However, German pathologist Friedrich Theodor von Frerichs argued that *Sarcina ventriculi* had no impact on digestion and were inactive inhabitants of the human body. Indeed, Frerichs' interpretation aligned with his pioneering description of microorganisms in the digestive tract, which he regarded as harmless commensals. Nevertheless, the debate concerning the mysterious presence of microorganisms in the human gut had already commenced.

In the latter half of the nineteenth century, several scientific publications confirmed the existence of bacteria in the human body. Specifically, British scientist Lionel Smith Beale reported the presence of microorganisms in the human stomach, intestines, and feces in 1867. German botanist Ernst Hallier and Army surgeon JJ Woodward asserted that millions of microorganisms comprised the feces of healthy individuals. In 1881, German physician Uffelmann found large numbers of microorganisms in the feces of breastfed infants. In 1885, Theodor Escherich was the first to systematically isolate and name *Escherichia coli* (initially termed *Bacterium coli commune*), who introduced a classification framework distinguishing between "obligate gut microbiota" (e.g., Bacterium lactis aërogenes) and "facultative gut microbiota" (opportunistic species such as Streptococcus), establishing the groundwork for modern gut microbial ecology. Through microscopy and pure-culture techniques, Escherich demonstrated how the gut microbiota of infants dynamically changes with dietary shifts (e.g., from sterile meconium to milk-dominated flora), pioneering research into diet-microbiome interactions (Biedert 1880).

For centuries, microorganisms inhabiting the human body were long relegated to the status of parasitic nuisances in medical discourse. While Louis Pasteur's groundbreaking germ theory of disease in the 1860s casted the microbes primarily as pathogenic villains, the visionary scientist harbored an unconventional intuition about their symbiotic potential. In 1885, Pasteur hypothesized that "animal life would not be possible in the absence of intestinal microorganisms." Elie Metchnikoff, a Nobel laureate, blamed decried gut flora as sources of "fermentations and putrefactions detrimental to the host." Metchnikoff's paradigm shift occurred in 1904 through his epidemiological study of Bulgarian centenarians. Observing their exceptional longevity coinciding with regular consumption of fermented yogurt, he isolated *Lactobacillus bulgaricus* as the active agent. This discovery led to his revolutionary proposal that "fermented milk prepared by lactic bacteria," marking the first scientific articulation of probiotic therapy. Parallel revelations emerged from germ-free animal research, where mice raised in the absence of microorganisms exhibited shrunken organs, weakened immune systems, and impaired digestion. These findings forced a fundamental reevaluation of host–microbe relationships, transforming our understanding from seventeenth-century "animalcules" to essential biological partners. Modern science now recognizes that these commensal organisms constitute not merely passive inhabitants, but rather active collaborators in maintaining human homeostasis.

11.2 Gut Microbiota and Human Health

The advent of molecular biology has led to significant evolutionary divergence, transcending traditional classifications. In 1990, Carl R. Woese, Otto Kandler, and Mark L. Wheelis proposed the Three-Domain System, dividing life into Bacteria, Archaea, and Eukaryotes. This framework revolutionized our comprehension of microbial diversity and evolutionary relationships. Although all body sites are colonized by microbes, the gut harbors the highest number of microbes and has been extensively studied.

The human gut microbiota represents a dynamic and complex ecosystem comprising trillions of microorganisms, which exhibit high diversity in composition. Microbiome development initiates in early life and is influenced by birth mode and infant feeding practices. Host genetics, epigenetic modifications, and environmental factors, such as nationality, lifestyle habits, and life events (e.g., puberty, ovarian cycle, pregnancy, and menopause), can impact the microbial population and its biological activity.

Over the past two decades, significant progress has been achieved to establish causal relationships between gut microbiota and health, transitioning from initial clinical observations to a mechanistic understanding. This chapter focuses on the fundamental characteristics of the three major domains: Bacteria, Eukaryotes (mainly fungi), and Archaea, and briefly discusses their interactions and the implications for human health (Fig. 11.1).

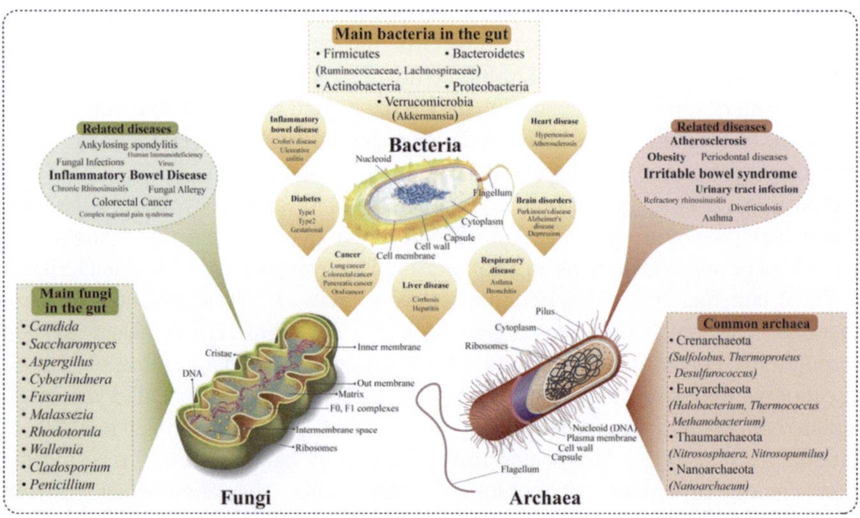

Fig. 11.1 Illustration of main bacteria, fungi and archaea prevalent in the gut. Current evidences supporting links between gut microbiota and multiple diseases

11.2.1 Bacteria: The Dominant Players

Bacteria, traditionally viewed as simple prokaryotic organisms devoid of nuclei and membrane-bound organelles, are increasingly recognized for their complex subcellular organization. Their membrane lipids primarily consist of diacylglycerol diesters, while their ribosomal RNA (rRNA) displays unique molecular characteristics. Bacteria play a crucial role in ecosystems, ranging from driving biogeochemical cycles to serving as pathogens and industrial catalysts. Recent advancements in imaging technologies have demonstrated that many bacteria possess intricate internal structures, which facilitate a variety of cellular processes. For example, bacterial micro-compartments (MCPs) are protein-bound organelles encapsulating metabolic enzymes, which optimize biochemical pathways by confining toxic or volatile intermediates. Additionally, bacteria exhibit a high degree of spatial organization, with proteins and nucleic acids forming dynamic, membrane-free organelles via liquid–liquid phase separation. This phenomenon is not only vital for compartmentalizing cellular components but also essential for regulating biochemical reactions in response to environmental alterations.

Bacteria constitute the most abundant and diverse group within the gut microbiota, with estimates suggesting that over 1000 species coexist in the human gastrointestinal tract. The gut microbiome is dominated by mainly anaerobic bacteria, including thousands of species and millions of genes, distributed among the major phyla of Firmicutes (predominantly Ruminococcaceae, Lachnospiraceae, and Clostridiaceae), Bacteroidetes (*Bacteroides*), Actinobacteria (*Bifidobacterium* spp.), Proteobacteria *(Enterobacteriaceae* and Desulfovibrionaceae), and Verrucomicrobia (*Akkermansia muciniphila*). Firmicutes, a gram-positive bacteria renowned for their carbohydrate-degrading capabilities, and Bacteroidetes (several species as masters of polysaccharide utilization) constitute approximately 90% of the total bacterial population, both playing pivotal roles in host metabolism, immune regulation, and disease pathogenesis. Firmicutes are enriched in individuals with high-fiber diets and are associated with efficient energy harvest, while Bacteroidetes populations decline in Western diets rich in fats and sugars, correlating with increased intestinal permeability and systemic inflammation.

Beyond their local intestinal functions, gut microbes produce bioactive metabolites that traverse the gut barrier or undergo systemic metabolism, influencing distant organs such as the liver, brain, and cardiovascular system (de Vos et al. 2022). Metabolic products of bacteria include lipopolysaccharide (LPS), trimethylamine-N-oxide (TMAO, a pro-inflammatory agent), short-chain fatty acids (SCFAs), bile acids, and more recently identified molecular actors, that is, tryptophan metabolites (AhR binding derivates) as well as immunomodulatory polysaccharides and sphingolipids produced specifically by *Bacteroides* spp. and muropeptides formed by *Enterococcus* spp., have shown to traverse the intestinal barrier or undergo systemic metabolism to impact distant organs.

The gut microbiome in human, dominated by anaerobic bacteria, functions as a dynamic metabolic organ with local and systemic impacts on human health. Advances in understanding microbial metabolites and their receptors will pave the way for gut microbiota derived precision medicine approaches for health promotion.

11.2.2 Fungi: The Understudied Kingdom

Eukaryotes represent a kingdom of eukaryotic organisms defined by their encapsulated nuclei and membrane-bound organelles. Fungi are microeukaryotes that have been identified in and on almost all human body sites, including the lungs, the vaginal tract, the urinary tract, the oral cavity, and the gut. The fungi community in gut, also called gut microbiota, although a smaller component of the gut microbiota compared to bacteria, has emerged as an important component of the gut environment and plays a crucial role in maintaining intestinal homeostasis and influencing systemic immunity. The dominant fungi found in gut include *Candida, saccharomyces, aspergillus, cyberlindnera*, etc. Despite being an anaerobic environment, the human gut does not seem to support the colonization of strictly anaerobic fungi, like *Neocallimastigomycota.*

Fungal dysbiosis has been increasingly recognized as a significant factor in colorectal diseases, including colorectal cancer (CRC). Fungal species, such as Candida and Malassezia, are more prevalent in the feces and mucosa of patients with Crohn's disease, while a lower fungal diversity is observed in patients with ulcerative colitis. *Candida albicans* (*C. albicans*), *Malassezia restricta, Cryptococcus neoformans* have been shown to bloom in the gut under inflammatory conditions. The intraneoplastic fungal community has been shown to differ between adenoma and CRC, with certain fungi such as *Malassezia* and *Aspergillus* playing a pivotal role in the progression from adenoma to CRC. These fungi exhibit increased abundance in CRC tissues, and their presence is correlated with host genetic mutations, such as KRAS-activating mutations and microsatellite instability, which are known to drive CRC progression (van de Haar et al. 2023). Moreover, *Mucor* genus and *Candida* spp. have shown to be predominated in obese subjects and diabetic patients compared to their healthy counterparts, respectively. *C. albicans and C. parapsilosis* transplantation evoked metabolism syndromes in mice models.

Emerging research highlights the therapeutic potential of cross-kingdom interactions between gut fungi and bacteria in maintaining intestinal health. For instance, the probiotic yeast *Saccharomyces boulardii* inhibits bacterial pathogens like *Clostridium* difficile by secreting proteases that degrade its toxins A and B. Both *Escherichia coli* (*E. coli*) and *Salmonella Typhimurium* (*S. Typhimurium*) bind to the surface of *S. boulardii*, potentially preventing adhesion to intestinal epithelial cells and thus allowing quicker excretion through fecal matter. Conversely, commensal bacteria antagonize pathogenic fungi, particularly *C. albicans*. Probiotic bacteria strains (*Lactobacillus acidophilus*, *Limosilactobacillus reuteri*, *Lacticaseibacillus rhamnosus GG*, and *Bifidobacterium animalis*) mitigate *C. albicans* infections in immunocompromised mice. Similarly, *Enterococcus faecalis* produces bacteriocin EntV to inhibit hyphal growth, reducing candidiasis severity. Gut bacteria like *Roseburia* and *Bacteroides ovatus* from healthy fecal samples could suppress *C. albicans* growth. Intriguingly, *C. albicans* reciprocally supports commensal anaerobes (e.g., *Bacteroides fragilis* and *B. vulgatus*) by providing mannan as a carbon source or reducing oxygen tension, illustrating bidirectional mutualism.

Cross-kingdom microbial interactions between fungi and bacteria are clinically relevant in diseases. However, challenges remain in translating these findings to human applications, necessitating further studies on strain-specific mechanisms and host-microbe dynamics.

11.2.3 Archaea: Methanogens and Beyond

Archaea, formally classified as a domain in 1990, are prokaryotes sharing morphological features with bacteria but exhibiting eukaryotic-like transcription/translation mechanisms (López-García et al. 2015). Pioneering rRNA analyses by Carl Woese in the 1970s resolved their phylogenetic ambiguity, establishing archaea as an evolutionarily distinct lineage with closer molecular affinity to eukaryotes than bacteria-fundamentally challenging the prokaryote-eukaryote paradigm (Zahn 2017; Zhou et al. 2022). Methanogenic archaea, are a unique domain of microorganisms that differ biochemically and genetically from bacteria. *Methanobrevibacter smithii*, a most prevalent methanogen in the human gastrointestinal tract (i.e., constituting up to 10% of gut anaerobes), consumes bacterial fermentation byproducts (e.g., H_2 and CO_2) to produce methane, which can be measured through clinical methane breath analyses. This methane production has been linked to various health conditions, although the precise role of Archaea in human health and disease remains largely unclear. Other archaeal taxa, such as *Methanosphaera* and ammonia-oxidizing archaea (AOA), are less common but contribute to nitrogen cycling.

The study of archaea in gut and their impact on human health is an emerging area of research in recent years. Archaea are resistant to most antibiotics due to their unique cell wall structures, enabling persistence during gut dysbiosis. Their role in human health remains under investigation, with emerging links between methanogenic archaea and cardiovascular disease, neurodegenerative diseases, and chronic kidney disease. Notably, methane production correlates with slower intestinal transit and conditions like constipation-predominant irritable bowel syndrome. Another archaeal community within the gastrointestinal tract, including *Methanobacterium* and *Woesearchaeota*, have been associated with various body sites such as the lung and skin. *Methanomassiliicoccales*, a novel order of methanogenic archaea, have been shown to utilize trimethylamine (TMA), a precursor to an atherosclerogenic metabolite, suggesting its role in modulating host metabolic pathways and influencing disease risk.

Despite these associations, Archaea are generally considered nonpathogenic, and their interactions within the human microbiome are not yet fully understood. Further research is needed to understand the biological significance of these microorganisms and their metabolites, which may reveal new therapeutic targets for maintaining gut health and preventing disease. The development of high-throughput sequencing technologies is crucial in advancing our understanding of the human archaeome and its role in health and disease.

In summary, the gut microbiota is a multispecies consortium where bacteria, fungi, and archaea collectively impact host health. While bacteria dominate in abundance and functional diversity, fungi and archaea contribute unique metabolic and immunological roles. The composition and activity of gut microbiota are shaped by diet, lifestyle, and disease states, offering actionable targets for therapeutic intervention. Future research must address gaps in culturing, strain-level analysis, and causal relationships to harness microbiota-based interventions fully. In-depth understanding this "hidden organ" promises breakthroughs in personalized medicine and disease prevention.

11.3 Physical Activity and the Lifespan Trajectory of the Gut Microbiome: Emerging Evidence and Implications from Population-Based Cohorts

Physical activity (PA) is any movement produced by skeletal muscles that requires energy expenditure. According to the World Health Organization (WHO), it includes leisure time, transport to and from places, and workdays. PA is widely recognized as a protective factor against numerous chronic diseases, such as obesity, type 2 diabetes, cardiovascular diseases, and inflammatory bowel disorders. Physical inactivity is responsible for more than 7% of all-cause mortality. Beyond its metabolic and cardiovascular benefits, PA offers a cost-effective and scalable strategy to enhance overall quality of life, with mounting evidence suggesting that exercise-induced microbial shifts contribute to improved metabolic regulation, immune function, while reduced systemic inflammation. This chapter summarizes the relationship between PA and gut microbiota, in particular the bacterial communities, in diverse populations (Fig. 11.2).

11.3.1 Associations Between PA and Gut Microbiota in Adults

Large-scale population-based studies (mostly cross-sectional and longitudinal studies) have provided critical insights into how PA influence gut microbial composition and function across diverse populations, offering potential pathways for microbiota-targeted interventions in chronic disease prevention and management. Specifically, positive associations were found between PA levels and greater gut microbial diversity—a hallmark of a healthy microbiome. Alpha diversity refers to the observed

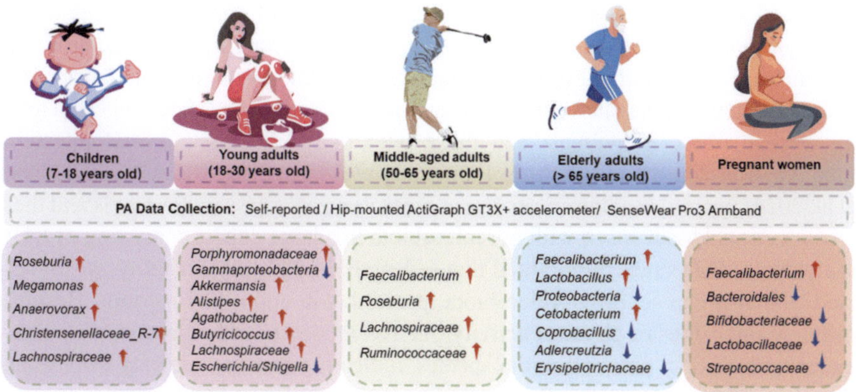

Fig. 11.2 Illustration of associations between physical activity and gut microbiota in different populations

richness (number of taxa) and the relative abundances of those taxa (also known as evenness) within a sample. Beta-diversity is defined as the variability in the microbial community composition among samples. Employing diversity indices allows for a comprehensive and nuanced exploration of associations between PA and microbial communities.

The moderate and vigorous PA levels are shown to be associated with a higher abundance of beneficial taxa such as SCFAs producers with anti-inflammatory properties (e.g., Lachnospiraceae, Ruminococcaceae, certain species *Faecalibacterium prausnitzii* and *Roseburia* spp.), and *Akkermansia muciniphila* (a next-generation probiotic for maintaining gut barrier integrity and improving immune responses and metabolic health) in young adults and middle-aged adults. Conversely, sedentary behavior has been linked to reduced microbial diversity and an overrepresentation of pro-inflammatory bacteria, such as Enterobacteriaceae (e.g., *Escherichia coli*) (Baldanzi et al. 2024), Desulfovibrionaceae, genera *Sutterella* and mucin-degrading *Ruminococcus gnavus*, which are implicated in metabolic disorders and insulin resistance. Notably, in 2025, the "macro-nutrient-to-physical activity ratio," a novel metric developed in a recent study—revealed that PA modulates the impact of dietary fat intake on gut microbiota composition. High saturated fat intake combined with low PA was associated with unfavorable microbial profiles (e.g., increased *Tyzzerella* and decreased *Roseburia*), whereas increased PA mitigated these effects, underscoring the interactive role of PA in shaping microbial ecosystems responding to the diet (Wang et al. 2025a).

Most studies investigating the relationship between PA and fecal microbiota composition in cohorts have used self-reported questionnaires to determine PA levels. A cross-sectional study enrolled 88 young adults aged 22.0 ± 2.3 years and objectively measured their PA at different intensities using a wrist-worn accelerometer for seven consecutive days (Ortiz-Alvarez et al. 2025). The study found that PA, particularly vigorous-intensity PA lasting 3–14 minutes per day, was positively correlated with increased fecal microbiota diversity and high abundances of the Porphyromonadaceae family and *Alistipes* genus (both belonging to the Bacteroidetes phylum) in young adults. It was also negatively correlated with the abundance of the Gammaproteobacteria class (Proteobacteria phylum), which has been reported to be elevated in obese individuals and those with metabolic diseases and intestinal inflammation. The Swedish CArdioPulmonary bioImage Study (SCAPIS) (Baldanzi et al. 2024) utilized accelerometer-based measurements to assess PA levels and their cross-sectional association with gut microbiota in a large cohort consisting of 8416 individuals. This study highlighted that sedentary and physical activity are associated with a similar set of gut microbiota species but in opposite directions. Besides, different intensities of physical activity, such as moderate and vigorous, were associated with distinct gut microbiota species, indicating that the intensity of physical activity might have specific effects on gut microbiota composition.

11.3.2 Associations Between PA and Gut Microbiota in Elderly People

In 2022, the global population aged 65 and older totaled 771 million, with projections indicating that this number will increase to 994 million by 2030 and 1.6 billion by 2050. The global population is aging, and concurrently becoming more sedentary. The Global Status Report on Physical Activity states that approximately one-quarter of adults do not meet the recommended levels of physical activity. Older adults spend, on average, 65–80% of their waking time sitting. PA, a crucial factor in promoting healthy aging, was defined by the WHO in 2020 as the process of developing and maintaining functional abilities that facilitate well-being in older adults. This includes the ability to meet basic needs, maintain mobility, make independent decisions, foster relationships, and contribute to society.

From infancy through old age, the gut microbiome undergoes dynamic transformations characterized by periods of rapid change and increasing individualization. While early life is marked by dramatic microbial shifts associated with dietary transitions and immune system development, adulthood brings relative stability, and old age is often accompanied by a gradual decline in microbial diversity and the emergence of age-associated dysbiosis. Emerging evidence suggests that PA may serve as a powerful modulator of these age-related microbial patterns, with even subtle but meaningful exercise-induced shifts potentially counteracting some of the detrimental changes observed in older adults. Although the associations between PA and gut microbiota across the lifespan have yet to be extensively studied, recent investigations have begun to illuminate these relationships.

In older populations, several cross-sectional studies have revealed intriguing connections between activity levels and microbial composition. One study of 39 older adults with insomnia demonstrated distinct microbial signatures associated with activity levels (Magzal et al. 2022), with more active individuals showing higher relative abundances of beneficial genera including *Bifidobacterium,*

Clostridium sensu stricto *1, Catenibacterium, Peptococcus, Holdemanella, and Butyricicoccus* (a butyrate producer implicated in gut barrier maintenance). Conversely, less active participants exhibited greater representation of *Barnesiella, Blautia, Lachnoclostridium*, Christensenellaceae_R-7_group, and UCG-005, some of which have been associated with metabolic dysfunction. *Faecalibacterium prausnitzii* (*F. prausnitzii*), *Barnesiella intestinihominis*, *Bacteroides caccae*, and *Clostridium citroniae* were reported to be higher in older adults with high fitness profile; meanwhile, a reduction in *Eubacterium biforme*, *Desulfovibrio* D168, and *Escherichia coli* was observed when compared to the Low Fitness group in a cross-sectional study where 29 older adults (mean age 70 years) (Fielding et al. 2019). Higher step counts were associated with a higher abundance of *Cetobacterium* and lower abundances of *Coprobacillus*, *Adlercreutzia*, and *Erysipelotrichaceae CC-115* (Langsetmo et al. 2019). The most comprehensive examination to date involved 101 community-dwelling older adults aged 65–85 years in the UK (Ramos et al. 2025). Researchers combined accelerometer-based PA monitoring with comprehensive anthropometric measures and a 6-minute walking test to objectively assess physical capacity. Participants who met WHO PA recommendations exhibited a gut microbial profile characterized by higher abundances of *Lactobacillus* (known for its probiotic effects), *F. prausnitzii*, and *Roseburia intestinalis* (both butyrate producers), along with reduced levels of potentially pathogenic organisms such as Enterobacterales.

These collective findings paint a compelling picture of PA as a potential modulator of aging-related microbial changes. As global populations continue to age, understanding and harnessing these PA-microbiome interactions could open new avenues for maintaining health and function in later life. The current evidence, while preliminary, strongly supports the notion that regular physical activity may help preserve a more youthful, health-associated gut microbial profile, potentially mitigating some of the detrimental effects of aging on gut ecosystem stability and function.

11.3.3 Associations Between PA and Gut Microbiota in Specific Populations

While most research on PA and gut microbiota has focused on adults or elderly populations, emerging studies suggest that exercise may influence microbial composition in *specific populations*, including children, adolescents, and pregnant women, with potentially profound implications for developmental and physiological health. Here, specific focuses are given to current evidence on associations between PA and gut microbiota across critical life stages such as childhood, adolescence, and pregnancy.

Specifically, children and adolescents can achieve substantial health benefits by engaging in moderate-to-vigorous PA for periods of time that add up to 60 minutes or more each day (current recommendations for adolescents aged 6–17 years). Based on the 2018 United States Report Card on PA for Children and Youth, only about 24% of children meet the recommended 60 minutes of PA every day. A birth cohort study of 321 children residing in the Detroit metropolitan area, conducted by Santarossa Sand colleagues, found that 189 (58.9%) met the PA recommendations and reported an association between meeting these recommendations and an increased abundance of *Megamonas*, *Anaerovorax*, and Christensenellaceae_R-7_group taxa (bacteria previously identified as part of a healthy gut microbiome) in children (Santarossa et al. 2021). Besides, a cross-sectional study of 267 individuals aged 7–18 years from the American Gut Project revealed positive associations between PA and increased abundances of *Roseburia* and Lachnospiraceae (Bai et al. 2019).

PA on gut microbial composition during pregnancy may exhibit profound implications for both maternal and fetal health outcomes, but there is a significant lack of research in humans. In a birth cohort study consisting of 23 pregnant women, 11 met or exceeded the PA recommendations (150 minutes per week of moderate-intensity exercise) during pregnancy, who exhibit markedly lower relative abundances of Bacteroidales, Bifidobacteriaceae, Lactobacillaceae, and Streptococcaceae compared to their less active counterparts (Santarossa et al. 2023). These microbial families crucial for nutrient metabolism, immune regulation, and mucosal barrier function, including specific species, exhibited positive associations with PA in adults. For instance, Bacteroidales contains numerous species involved in complex carbohydrate metabolism and SCFAs production that are generally considered beneficial for gut health. However, a higher relative abundance of Bacteroidales was reported in pregnant women with pre-eclampsia (Wang et al. 2020) and overweight pregnant women (Collado et al. 2008); their relative decrease in physically active pregnant women may reflect an optimized balance between energy harvest and immune modulation during this physiologically demanding period.

The timing and intensity of PA during pregnancy appear crucial in shaping these microbial responses. Microbial changes are most pronounced when exercise is maintained consistently throughout pregnancy rather than initiated at later stages. The moderate-intensity activities (such as brisk walking or swimming) showing the most consistent associations with beneficial microbial profiles, while vigorous exercise may have more variable effects depending on the individual's pre-pregnancy fitness level. These findings underscore the importance of personalized exercise prescriptions during pregnancy that consider baseline fitness, microbiome composition, and pregnancy trimester. PA exerts benefits in pregnancy, including reduced risks of preeclampsia and excessive gestational weight gain. Additionally, these maternal microbial shifts may have lasting effects on offspring health through vertical transmission during birth and breastfeeding, representing a potential mechanism for the intergenerational benefits of maternal exercise. Future studies that integrate deep microbial sequencing with detailed metabolic and immunological profiling will help unravel the complex interplay between PA, gut microbiota, and pregnancy physiology, potentially leading to more effective strategies for promoting maternal and fetal well-being.

Collectively, current evidence from population-based cohort studies collectively underscore PA as a potent modulator of gut microbiota in diverse populations, with far-reaching effects on metabolic, immune, and inflammatory pathways. Integrating PA into public health recommendations, tailored to individual microbial and metabolic profiles, could revolutionize preventive medicine in the era of precision health.

11.4 Exercise as a Multifaceted Modulator of Gut Microbiota: Evidence from Human Trials and Animal Interventions

A systematic review and meta-analysis consisting of 25 human trials conducted between 2016 and 2023, revealed the effect of exercise training on gut microbiota in healthy and unhealthy participants. It reported a significant increase in the gut microbiota diversity after exercise in adults, particularly reflected in the Shannon index, along with a significant impact on the Bacteroidetes and Firmicutes phyla (specifically, *Faecalibacterium prausnitzii* and species from the genus *Oscillospira, Lachnospira, and Coprococcus*). However, these effects are not uniformly observed across all bacterial groups, hinting at intricate relationships influenced by potential variations in types of exercise and/or the exercise intensities.

The realm of exercise encompasses a broad spectrum of modalities and intensity gradients, including aerobic exercise, resistance exercise, high-intensity interval training (HIIT), and flexibility/balance regimens, each demonstrating unique potential to influence gut microbial ecology. Thus, this chapter focuses on critical findings from human clinical trials to systematically elucidate the complex interplay between various exercise paradigms and gut microbiome remodeling, with particular emphasis on distinct microbial responses to specific exercise types and their translational clinical significance (Fig. 11.3).

This chapter further summarizes recent findings on the interaction between exercise and individual characteristics, including baseline fitness, age, and dietary patterns, in shaping the microbial response to exercise. Current evidence supports the concept of personalized exercise prescriptions tailored to microbial profiles. However, significant knowledge gaps remain regarding the durability of exercise-induced microbial alterations and their causal impacts on health outcomes.

11.4.1 Aerobic Exercise Modulates Gut Microbiota

Aerobic exercise (e.g., running, cycling, and swimming) involves sustained, rhythmic activity that increases cardiovascular endurance. It is typically performed at moderate intensity for extended durations (>30 minutes). Aerobic exercise has emerged as a potent lifestyle intervention capable of reshaping gut microbial

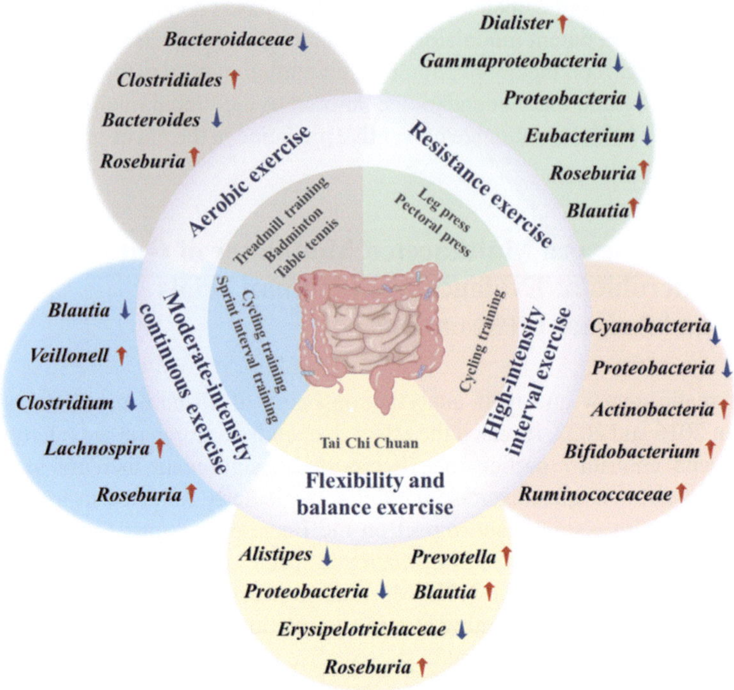

Fig. 11.3 Illustration on the modulatory effects of exercise on gut microbiota. Evidence is derived from randomized controlled trails

composition and function, with implications for metabolic health, inflammation, and disease prevention. Evidences from randomized controlled trials reveal the promoting effects of aerobic exercise on the growth of SCFAs-producing bacteria (e.g., *Faecalibacterium prausnitzii*, *Roseburia, Bacteroides, Blautia*). These bacteria have shown to enhance gut barrier function and reduce inflammation. By contrast, *Proteobacteria* phylum and *Gammaproteobacteria* class were shown to reduce upon aerobic exercise. These changes were most notable at 3 months into the exercise regimen, highlighting the importance of sustained physical activity for gut health.

Consistent with human trials, research using mouse models demonstrates that aerobic exercise increases the abundance of *Faecalibacterium prausnitzii*, a key butyrate producer essential for maintaining intestinal epithelial integrity, and *Akkermansia muciniphila*, a mucin-degrading bacterium associated with improved metabolic health and reduced obesity risk. Additionally, aerobic exercise enhances SCFAs production, ultimately improving gut health and energy metabolism. Aerobic exercise not only enriches beneficial microbes but also suppresses pro-inflammatory bacteria: a notable decrease in Proteobacteria, a phylum that includes pathogens such as *Escherichia* and *Salmonella*, has been observed in mice subjected to aerobic exercise, leading to reduced systemic inflammation.

11.4.2 Resistance Exercise Modulates Gut Microbiota

Resistance exercise, encompassing exercises such as weightlifting, leg presses, and pectoral presses, is a structured form of physical activity designed to enhance muscle strength and hypertrophy through repetitive contractions against external resistance. Resistance exercise has been extensively studied and is widely regarded as one of the most effective interventions to combat age-related sarcopenia, a progressive loss of muscle mass and strength functioning, thereby contributing to better physical performance and functional independence in elderly adults.

Beyond its role in mitigating sarcopenia, resistance exercise could offer various health benefits, such as increased bone density, improved glucose metabolism, decreased risk of type 2 diabetes and cardiovascular diseases, enhanced recovery from injuries and surgeries through restored muscular function and joint stability in rehabilitation, ultimately enhancing quality of life throughout one's lifespan. Nevertheless, a limited number of human studies have evaluated the modulatory impact of resistance exercise on gut microbiota.

Cullen et al. conducted a 6-week randomized controlled trial in young adults with overweight or obesity. The study revealed that resistance exercise (i.e., three sessions per week, three sets of 8–12 repetitions, at an intensity of up to 70% 1RM) selectively increased the abundance of *Roseburia*, a butyrate-producing bacterium crucial for enhancing intestinal barrier integrity, reducing systemic inflammation, and improving insulin sensitivity. A 12-week training combined endurance and strength protocol in obese children demonstrated a reduction in obesity-associated deleterious microbiota within the Proteobacteria phylum and Gammaproteobacteria class. Additionally, it enriched metabolic health-associated bacteria, including *Blautia*, *Dialister*, and *Roseburia*.

Evidence derived from mice study showed that resistance exercise elevated *Lactobacillus* and *Bifidobacterium*, genera known for their probiotic properties and ability to enhance nitrogen recycling, a process vital for muscle protein synthesis (Yoon et al. 2024). Besides, these microbes may synergize with myokines, such as irisin, released during muscle contraction, to enhance gut barrier function and promote systemic metabolic health. When combined with high-protein diets, resistance training augments the abundance of *Bacteroides*, thereby optimizing nutrient uptake and facilitating muscle growth.

11.4.3 HIIT, MICT, and Other Types of Exercises Training

High-intensity interval training (HIIT), characterized by short bursts of vigorous activity interspersed with recovery periods, has gained widespread recognition for its effectiveness in enhancing cardiovascular fitness, improving insulin sensitivity, and optimizing body composition. As a popular exercise modality, HIIT exerts pulsatile modulation on gut microbiota in relation to host metabolism. HIIT may elicit distinct microbial responses compared to moderate-intensity continuous training, due to its unique metabolic demands, characterized by rapid transitions between anaerobic and aerobic energy systems.

Specifically, swimmers who participated in a 7-week HIIT program exhibited a significant increase in alpha diversity of microbiota (Bielik et al. 2022). A 9-week HIIT program (Donati Zeppa et al. 2021) increased *Bifidobacterium* levels and decreased *Parabacteroides* and *Prevotella* (both belonging to the Bacteroidetes phylum) in healthy male college students, accompanied by reduced fat mass, increased muscle mass, and enhanced fecal concentration of SCFAs. HIIT showed enrichment in relative abundance of *Parabacteroides* and Defluviitaleaceae_UCG_011 among inactive adults with celiac disease (Warbeck et al. 2021). In post-menopausal women with overweight/obesity, a combination of HIIT and resistance exercise increased the relative abundances of Bifidobacteriaceae and Prevotellaceae, in association with significant fat loss and muscle mass gain (Nechalová et al. 2024).

Moderate-intensity continuous training (MICT) can enhance gut microbial diversity, enrich beneficial microbial species, optimize gut functionality, and modify gut physiology favorably. A randomized controlled trial on insulin-resistant subjects demonstrated, for the first time, that both MICT and sprint interval training (SIT) reduce systemic inflammation, increase Bacteroidetes and *Veillonella*, decrease the Firmicutes/Bacteroidetes ratio (i.e., a hallmark of obesity) as well as the relative abundances of the genera *Clostridium* and *Blautia* (Motiani et al. 2020). A 6-week endurance exercise regimen significantly increased fecal SCFAs post-exercise in lean subjects, but had no effect in obese subjects (Allen et al. 2018).

Most recently, a RCT study conducted in 2025 comparing the gut microbiota profiles of participants engaged in MICT, HIIT, and high-intensity functional training (HIFT). HIFT is characterized by functional, multi-joint movements that can be tailored to various fitness levels and are associated with greater muscle recruitment compared to traditional exercise forms. The findings indicated distinct gut microbiome profiles among participants following different exercise regimens. Specifically, MICT was associated with a significant increase in gut microbial diversity. HIFT uniquely enriched the presence of Faecalibacterium, Lactobacillus, and Limosilactobacillus, while reducing the abundance of Actinomyces and Anaeromassilibacillus, promoting a more favorable microbial profile compared to traditional endurance training (Wang et al. 2025b).

Flexibility and balance training, encompassing practices such as yoga and tai chi, has long been recognized for its profound benefits on physical and mental well-being. These ancient disciplines, rooted in mind–body integration, enhance muscular elasticity, joint mobility, postural stability, and neuromuscular coordination, thereby reducing the risk of falls, alleviating chronic pain, and improving overall functional fitness. Although the direct impacts on gut microbiota have been largely unexplored, a randomized controlled trial involving 90 college students with problematic mobile phone use found that 8-week Tai Chi Chuan particularly reduced Erysipelotrichaceae and *Alistipes* abundance (Zhang et al. 2024). The intersection of flexibility and balance training with gut microbiota presents a compelling area of exploration, suggesting that these time-honored practices may offer a holistic approach to health by harmonizing both physical and microbial well-being. As modern science continues to unravel the intricate connections between movement, stress, and the microbiome, the integration of flexibility and balance training, for example, yoga and tai chi into preventive and therapeutic health strategies may gain further validation, bridging ancient wisdom with contemporary biomedical understanding.

In summary, exercise is beneficial for overall health via multiple mechanisms, including modulating the gut microbiome. Notably, combining various exercise modalities (e.g., aerobic exercises and resistance training) has a more pronounced effect on bacterial diversity than engaging in a single type of exercise. Nonetheless, intense and prolonged physical activities may increase inflammation, indicating the need for a delicate balance between exercise intensity and its health benefits.

11.5 Gut Microbiota: The New Frontier of Precision Sports Medicine of Athletes

High-level athletes demonstrate remarkable physiological and metabolic adaptations, such as enhanced muscular strength and power, exceptional aerobic capacity, increased energy expenditure, and advanced thermoregulation, distinguishing them from sedentary individuals. Gut microbiome plays multifaceted roles in energy harvesting, immune modulation, and gastrointestinal homeostasis, which exhibit influences on various indices of exercise performance, recovery, and patterns of illness, such as signaling through myokines and other cytokines, modulating activation of the hypothalamic–pituitary–adrenal (HPA) axis, and affecting performance-associated metabolic pathways. While the current evidence of this comparative analysis is limited, collectively it provides important insight and highlights key areas of future study.

This chapter summarizes unique microbiota observed in athletes, coupled with their distinct dietary and training regimens, with specific focus on the current understanding of the gut microbiome's contribution to athletic performance, examining how microbial communities may support (or impair) energy metabolism, immune resilience, and recovery processes in high-level athletes. By synthesizing findings from human studies, this chapter further clarifies the potential for microbiome-targeted strategies to enhance athletic achievement.

In 2014, a landmark study by Professor Fergus Shanahan and his team at University College Cork unveiled groundbreaking differences in gut microbial ecology between elite rugby players during a preseason camp and sedentary individuals as control groups (i.e., one matched for size and Body Mass Index (BMI), and another representing the general age- and gender-matched population) (Clarke et al. 2014). Rugby is a vigorous contact sport requiring considerable fitness and increased dietary requirements. Results indicate that rugby athletes exhibit higher α-diversity and increased abundances of *Firmicutes* and *Akkermansia*. Differences in gut microbiota between athletes and sedentary individuals were associated with more favorable metabolic and inflammatory profiles, characterized by improved intestinal barrier integrity and enhanced protection against pathogen invasion and systemic inflammation. This pioneering work represented the first systematic comparative analysis of athlete microbiota, establishing foundational knowledge for what would later become the field of "exercise microbiology."

In 2019, the research team from Harvard Medical School, Harvard University and Joslin Diabetes Center etc., elucidated the direct impact of gut microbiota on exercise capacity. Most importantly, it marks the first isolation and validation of a performance-enhancing bacteria from elite athletes—underscores the potential of microbiota-targeted interventions to optimize athletic performance. Specifically, by analyzing marathon runners' gut microbiota before and after races, this study revolutionizes our understanding of exercise biology by identifying *Veillonella atypica,* as the first gut microbiome-derived biological catalyst for human athletic performance enhancement. Inoculation of this strain into mice significantly increased exhaustive treadmill runtime. Mechanistically, *Veillonella atypica* metabolize lactate into the propionate via the methylmalonyl-CoA pathway. Propionate has been shown to increase heart rate, VO_2 max, and affect blood pressure, and introduction of propionate into the colon is sufficient to result in an enhanced exercise phenotype. The translational impact of *Veillonella*'s lactate-catabolizing capacity is profound, underpinning the development of next-generation probiotics and microbiome-based performance diagnostics (Fig. 11.4). These advancements pave the way for precision sports medicine, where microbiota-targeted interventions, such as phage-guided bacterial modulation and metabolite supplementation, are pushing the limits of human athletic potential.

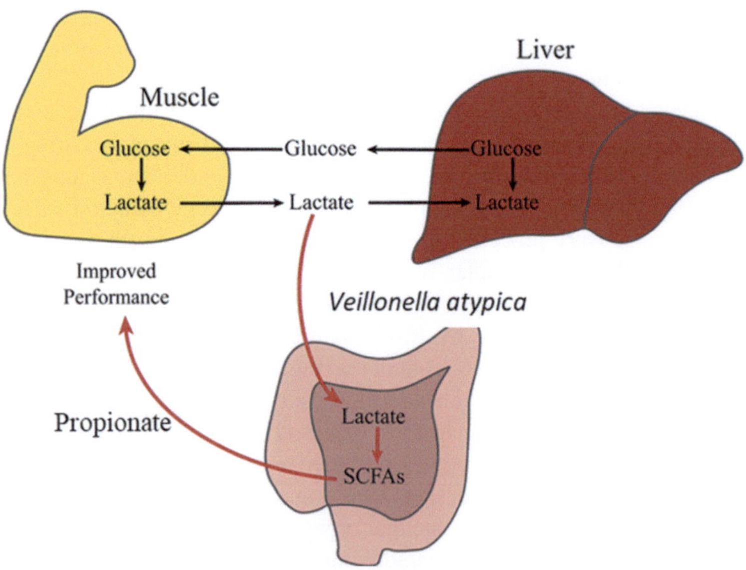

Fig. 11.4 Proposed model of the microbiome-exercise interaction induced by the *Veillonella atypica*

Previous studies have shown that athletes possess greater gut microbiota diversity compared to the general population, particularly endurance athletes who exhibit microbiota profiles correlated with stamina. Recently, research has begun to identify microbial signatures associated with strength training. A promising example is the identification of *Lactiplantibacillus plantarum Tana* (*L. plantarum Tana*) (Lee et al. 2022). *L. plantarum Tana* was isolated from the feces of Hsing-Chun Kuo, a gold-medal-winning weightlifter who has participated in numerous international competitions, including the 2020 World Championships and Olympic Games in Tokyo. A 5-week oral administration of *L. plantarum Tana* was found to enhance exercise performance and exhibit anti-fatigue effects in mouse models. Additionally, it increased the abundance of strength-oriented microbiota, such as Prevotellaceae, and significantly improved hepatic and muscular glycogen storage, prioritizing glycogen synthesis and amino acid utilization, while decreasing levels of fatigue markers such as lactate, blood urinary nitrogen, and creatine kinase. These findings provide a functional framework linking biomechanical demands to microbial ecosystem services: endurance sports cultivate microbiota specializing in oxidative phosphorylation substrates, whereas strength disciplines select for communities proficient in creatine recycling and myofibrillar repair.

Building on strength-specific microbiota research, Irish investigators pioneered a novel sports classification system based on dynamic/static component groups (SCGs) (O'Donovan et al. 2020), uncovering distinct microbial signatures across athletic disciplines. Moderate-dynamic sports (e.g., fencing) enriched *Streptococcus suis*, *Clostridium bolteae*, *Lactobacillus* phage LfeInf, and *Anaerostipes hadrus*, specific taxa implicated in rapid energy turnover and neuromuscular coordination. High-dynamic/low-static sports (e.g., field hockey) demonstrated preferential colonization of *Bifidobacterium animalis*, *Lactobacillus acidophilus, Prevotella intermedia*, and *Faecalibacterium prausnitzii*, reflecting adaptations to sustained aerobic demand through enhanced carbohydrate metabolism and anti-inflammatory modulation. High-dynamic/high-static disciplines (e.g., rowing) uniquely harbored *Bacteroides caccae*, suggesting evolutionary pressure toward concurrent glycogen storage and amino acid catabolism for explosive-endurance hybrid performance.

A recent multi-cohort study (Li et al. 2023), involving 543 fecal samples from athletes who regularly engage in high-intensity exercise (aerobics, $n = 316$; wrestling, $n = 53$; and rowing, $n = 174$), investigated whether distinct gut microbiome profiles exist among different athlete types and whether these profiles can be used to effectively monitor inflammatory risk. Using latent Dirichlet allocation, 10 microbial subgroups of athletes' gut microbiota were identified, each of which had specific correlations with inflammation, diet, and anaerobic performance in different types of athletes. Rowing athletes had significantly higher Shannon diversity compared to other athletes *in a multi-cohort study across different types of sports*. Professional football players, undergoing high-intensity mixed aerobic/anaerobic training, displayed greater alpha diversity and a higher abundance of butyrate-producing *Phascolarctobacterium* and *Streptococcus salivarius* (Urban et al. 2024). Elite soccer players, who trained over 6 hours weekly, had increased *Prevotella albensis, Roseburia* and *Phascolarctobacterium*, a propionate producer linked to metabolic health (Si et al. 2022). These findings shed light on the complex associations between physical factors, gut microbiota, and metabolic status in athletes of different sports types and could have significant implications for monitoring health status and developing personalized exercise programs.

The convergence of microbiome science and athletic performance has unveiled gut microbiota as the next paradigm in precision sports medicine. The discovery of performance-enhancing taxa like *Veillonella atypica*, which converts exercise-induced lactate into ergogenic propionate, has emerged as a keystone species, enhancing endurance capacity through improved gut barrier function and systemic metabolic efficiency. The implications extend beyond elite sports, offering novel strategies for metabolic rehabilitation and healthy aging. As metabolomic and multi-omics technologies mature, the sports microbiome is poised to redefine human performance boundaries, establishing microbiota-targeted therapies as indispensable tools in the future of athletic training, recovery, and resilience.

11.6 Individual Characteristics Shape the Beneficial Responses to Exercise: Insights for Personalized Exercise Programs Tailored by Gut Microbiota

Recent research emphasizes that exercise has a non-uniform impact on the gut microbiome, inducing highly individualized alterations influenced by both intrinsic and extrinsic factors. The interplay between exercise and individual factors, including baseline fitness, age, and dietary habits, can influence the pre-exercise microbiota profile and subsequently modify the microbial response to exercise training. Recognizing the complex interplay between exercise and host-specific factors is crucial for enhancing our understanding of personalized exercise interventions, which have significant implications for microbiome-related health outcomes.

Baseline fitness level, for instance, modulates microbial diversity, with trained individuals often harboring a more resilient and metabolically active microbiota compared to sedentary counterparts. Age further complicates this relationship, as the gut microbiome undergoes natural shifts over the lifespan, altering its responsiveness to exercise-induced perturbations. Younger individuals might experience different microbial responses to exercise compared to older adults, who may have a less diverse microbiome and different metabolic needs. Dietary patterns, perhaps the most malleable factor, interact synergistically with exercise to either amplify or dampen microbial adaptations, with high-fiber diets promoting the growth of beneficial SCFAs-producing bacteria, while high-fat or low-nutrient diets may counteract the positive effects of exercise. Moreover, dietary strategies targeting the gut microbiota, such as probiotic supplementation, have been shown to improve exercise performance by enhancing nutrient absorption and skeletal muscle function while reducing exercise-induced fatigue. Additionally, genetic predispositions, body composition, and even psychological stress levels contribute to the heterogeneity in microbial responses, underscoring the need for a precision-based approach to exercise recommendations.

This bidirectional relationship suggests that not only regular exercise promotes a more favorable gut microbiota composition, but a healthy gut microbiome (young adults or people who consumed high-fiber diets) can potentially enhance exercise ability through various mechanisms. Gut microbiota plays a crucial role in regulating brain function through the gut–brain axis, which has significant implications for exercise motivation and performance. This intricate communication network between the gut and the brain involves various pathways, including neural, endocrine, and immune mechanisms. For instance, gut microbiota can influence neurotransmitter metabolism and gut-to-brain signaling, which are essential for maintaining brain function and can impact neuropsychiatric conditions such as anxiety and depression, potentially affecting exercise motivation (Qu et al. 2024).

Intriguingly, a study published in *Nature* (Dohnalová et al. 2022) in 2022 reveals a novel gut–spinal cord–striatal axis (Fig. 11.5), through which microbial metabolites regulate dopamine dynamics to influence exercise persistence, with *Eubacterium rectale* and *Coprococcus eutactus* being identified as key taxa driving this phenotypic rescue. The study found that microbiome-dependent production of endocannabinoid metabolites in the gut stimulates CB1-expressing TRPV1+ sensory neurons, leading to downregulation of MAO expression in the striatum and consequently elevated dopamine levels in the ventral striatum during exercise. Stimulation of this pathway enhances running performance. Conversely, microbiome depletion, inhibition of peripheral endocannabinoid receptors, ablation of spinal afferent neurons, or dopamine blockade impair exercise capacity.

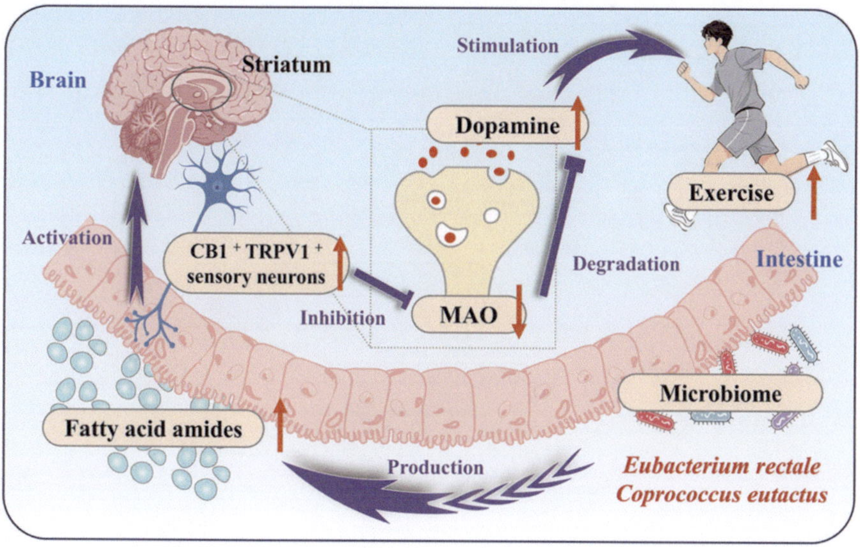

Fig. 11.5 Illustration of a microbiome-dependent gut–brain pathway regulates exercise motivation

The intricate bidirectional crosstalk between gut microbiota, brain function, and musculoskeletal systems forms a sophisticated biological network that underpins exercise-induced enhancements across metabolic, neurocognitive, and immunological domains. This tripartite communication framework, encompassing the gut–brain–muscle axis, operates through an orchestra of microbial-derived metabolites (e.g., SCFAs, tryptophan derivatives), neuroendocrine signaling (HPA axis modulation, catecholamine release), and immunoregulatory pathways (cytokine balance, Toll-like receptor activation) that collectively maintain organismal homeostasis while adapting to physical activity demands.

This conceptual framework (illustrated in Fig. 11.6) not only redefines our understanding of exercise physiology as a microbiome-influenced phenomenon but also opens therapeutic avenues for manipulating microbial networks to amplify training adaptations, mitigate overtraining syndrome, and potentially decelerate age-related declines in musculoskeletal and neurological function through targeted microbial ecosystem engineering.

Specifically, central to exercise-induced microbiota remodeling, the gut–brain axis orchestrates neurotransmitter regulation (i.e., serotonin, GABA, BDNF) through microbial-derived SCFAs. Aerobic exercise enhances intestinal barrier function via butyrate—a histone deacetylase inhibitor produced by commensals like *Faecalibacterium prausnitzii*. Butyrate upregulates tight junction proteins

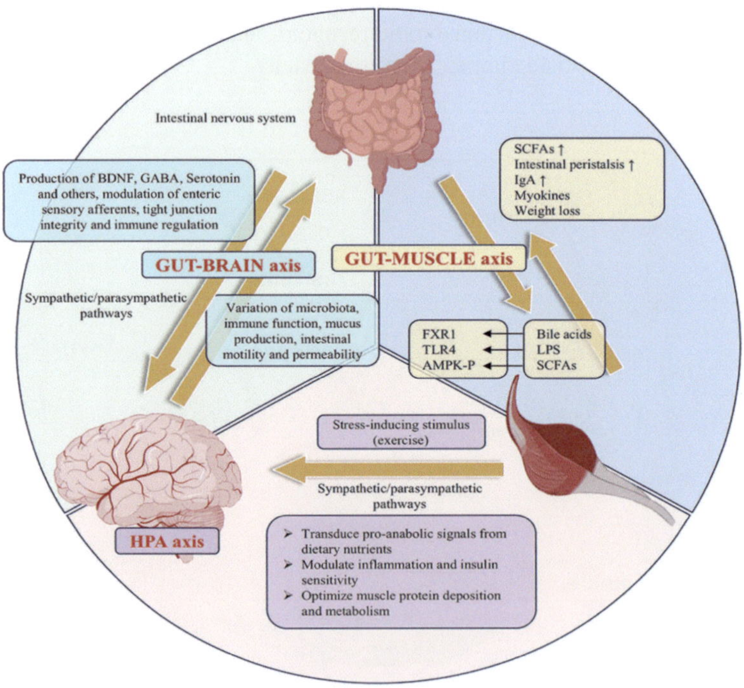

Fig. 11.6 Illustration of the gut–brain/muscle axis. Main interactions during exercise are presented

(occludin, ZO-1), suppresses NF-κB-mediated inflammation, and potentiates mucosal immunity through secretory IgA amplification. Paradoxically, high-intensity exercise (>60% VO_2max) activates HPA-axis-driven cortisol release, inducing transient gut hyperpermeability and microbial shifts. Chronic HPA overactivation may precipitate dysbiosis, underscoring the hormetic nature of exercise stress. SCFAs and bile acids mediate microbiota-muscle communication through AMPK activation, enhancing glucose utilization and lipid β-oxidation. Butyrate and propionate stimulate enteroendocrine L-cells to secrete peptide YY (PYY), improving insulin sensitivity and satiety. Exercise-induced myokines may exhibit dual anti-inflammatory roles: IL-6 triggers IL-10 production while suppressing TNF-α, and promotes AMPK-dependent fat catabolism. Notably, exercise mitigates TLR4-mediated muscle atrophy by reducing circulating LPS, thus countering metabolic endotoxemia. Catecholamine surges during acute stress or exhaustive exercise increase virulence gene expression in Salmonella and E. coli, while cortisol dysregulation depletes immunomodulatory taxa (e.g., *Lactobacillus*). Contrastingly, moderate exercise enriches butyrogenic microbiota (e.g., *F. prausnitzii*), stabilizes HPA reactivity, and enhances stress resilience—as evidenced by athletes' elevated SCFA levels and attenuated cortisol responses. Microbiota adaptations exhibit exercise-intensity gradients: moderate regimens increase SCFAs producers (e.g., *Bacteroidota*) and reduce bile acid hydrophobicity, whereas overtraining depletes mucin-degraders (e.g., *Akkermansia*). Bile acids (e.g., cholic and deoxycholic acid) could modulate microbial composition through FXR signaling, favoring *Bacillota* over *Bacteroidota* to regulate lipid metabolism.

In conclusion, a healthy gut microbiome can enhance exercise capacity via diverse mechanisms, such as energy metabolism, immune modulation, and the gut–brain–muscle axis (Dohnalová et al. 2022). These findings emphasize the significance of maintaining a balanced gut microbiota for achieving optimal physical performance and overall health. The implications of these findings extend beyond elite sports, suggesting novel approaches for metabolic rehabilitation and promoting healthy aging.

Parallel breakthroughs across global research consortia across the U.S., Japan, the U.K., and China, have crystallized a revolutionary "sports microbiome" framework, positioning microbial communities as master regulators of energy homeostasis, recovery, and peak performance. This collaborative effort underscores the importance of international partnerships in microbiome research, as highlighted in various studies and symposiums. For instance, the American Society for Microbiology Global Research Symposium on Microbes in Human Health, hosted in partnership with Tsinghua University, was held in Beijing, China, from 25 to 27 September 2024. The conference provided an international forum for microbiologists from different disciplines to present and discuss new findings, emphasizing the global nature of microbiome research. Moreover, the Tri-Service Microbiome Consortium symposiums have played a crucial role in enhancing collaboration and communication of microbiome research among U.S. Department of Defense organizations, facilitating the development of new collaborative research opportunities.

These symposiums have highlighted the potential for microbiome research to transition into practical solutions, including those applicable to sports and performance. Additionally, the National Microbiome Data Collaborative has been instrumental in enabling microbiome science by providing a platform for data sharing and collaboration, which is crucial for advancing research in the sports microbiome domain. The collaborative efforts between China and the U.S. in antimicrobial resistance research also highlight the potential for international partnerships to improve health outcomes (van Duin et al. 2018), which contribute to the development and refinement of the sports microbiome framework, paving the way for innovative applications in sports science and health.

11.7 Exercise, Gut Microbiota and Epigenetics: Novel Insights into Mechanisms of Health Promotion

Epigenetics is the study of changes in organisms caused by the modification of gene expression rather than alteration of the genetic code itself, according to the Oxford dictionary. Epigenetic modifications represent a fundamental mechanism by which mammalian cells can adapt their transcriptional program to environmental cues, including exercise. The predominant epigenetic mechanisms include DNA methylation, histone modifications such as acetylation, methylation, phosphorylation, SUMOylation and ubiquitination, non-coding RNAs (i.e., microRNAs, long non-coding RNAs, and piwi-interacting RNAs), as well as modifications to chromatin (McGee and Hargreaves 2019).

Beyond established bacterial pathways in metabolic regulations, the gut microbiota emerges as a master regulator of host epigenetic modifications (Fig. 11.7). These modifications, encompassing DNA and histone alterations, as well as microRNA regulation, have shown to directly calibrate host cell transcriptional programs without altering the genetic code (Woo and Alenghat 2022). This gut microbiota-driven epigenetic modifications have become a pervasive and fundamental regulatory mechanism, particularly critical in regulating host physiology, for example, aging, metabolic homeostasis, inflammation, and disease processes.

This chapter delves into the intricate relationships among gut microbiota, particularly those have shown to be modulated by exercise, and epigenetics. Beyond merely cataloging microbiota composition, we dissect how microbial metabolites such as SCFAs, lactate, and tryptophan metabolites that are remodeled by exercise, orchestrate host epigenetic reprogramming. The clinical implications of this tripartite relationship are profound. We collect evidence demonstrating how the exercise-microbiota-epigenetics nexus plays roles in maintaining immunoregulatory functions, regulating metabolic homeostasis, and affecting diseases progressions, establishing the gut microbiome as an epigenetic interface that translates lifestyle interventions into sustained physiological adaptations for health promotion.

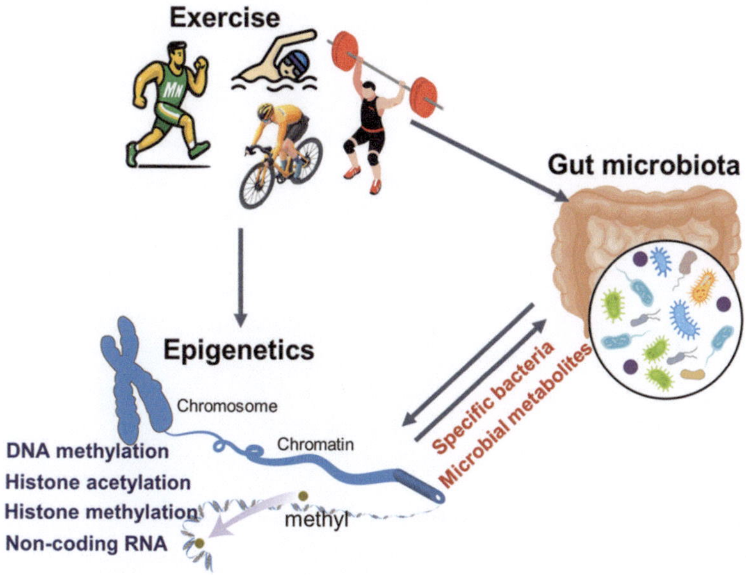

Fig. 11.7 Illustration of interplay between exercise, gut microbiota and epigenetics

11.7.1 Effects of Exercise-Associated Gut Microbiota on Epigenetic Regulations

The gut microbiota could induce specific epigenetic modifications that regulate host physiological functions through multiple mechanisms (Fig. 11.8, Table 11.1). Recent studies demonstrate that gut microbiota dysbiosis contributes to metabolic disorders by modulating organ-specific epigenetic mechanisms such as DNA methylation, histone modifications (e.g., H3 acetylation), and microRNA (miRNA) expressions. For example, disruption of the gut microbiota–miR-181 axis has been shown to drive white adipose tissue inflammation, obesity, and insulin resistance (Virtue et al. 2019). Circadian metabolic rhythms, which synchronize physiological processes with light–dark cycles, are also influenced by the gut microbiota. The histone deacetylase 3 (HDAC3) mediates microbiota-dependent regulation of circadian metabolism. Beyond its canonical role in deacetylation, HDAC3 coactivates estrogen-related receptor α (ERRα) and lipid transporter gene Cd36, thereafter enhancing lipid absorption and exacerbating diet-induced obesity (Kuang et al. 2019). Intestinal epithelial cells (IECs), as key players mediating nutrient absorption and maintain gut barrier integrity, are regulated by microbiota-dependent epigenetic mechanisms (Roostaee et al. 2016; Kraiczy and Zilbauer 2019). For instance, commensal bacteria such as *Bacteroides acidifaciens* and *Lactobacillus johnsonii* modulate miR-21-5p expression in IECs, thereby influencing intestinal permeability through ARF4 signaling (Nakata et al. 2017).

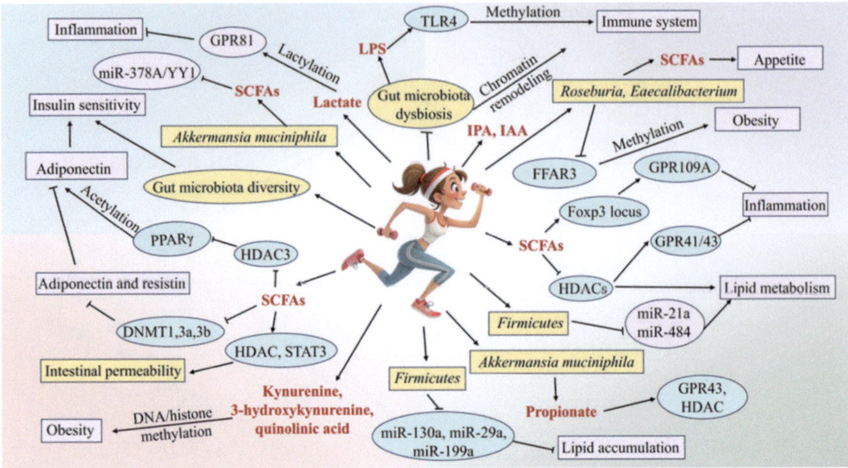

Fig. 11.8 Illustration of exercise associated gut microbiota and microbial metabolites on epigenetic factors

Table 11.1 Effects of gut microbiota and microbial metabolites on epigenetic regulations

Gut microbiota or its metabolites	Microbiota/metabolites-associated epigenetic changes	Effects on health or diseases	Reference
Gut dysbacteriasis	Affect DNA methylation	Promote the maturation of epithelial cells	Yu et al. (2015)
	Affect HDAC3	Control metabolic rhythms and promote obesity under high-fat diet condition	Kuang et al. (2019)
Gut dysbacteriasis	Promote miR-181	Contribute to the progression of obesity, insulin resistance, and inflammation	Virtue et al. (2019)
Aging-related gut dysbacteriasis	Inhibit M6A methylation	Accelerate intestinal aging and exacerbates chronic inflammation and metabolic disease	Ding et al. (2024)
Obesity-associated microbiota	Inhibit binding to HNF4α; Altered H3K27ac and H3K4me1 levels	Promote metabolic disorders (e.g., obesity, hyperlipidemia)	Qin et al. (2018)
Roseburia	Promote DNA methylation	Enhance anti-inflammatory, neuroprotective and antidepressant function; Alleviate Parkinson's disease	Xie et al. (2022)
	Correlated miR-425-3p and miR-638 positively	Maintain the metabolism and intestinal health	Tarallo et al. (2022)

(continued)

Table 11.1 (continued)

Gut microbiota or its metabolites	Microbiota/metabolites-associated epigenetic changes	Effects on health or diseases	Reference
Faecalibacterium prausnitzii	Inhibit HDAC Downregulate LCN2	Alleviate age-related heart failure	Zhang et al. (2025)
Akkermansia muciniphila and *Lactobacillus plantarum*	Enhance modification of N6-methyladenosine (m6A)	Maintain the metabolism and reduce inflammation of the host	Jabs et al. (2020)
Limosilactobacillus fermentum MTCC 5898 and *Lacticaseibacillus rhamnosus* MTCC 5897	Activate H3 histone acetylation	Maintain immunoregulatory functions in intestinal cells	Wieërs et al. (2019), Kumari et al. (2023)
Limosilactobacillus fermentum MTCC 5898	Enhance the transcriptional expression of DNA and histone modifiers (DNMT1, TET2, p300, HDAC1, KMT2A, KDM5B, EzH2 and JMJD3)	Maintain intestinal epithelial functions	Kumari et al. (2022)
Commensal microbiome-dependent (*Bacteroides acidifaciens* type A43 (BA) and *Lactobacillus johnsonii* 129 (LJ), etc.)	Promote miR-21-5p	Enhance intestinal epithelial permeability	Nakata et al. (2017)
Bacteroides, Bifidobacterium	Promote DNA methylation	Alleviate thyroid disorders	Ludgate et al. (2024)
Increased *Proteobacteria* phylum and Enterobacteriaceae family, decreased Ruminococcaceae family and Faecalibacterium genus	Modulate NAFLD-related gene expression through DNA methylation	Promote NAFLD development and progression	Tang et al. (2024)
Clostridiales and butyrate	Inhibit HDAC	Promote colonic Treg differentiation; Enhance anti-inflammatory function; Ameliorate experimental colitis; Maintain gut immune homeostasis	Furusawa et al. (2013)

(continued)

Table 11.1 (continued)

Gut microbiota or its metabolites	Microbiota/metabolites-associated epigenetic changes	Effects on health or diseases	Reference
Bacteroides thetaiotamicron	Activate HDAC3	Dampen diurnal rhythm	Kuang et al. (2019)
Bacteroidetes-to-Firmicutes ratio	Promote DNA methylation	Protect against obesity	Ramos-Molina et al. (2019)
Fusobacterium	Promote DNA methylation	Promote ulcerative colitis	Tahara et al. (2017)
Clostridium scindens	Reduce H3K27me3 at the C/EBPα promoter and increasing H3K4me3 at the C/EBPβ promoter	Enhance innate immunity via intestinal neutrophil recruitment; Protect against *Entamoeba histolytica* colitis.	Burgess et al. (2020)
	Inhibit histone deacetylase activity Promote the differentiation and anti-inflammatory function of regulatory T cells	Maintain immune homeostasis and intestinal health	Brown et al. (2003), Arpaia et al. (2013), Smith et al. (2013), Zeng et al. (2013), Park et al. (2015), and Kumar et al. (2017)
SCFAs	Inhibit HDAC3	Improve glycolipid metabolism and protect nerves	Kuang et al. (2019)
	Activate GPR43 (FFAR2) Inhibit H3K27me3 and H3K4me3 histone	Regulate colonic inflammation; Protect against colon carcinogenesis	Pan et al. (2018)
	Activate SLC5A8	Promote apoptosis in cancer cells; Maintain colonic energy metabolism and suppress tumorigenesis	Gopal et al. (2004), Miyauchi et al. (2004)
	SIRT1 influences DNA methylation by regulating DNMTs (DNA methyltransferases) and TET enzymes	Regulate mitochondrial biogenesis, metabolism, stress response, genome stability, and aging	Borra et al. (2005), Clark and Mach (2017)
	Activate Stat3 Inhibit HDACs	Promote the formation of epithelial barrier	Zheng et al. (2017)
	Promote histone crotonylation through histone deacetylases	Maintain intestinal homeostasis	Fellows et al. (2018)
	Inhibit HDAC1 and HDAC2	Inhibit cell proliferation, promote differentiation, treat cancer	Hamer et al. (2008), Poupeau et al. (2019)
	Inhibit HDAC3	Enhance the intestinal resistance to pathogens; Prevent obesity	Whitt et al. (2018), Schulthess et al. (2019)

(continued)

Table 11.1 (continued)

Gut microbiota or its metabolites	Microbiota/metabolites-associated epigenetic changes	Effects on health or diseases	Reference
Butyrate	Downregulate miR-24	Reduce the apoptosis in cancer cells	Bartley et al. (2011), Xie et al. (2013)
	FFAR3 and LINE1 gene hypomethylation	Affect metabolic diseases (e.g., obesity and type 2 diabetes)	Remely et al. (2014)
	Reduce the deposition of H3K27me3	Increase H3K27me3 of NFκB1 promoter in colon tissue; Reduce intestinal inflammation	Liu et al. (2016)
	Reduce the levels of the miR-17-92a (oncomiR-1)	Reduce the proliferation and apoptosis of colon cancer cells; Stimulate apoptosis	Hu et al. (2015)
	Inhibit HDAC	Promote pancreatic cancer development and immunosuppressive microenvironment formation	Elango et al. (2024)
Acetate and propionate	Inhibit HDAC2 and HDAC3	Reduce colon inflammation, hepatic lipid metabolism disorders; Protect against neurodegenerative diseases	Davie (2003)
Butyrate and propionate	Inhibit HDAC	Inhibit the production of autoantibodies and autoimmunity in mice with lupus erythematosus	Sanchez et al. (2020)
Choline metabolites	Methyl donors for DNA methylation	Affect the susceptibility to colitis and metabolic diseases	Kok et al. (2018), Wu et al. (2020)
Indole propionic acid, indole acetic acid	Activation of AhR receptors regulate DNA methylation and histone modification	Inhibit intestinal inflammation; Protect against IBD	Li et al. (2021), Akhtar et al. (2022)
	DNA methylation	Induce apoptosis; Reduce mitochondrial respiration as well as modify cell morphology, and mitochondrial dynamics	Ilha et al. (2025)

(continued)

Table 11.1 (continued)

Gut microbiota or its metabolites	Microbiota/metabolites-associated epigenetic changes	Effects on health or diseases	Reference
Indole-3-propionic acid	Enhance mitochondrial transcription factor A (Tfam) expression Promote Kdm6b/Jmjd3 histone demethylase Inhibit H3K27me3 epigenetic methylation at the Tfam promoter	Stimulate bone formation in obese mice via histone methylations	Behera et al. (2021)
Kynurenine	Activate AHR and indoleamine 2,3-dioxygenase (IDO) methylation	Promote maternal inflammation; Increase risk of obesity in offspring	Gharipour et al. (2025)
3-OH-Kynurenine	Activate H3K4me3 and H2AS40Gc	Regulate metabolism and may contribute to neurodegenerative diseases	Hayakawa et al. (2019)
	Activate H3K9 acetylation, H3K18 lactylation	Mitigate colitis	Sun et al. (2021)
	Activate H3K18 lactylation	Alleviate intestinal inflammation and autoimmune diseases	Lopez Krol et al. (2022)
Lactate	Activate histone lactylation (H3K18la, H4K12la, H3K9la) and non-histone lactylation (PKM2, P4HB, SNAP91)	Enhance insulin sensitivity, reduce chronic inflammation, improve metabolic flexibility; Lead to metabolic disorders and organ damage	Chen et al. (2025)

Moreover, N6-methyladenosine (m6A), the most abundant RNA modification in eukaryotes, affects approximately 30–50% of mammalian mRNAs and plays critical roles in RNA metabolism. Given its crucial role in essential biological processes, including stem cell differentiation, cell division, and biorhythms, m6A methylation is intimately associated with numerous human diseases, garnering significant attention over the past two decades. A study by Jabs et al. (2020) (Jabs et al. 2020) revealed that the gut microbiota directly shapes the m6A epitranscriptome in murine cecum and liver. Importantly, *Akkermansia muciniphila* and *Lactobacillus plantarum* were found to modulate host m6A modifications, impacting cellular proliferation, growth, and apoptosis. This highlights epitranscriptomic regulation as a novel axis of host-microbiota crosstalk. M6A methylation is also implicated in the regulation of gut aging and may serve as a signaling molecule facilitating communication between the host and microbial communities. Besides, emerging research has revealed a sophisticated bidirectional relationship between RNA m6A methylation and gut microbiota homeostasis, with profound implications for intestinal aging and systemic age-related conditions, including chronic infections, gastrointestinal malignancies, nutritional deficiencies, chronic constipation, and metabolic dysfunctions (Ding et al. 2024). With 901 million individuals aged 60 or older, accounting for 12.3% of the global population and this proportion expected to rise sharply, targeting the gut microbiome-m6A methylation axis is being considered as a strategy to alleviate the detrimental effects of flora-related gut aging, presenting a novel approach to promoting longevity.

Beyond the pathways above-mentioned, current research has identified specific commensal bacteria that exert profound epigenetic influences on host physiology (Table 11.1). They include *Bacteroides, Bifidobacterium, Faecalibacterium prausnitzii, Roseburia, Fusobacterium,* and several species belonging to the *Limosilactobacillus*. Mechanisms involve the regulation of DNA methylation, HDAC3, H3K27me3 and H3K4me3 histone, as well as modifying microRNAs. These microbiota-epigenetic interactions occur in a tissue-specific manner, with particularly strong effects observed in the intestinal epithelium, liver, and adipose tissue, which represent promising therapeutic targets for various metabolic diseases, although further research is needed to fully elucidate the complex interactions and develop targeted interventions.

11.7.2 Effects of Exercise-Associated Microbial Metabolites on Epigenetic Regulations

The gut microbiota plays a pivotal role in shaping the host's epigenetic landscape through metabolite-mediated mechanisms. Notably, SCFAs that are produced by commensal bacteria such as Clostridia, Bacteroides and many Negativicutes (such as *Veilonella*, *Roseburia*, and *Ruminococcus*) are well-known to adjust the host's homeostasis by regulating epigenetic pathways (Wu et al. 2021; Li et al. 2022) (Fig. 11.8, Table 11.1). The most abundant SCFAs in the colon are acetate, butyrate, and propionate. SCFAs are enriched by exercise which could help to maintain skeletal muscle homeostasis and improve exercise performance (Liu et al. 2025). Briefly, acetate serves as a potent HDAC inhibitor, which has shown to effectively elevate global histone acetylation levels and enhancing chromatin accessibility. Butyrate can induce the differentiation of colonic Treg cells in mice, and it enhances histone H3 acetylation within the promoter and conserved non-coding sequences of the Foxp3 gene. Propionate and butyrate both inhibit HDAC activity in IECs and immune cells by promoting hyperacetylation of histones and specific transcription factors involved in signal transduction, thereby playing a crucial role in cancer development. IECs absorb acetate and propionate through passive diffusion, electroneutralization, or active transport, and these short-chain fatty acids are subsequently transported to peripheral tissues. SCFAs also downregulate the expression of DNMTs (DNMT1, DNMT3A, and DNMT3B) and methyl-CpG-binding domain protein 2 (MBD2), inhibiting their binding to the promoters of adiponectin and resistin genes, thereby mitigating obesity and associated comorbidities. Although their mechanism of action is not fully understood, SCFAs are thought to modulate host cellular processes through activation of G-protein-coupled-receptors (GPCRs), in particular GPR41 and GPR43. These receptors are integral to various physiological functions, including immune regulation and metabolic processes. The interaction between SCFAs and GPCRs represents a critical mechanism by which gut microbiota can influence host physiology, offering potential therapeutic avenues for a range of diseases.

Lactate, traditionally dismissed as a glycolytic waste product, has been redefined as a dynamic signaling molecule and master regulator of epigenetic adaptations. Emerging research reveals its central role in coordinating systemic metabolic responses through histone lactylation, a novel post-translational modification directly linking cellular metabolism to gene regulation, which has intrinsic connections with cell metabolism that represents a novel epigenetic code affecting the fate of cells including carcinogenesis (Lopez Krol et al. 2022; Wang et al. 2023).

Exercise effectively remodels lactate flux and lactylation dynamics. The dynamic interplay among exercise, lactate metabolism, and lactylation modifications serves as a pivotal axis in understanding exercise-induced protection against metabolic disorders. For instance, during high-intensity exercise, rapid glycolysis generates the accumulation of transient lactate (5–10 mM). Conversely, moderate-intensity exercise improves the mitochondrial oxidative capacity and reduces chronic lactate overproduction, thereby enhancing metabolic flexibility and lactate clearance. Regular moderate-intensity exercise lowers baseline blood lactate levels in type 2 diabetes patients, improving their exercise tolerance and glycemic control (Huebschmann et al. 2024). Lactate has also emerged as a multifaceted immunomodulator through both GPR81-dependent and -independent mechanisms (Bartoloni et al. 2024). GPR81 is expressed in the skeletal muscle, liver, spleen, kidney, and brain. A lactate concentration ranging from 0.2 to 1.0 mM activates GPR81, leading to downregulation of cyclic AMP (cAMP) and inhibition of protein kinase A (PKA)-mediated signaling. Recent research demonstrated that lactate-induced activation of GPR81 signals via a non-canonical cAMP/PKA-independent pathway, involving beta-arrestin, an adaptor protein for GPR81. This results in the inhibition of TLR-4 and NLRP3 inflammasome, thereby downregulating production of proinflammatory cytokines, including IL-1β and IL-18. Besides, lactate enhances the production of anti-inflammatory IL-10 in dendritic cells and reduces the production of proinflammatory IL-12 in response to TLR agonists. Interplay between lactate metabolism and immune regulation underscores its evolution from a simple glycolytic byproduct to a central coordinator of inflammatory responses with far-reaching clinical implications.

Moreover, activation of the aryl hydrocarbon receptor (AhR) /CYP1A1 pathway has shown to affect tumorigenesis through modulation of DNA methylation, histone acetylation/deacetylation, and expression of several miRNAs (Akhtar et al. 2022). The gut microbiota-derived tryptophan metabolites, such as indole-3-propionic acid

(IPA) and indole acetic acid (IAA), serve as potent endogenous ligands for AhR, bridging microbial metabolism with host epigenetic regulation. Administration of IPA has shown to upregulate mitochondrial transcription factor A (Tfam) in osteoblasts by activating the histone demethylase Kdm6b/Jmjd3 and inhibiting H3K27 trimethylation (H3K27me3) at the Tfam promoter. This, in turn, ameliorated obesity-induced impairments in osteoblast function, ensuring normal glucose uptake, mitochondrial biogenesis, and mineralization. Most recently, it is reported that serum IPA levels were lower in individuals with liver fibrosis, while had positive association with expression of liver genes enriched for apoptosis, mitophagy, and longevity pathways (Ilha et al. 2025). Specifically, AKT serine/threonine kinase 1 (AKT1) was identified as a key interactive gene in both liver transcriptomic and DNA methylation profiles. In vitro studies using LX-2 cell models suggest that IPA may inhibit hepatic fibrogenesis by modulating hepatic stellate cell activation and mitochondrial metabolism. Besides, in a recent review focusing on the impact of maternal inflammation on childhood obesity risk, kynurenine, another critical tryptophan metabolite, has been highlighted as a novel epigenetic biomarker and a promising therapeutic target for mitigating maternal inflammation and preventing offspring obesity (Gharipour et al. 2025). This might be attributed to its ability to activate AHR and promote methylation of indoleamine 2,3-dioxygenase (IDO). Collectively, these findings highlight the critical role of microbiota-derived tryptophan metabolites in modulating epigenetic pathways to preserve metabolic health under diverse physiological conditions.

Additionally, certain gut bacteria can synthesize bioactive compounds that serve as essential substrates for epigenetic regulations, such as methyl or acetyl groups, which are critical for DNA methylation and histone modifications in relation with both physiological and pathological processes in the host. Epigenetic-modifying enzymes depend on the availability of specific substrates to mediate chromatin remodeling. For instance, DNA methyltransferases (DNMTs) and histone methyltransferases (HMTs) require methyl donors, while histone acetyltransferases (HATs) rely on acetyl-CoA for their catalytic activity. Emerging evidence highlights the gut microbiota as a key contributor to the pool of methyl donors, as microbial metabolism of dietary and endogenous compounds directly modulates the methionine cycle and the bioavailability of S-adenosylmethionine (SAM), the universal methyl donor. Additionally, certain commensal microbes, including probiotic *Bifidobacterium* and *Lactobacillus* species, produce folate, a vital nutrient involved in one-carbon metabolism. This pathway is essential for SAM synthesis, which in turn serves as the primary methyl donor for both DNA and histone methylation.

The interplay between exercise, gut microbiota, epigenetics, and metabolic health is complex, and remains incompletely understood. Emerging evidence indicates that exercise induces significant remodeling of the gut microbiome, with cascading effects on host metabolism. Despite direct influence induced by exercise on epigenetics factors, the gut microbiota and microbial metabolites, such as SCFAs, LPS, and tryptophan metabolites, could modify epigenetic marks (e.g., DNA methylation, histone modifications) that govern key metabolic processes, such as inflammation, immune function, glucose/lipid homeostasis, and intestinal barrier integrity.

Collectively, this microbiota-epigenetics crosstalk mediates, at least in part, the beneficial effects of exercise on health promotion and disease prevention (Fig. 11.8).

Notably, the profound influence of exercise on the gut microbiota and epigenetic modifications presents a compelling rationale for developing precision exercise interventions tailored to an individual's microbial and epigenetic profile, which holds great potentials to optimize metabolic health, training adaptation, and disease prevention. For instance, a miRNA-based fitness score was proposed to assess an individual's response to nutrition, exercise, and metabolic influence, offering insights into training efficacy, body composition, recovery needs, and nutritional requirements. In metabolic disease management, integrating multi-omics data (i.e., genomics, epigenetics, and gut microbiota) might revolutionize intervention strategies by enabling personalized exercise prescriptions. Moreover, the Developmental Origins of Health and Disease program highlights the critical role of maternal exercise in shaping offspring metabolism through gut microbiota-metabolite-epigenetic crosstalk, suggesting early-life interventions may have long-term health benefits (Ren et al. 2024). Another key perspective lies in early disease prevention. Deciphering how exercise modulates the microbiota-epigenetics axis during predisease stages could uncover novel biomarkers for risk stratification and facilitate timely and effective interventions. By bridging these research domains, we can advance toward precision exercise medicine, where interventions are dynamically adjusted based on real-time microbial and epigenetic feedback, ultimately enabling more effective, individualized healthcare strategies for chronic disease prevention and health optimization.

Compliance with Ethical Standards

Competing Interests All authors declare no conflict of interest.

Ethical Approval This chapter is a review of previously published accounts; as such, no animal or human studies were performed.

References

Akhtar S, Hourani S, Therachiyil L, Al-Dhfyan A, Agouni A, Zeidan A, Uddin S, Korashy HM (2022) Epigenetic regulation of cancer stem cells by the aryl hydrocarbon receptor pathway. Semin Cancer Biol 83:177–196. https://doi.org/10.1016/j.semcancer.2020.08.014

Allen JM, Mailing LJ, Niemiro GM, Moore R, Cook MD, White BA, Holscher HD, Woods JA (2018) Exercise alters gut microbiota composition and function in lean and obese humans. Med Sci Sports Exerc 50(4):747–757. https://doi.org/10.1249/mss.0000000000001495

Arpaia N, Campbell C, Fan X, Dikiy S, van der Veeken J, deRoos P, Liu H, Cross JR, Pfeffer K, Coffer PJ, Rudensky AY (2013) Metabolites produced by commensal bacteria promote peripheral regulatory T-cell generation. Nature 504(7480):451–455. https://doi.org/10.1038/nature12726

Bai J, Hu Y, Bruner DW (2019) Composition of gut microbiota and its association with body mass index and lifestyle factors in a cohort of 7-18 years old children from the American gut project. Pediatr Obes 14(4):e12480. https://doi.org/10.1111/ijpo.12480

Baldanzi G, Sayols-Baixeras S, Ekblom-Bak E, Ekblom Ö, Dekkers KF, Hammar U, Nguyen D, Ahmad S, Ericson U, Arvidsson D, Börjesson M, Johanson PJ, Smith JG, Bergström G, Lind L, Engström G, Ärnlöv J, Kennedy B, Orho-Melander M, Fall T (2024) Accelerometer-based physical activity is associated with the gut microbiota in 8416 individuals in SCAPIS. EBioMedicine 100:104989. https://doi.org/10.1016/j.ebiom.2024.104989

Bartley AN, Yao H, Barkoh BA, Ivan C, Mishra BM, Rashid A, Calin GA, Luthra R, Hamilton SR (2011) Complex patterns of altered MicroRNA expression during the adenoma-adenocarcinoma sequence for microsatellite-stable colorectal cancer. Clin Cancer Res 17(23):7283–7293. https://doi.org/10.1158/1078-0432.Ccr-11-1452

Bartoloni B, Mannelli M, Gamberi T, Fiaschi T (2024) The multiple roles of lactate in the skeletal muscle. Cells 13(14). https://doi.org/10.3390/cells13141177

Behera J, Ison J, Voor MJ, Tyagi N (2021) Probiotics stimulate bone formation in obese mice via histone methylations. Theranostics 11(17):8605–8623. https://doi.org/10.7150/thno.63749

Biedert P (1880) Die Kinderernährung im Säuglingsalter. Verlag von Ferdinand Enke

Bielik V, Hric I, Ugrayová S, Kubáňová L, Putala M, Grznár Ľ, Penesová A, Havranová A, Šardzíková S, Grendar M, Baranovičová E, Šoltys K, Kolisek M (2022) Effect of high-intensity training and probiotics on gut microbiota diversity in competitive swimmers: randomized controlled trial. Sports Med Open 8(1):64. https://doi.org/10.1186/s40798-022-00453-8

Borra MT, Smith BC, Denu JM (2005) Mechanism of human SIRT1 activation by resveratrol. J Biol Chem 280(17):17187–17195. https://doi.org/10.1074/jbc.M501250200

Brown AJ, Goldsworthy SM, Barnes AA, Eilert MM, Tcheang L, Daniels D, Muir AI, Wigglesworth MJ, Kinghorn I, Fraser NJ, Pike NB, Strum JC, Steplewski KM, Murdock PR, Holder JC, Marshall FH, Szekeres PG, Wilson S, Ignar DM, Foord SM, Wise A, Dowell SJ (2003) The Orphan G protein-coupled receptors GPR41 and GPR43 are activated by propionate and other short chain carboxylic acids. J Biol Chem 278(13):11312–11319. https://doi.org/10.1074/jbc.M211609200

Burgess SL, Leslie JL, Uddin J, Oakland DN, Gilchrist C, Moreau GB, Watanabe K, Saleh M, Simpson M, Thompson BA, Auble DT, Turner SD, Giallourou N, Swann J, Pu Z, Ma JZ, Haque R, Petri WA Jr (2020) Gut microbiome communication with bone marrow regulates susceptibility to amebiasis. J Clin Invest 130(8):4019–4024. https://doi.org/10.1172/jci133605

Chen G, Liu J, Guo Y, Sun P (2025) Mechanisms for regulatory effects of exercise on metabolic diseases from the lactate-lactylation perspective. Int J Mol Sci 26(8). https://doi.org/10.3390/ijms26083469

Clark A, Mach N (2017) The crosstalk between the gut microbiota and mitochondria during exercise. Front Physiol 8:319. https://doi.org/10.3389/fphys.2017.00319

Clarke SF, Murphy EF, O'Sullivan O, Lucey AJ, Humphreys M, Hogan A, Hayes P, O'Reilly M, Jeffery IB, Wood-Martin R, Kerins DM, Quigley E, Ross RP, O'Toole PW, Molloy MG, Falvey E, Shanahan F, Cotter PD (2014) Exercise and associated dietary extremes impact on gut microbial diversity. Gut 63(12):1913–1920. https://doi.org/10.1136/gutjnl-2013-306541. Available at: https://gut.bmj.com/content/gutjnl/63/12/1913.full.pdf

Collado MC, Isolauri E, Laitinen K, Salminen S (2008) Distinct composition of gut microbiota during pregnancy in overweight and normal-weight women. Am J Clin Nutr 88(4):894–899. https://doi.org/10.1093/ajcn/88.4.894

Davie JR (2003) Inhibition of histone deacetylase activity by butyrate. J Nutr 133(7 Suppl):2485s–2493s. https://doi.org/10.1093/jn/133.7.2485S

Ding M, Yan J, Chen Y, Liu J, Chao G, Zhang S (2024) Changes in M6A methylation: a key factor in the vicious cycle of flora -gut aging. Ageing Res Rev 98:102351. https://doi.org/10.1016/j.arr.2024.102351

Dohnalová L, Lundgren P, Carty JRE, Goldstein N, Wenski SL, Nanudorn P, Thiengmag S, Huang KP, Litichevskiy L, Descamps HC, Chellappa K, Glassman A, Kessler S, Kim J, Cox TO,

Dmitrieva-Posocco O, Wong AC, Allman EL, Ghosh S, Sharma N, Sengupta K, Cornes B, Dean N, Churchill GA, Khurana TS, Sellmyer MA, FitzGerald GA, Patterson AD, Baur JA, Alhadeff AL, Helfrich EJN, Levy M, Betley JN, Thaiss CA (2022) A microbiome-dependent gut-brain pathway regulates motivation for exercise. Nature 612(7941):739–747. https://doi.org/10.1038/s41586-022-05525-z

Donati Zeppa S, Amatori S, Sisti D, Gervasi M, Agostini D, Piccoli G, Pazienza V, Gobbi P, Rocchi MBL, Sestili P, Stocchi V (2021) Nine weeks of high-intensity indoor cycling training induced changes in the microbiota composition in non-athlete healthy male college students. J Int Soc Sports Nutr 18(1):74. https://doi.org/10.1186/s12970-021-00471-z

van Duin D, Gu P, Dong J, Paff M, Arias RM, Evans B, Yu Y, Li L, Zhang F, Liu Z, Cao B, Fowler VG Jr, Wang M (2018) China-United States research collaborations in antimicrobial resistance. Clin Infect Dis 67(suppl_2):S142–s145. https://doi.org/10.1093/cid/ciy694

Elango A, Nesam VD, Sukumar P, Lawrence I, Radhakrishnan A (2024) Postbiotic butyrate: role and its effects for being a potential drug and biomarker to pancreatic cancer. Arch Microbiol 206(4):156. https://doi.org/10.1007/s00203-024-03914-8

Fellows R, Denizot J, Stellato C, Cuomo A, Jain P, Stoyanova E, Balázsi S, Hajnády Z, Liebert A, Kazakevych J, Blackburn H, Corrêa RO, Fachi JL, Sato FT, Ribeiro WR, Ferreira CM, Perée H, Spagnuolo M, Mattiuz R, Matolcsi C, Guedes J, Clark J, Veldhoen M, Bonaldi T, Vinolo MAR, Varga-Weisz P (2018) Microbiota derived short chain fatty acids promote histone crotonylation in the colon through histone deacetylases. Nat Commun 9(1):105. https://doi.org/10.1038/s41467-017-02651-5

Fielding RA, Reeves AR, Jasuja R, Liu C, Barrett BB, Lustgarten MS (2019) Muscle strength is increased in mice that are colonized with microbiota from high-functioning older adults. Exp Gerontol 127:110722. https://doi.org/10.1016/j.exger.2019.110722

Furusawa Y, Obata Y, Fukuda S, Endo TA, Nakato G, Takahashi D, Nakanishi Y, Uetake C, Kato K, Kato T, Takahashi M, Fukuda NN, Murakami S, Miyauchi E, Hino S, Atarashi K, Onawa S, Fujimura Y, Lockett T, Clarke JM, Topping DL, Tomita M, Hori S, Ohara O, Morita T, Koseki H, Kikuchi J, Honda K, Hase K, Ohno H (2013) Commensal microbe-derived butyrate induces the differentiation of colonic regulatory T cells. Nature 504(7480):446–450. https://doi.org/10.1038/nature12721

Gharipour M, Craig JM, Stephenson G (2025) Epigenetic programming of obesity in early life through modulation of the kynurenine pathway. Int J Obes 49(1):49–53. https://doi.org/10.1038/s41366-024-01647-8

Goodsir J, Wilson G (1842) History of a case in which a fluid periodically ejected from the stomach contained vegetable organisms of an undescribed form. Edinb Med Surg J 57(151):430–443

Gopal E, Fei YJ, Sugawara M, Miyauchi S, Zhuang L, Martin P, Smith SB, Prasad PD, Ganapathy V (2004) Expression of slc5a8 in kidney and its role in Na(+)-coupled transport of lactate. J Biol Chem 279(43):44522–44532. https://doi.org/10.1074/jbc.M405365200

van de Haar J, Ma X, Ooft SN, van der Helm PW, Hoes LR, Mainardi S, Pinato DJ, Sun K, Salvatore L, Tortora G, Zurlo IV, Leo S, Giampieri R, Berardi R, Gelsomino F, Merz V, Mazzuca F, Antonuzzo L, Rosati G, Stavraka C, Ross P, Rodriquenz MG, Pavarana M, Messina C, Iveson T, Zoratto F, Thomas A, Fenocchio E, Ratti M, Depetris I, Cergnul M, Morelli C, Libertini M, Parisi A, De Tursi M, Zanaletti N, Garrone O, Graham J, Longarini R, Gobba SM, Petrillo A, Tamburini E, La Verde N, Petrelli F, Ricci V, Wessels LFA, Ghidini M, Cortellini A, Voest EE, Valeri N (2023) Codon-specific KRAS mutations predict survival benefit of trifluridine/tipiracil in metastatic colorectal cancer. Nat Med 29(3):605–614. https://doi.org/10.1038/s41591-023-02240-8

Hamer HM, Jonkers D, Venema K, Vanhoutvin S, Troost FJ, Brummer RJ (2008) Review article: the role of butyrate on colonic function. Aliment Pharmacol Ther 27(2):104–119. https://doi.org/10.1111/j.1365-2036.2007.03562.x

Hayakawa K, Nishitani K, Tanaka S (2019) Kynurenine, 3-OH-kynurenine, and anthranilate are nutrient metabolites that alter H3K4 trimethylation and H2AS40 O-GlcNAcylation at hypothalamus-related loci. Sci Rep 9(1):19768. https://doi.org/10.1038/s41598-019-56341-x

Hu S, Liu L, Chang EB, Wang JY, Raufman JP (2015) Butyrate inhibits pro-proliferative miR-92a by diminishing c-Myc-induced miR-17-92a cluster transcription in human colon cancer cells. Mol Cancer 14:180. https://doi.org/10.1186/s12943-015-0450-x

Huebschmann AG, Scalzo RL, Yang X, Schmiege SJ, Reusch JEB, Dunn AL, Chapman K, Regensteiner JG (2024) Type 2 diabetes is linked to higher physiologic markers of effort during exercise. Front Clin Diabetes Healthc 5:1346716. https://doi.org/10.3389/fcdhc.2024.1346716

Ilha M, Sehgal R, Matilainen J, Rilla K, Kaminska D, Gandhi S, Männistö V, Ling C, Romeo S, Pajukanta P, Pirinen E, Virtanen KA, Pietiläinen KH, Vaittinen M, Pihlajamäki J (2025) Indole-3-propionic acid promotes hepatic stellate cells inactivation. J Transl Med 23(1):253. https://doi.org/10.1186/s12967-025-06266-z

Jabs S, Biton A, Bécavin C, Nahori MA, Ghozlane A, Pagliuso A, Spanò G, Guérineau V, Touboul D, Giai Gianetto Q, Chaze T, Matondo M, Dillies MA, Cossart P (2020) Impact of the gut microbiota on the m(6)A epitranscriptome of mouse cecum and liver. Nat Commun 11(1):1344. https://doi.org/10.1038/s41467-020-15126-x

Kok DE, Steegenga WT, McKay JA (2018) Folate and epigenetics: why we should not forget bacterial biosynthesis. Epigenomics 10(9):1147–1150. https://doi.org/10.2217/epi-2018-0117

Kraiczy J, Zilbauer M (2019) Intestinal epithelial organoids as tools to study epigenetics in gut health and disease. Stem Cells Int 2019:7242415. https://doi.org/10.1155/2019/7242415

Kuang Z, Wang Y, Li Y, Ye C, Ruhn KA, Behrendt CL, Olson EN, Hooper LV (2019) The intestinal microbiota programs diurnal rhythms in host metabolism through histone deacetylase 3. Science 365(6460):1428–1434. https://doi.org/10.1126/science.aaw3134

Kumar P, Gogulamudi VR, Periasamy R, Raghavaraju G, Subramanian U, Pandey KN (2017) Inhibition of HDAC enhances STAT acetylation, blocks NF-κB, and suppresses the renal inflammation and fibrosis in Npr1 haplotype male mice. Am J Physiol Renal Physiol 313(3):F781–f795. https://doi.org/10.1152/ajprenal.00166.2017

Kumari A, Bhawal S, Kapila S, Kapila R (2022) Strain-specific effects of probiotic lactobacilli on mRNA expression of epigenetic modifiers in intestinal epithelial cells. Arch Microbiol 204(7):411. https://doi.org/10.1007/s00203-022-03027-0

Kumari A, Bhawal S, Kapila S, Kapila R (2023) Probiotic lactobacilli mediate their immunoregulatory functions in intestinal cells via modulation of H3 histone acetylation. J Appl Microbiol 134(2). https://doi.org/10.1093/jambio/lxac045

Langsetmo L, Johnson A, Demmer RT, Fino N, Orwoll ES, Ensrud KE, Hoffman AR, Cauley JA, Shmagel A, Meyer K, Shikany JM (2019) The association between objectively measured physical activity and the gut microbiome among older community dwelling men. J Nutr Health Aging 23(6):538–546. https://doi.org/10.1007/s12603-019-1194-x

Lee MC, Chen MJ, Huang HW, Wu WK, Lee YW, Kuo HC, Huang CC (2022) Probiotic Lactiplantibacillus plantarum Tana isolated from an international weightlifter enhances exercise performance and promotes antifatigue effects in mice. Nutrients 14(16). https://doi.org/10.3390/nu14163308

Li K, Hao Z, Du J, Gao Y, Yang S, Zhou Y (2021) Bacteroides thetaiotaomicron relieves colon inflammation by activating aryl hydrocarbon receptor and modulating CD4(+)T cell homeostasis. Int Immunopharmacol 90:107183. https://doi.org/10.1016/j.intimp.2020.107183

Li D, Li Y, Yang S, Lu J, Jin X, Wu M (2022) Diet-gut microbiota-epigenetics in metabolic diseases: from mechanisms to therapeutics. Biomed Pharmacother 153:113290. https://doi.org/10.1016/j.biopha.2022.113290

Li Y, Cheng M, Zha Y, Yang K, Tong Y, Wang S, Lu Q, Ning K (2023) Gut microbiota and inflammation patterns for specialized athletes: a multi-cohort study across different types of sports. mSystems 8(4):e0025923. https://doi.org/10.1128/msystems.00259-23

Liu Y, Upadhyaya B, Fardin-Kia AR, Juenemann RM, Dey M (2016) Dietary resistant starch type 4-derived butyrate attenuates nuclear factor-kappa-B1 through modulation of histone H3 trimethylation at lysine 27. Food Funct 7(9):3772–3781. https://doi.org/10.1039/c6fo00856a

Liu X, Xu M, Wang H, Zhu L (2025) Role and mechanism of short-chain fatty acids in skeletal muscle homeostasis and exercise performance. Nutrients 17(9). https://doi.org/10.3390/nu17091463

Lopez Krol A, Nehring HP, Krause FF, Wempe A, Raifer H, Nist A, Stiewe T, Bertrams W, Schmeck B, Luu M, Leister H, Chung HR, Bauer UM, Adhikary T, Visekruna A (2022) Lactate induces metabolic and epigenetic reprogramming of pro-inflammatory Th17 cells. EMBO Rep 23(12):e54685. https://doi.org/10.15252/embr.202254685

López-García P, Zivanovic Y, Deschamps P, Moreira D (2015) Bacterial gene import and mesophilic adaptation in archaea. Nat Rev Microbiol 13(7):447–456

Ludgate ME, Masetti G, Soares P (2024) The relationship between the gut microbiota and thyroid disorders. Nat Rev Endocrinol 20(9):511–525. https://doi.org/10.1038/s41574-024-01003-w

Magzal F, Shochat T, Haimov I, Tamir S, Asraf K, Tuchner-Arieli M, Even C, Agmon M (2022) Increased physical activity improves gut microbiota composition and reduces short-chain fatty acid concentrations in older adults with insomnia. Sci Rep 12(1):2265. https://doi.org/10.1038/s41598-022-05099-w

McGee SL, Hargreaves M (2019) Epigenetics and exercise. Trends Endocrinol Metab 30(9):636–645. https://doi.org/10.1016/j.tem.2019.06.002

Miyauchi S, Gopal E, Fei YJ, Ganapathy V (2004) Functional identification of SLC5A8, a tumor suppressor down-regulated in colon cancer, as a Na(+)-coupled transporter for short-chain fatty acids. J Biol Chem 279(14):13293–13296. https://doi.org/10.1074/jbc.C400059200

Motiani KK, Collado MC, Eskelinen JJ, Virtanen KA, Löyttyniemi E, Salminen S, Nuutila P, Kalliokoski KK, Hannukainen JC (2020) Exercise training modulates gut microbiota profile and improves endotoxemia. Med Sci Sports Exerc 52(1):94–104. https://doi.org/10.1249/mss.0000000000002112

Nakata K, Sugi Y, Narabayashi H, Kobayakawa T, Nakanishi Y, Tsuda M, Hosono A, Kaminogawa S, Hanazawa S, Takahashi K (2017) Commensal microbiota-induced microRNA modulates intestinal epithelial permeability through the small GTPase ARF4. J Biol Chem 292(37):15426–15433. https://doi.org/10.1074/jbc.M117.788596

Nechalová L, Bielik V, Hric I, Babicová M, Baranovičová E, Grendár M, Koška J, Penesová A (2024) Gut microbiota and metabolic responses to a 12-week caloric restriction combined with strength and HIIT training in patients with obesity: a randomized trial. BMC Sports Sci Med Rehabil 16(1):239. https://doi.org/10.1186/s13102-024-01029-7

O'Donovan CM, Madigan SM, Garcia-Perez I, Rankin A, O'Sullivan O, Cotter PD (2020) Distinct microbiome composition and metabolome exists across subgroups of elite Irish athletes. J Sci Med Sport 23(1):63–68. https://doi.org/10.1016/j.jsams.2019.08.290. Accessed 2025/03/17

Ortiz-Alvarez L, Xu H, Ruiz-Campos S, Acosta FM, Migueles JH, Vilchez-Vargas R, Link A, Plaza-Díaz J, Gil A, Labayen I, Ruiz JR, Martinez-Tellez B (2025) Higher physical activity levels are related to faecal microbiota diversity and composition in young adults. Biol Sport 42(1):123–135. https://doi.org/10.5114/biolsport.2025.139850

Pan P, Oshima K, Huang YW, Agle KA, Drobyski WR, Chen X, Zhang J, Yearsley MM, Yu J, Wang LS (2018) Loss of FFAR2 promotes colon cancer by epigenetic dysregulation of inflammation suppressors. Int J Cancer 143(4):886–896. https://doi.org/10.1002/ijc.31366

Park J, Kim M, Kang SG, Jannasch AH, Cooper B, Patterson J, Kim CH (2015) Short-chain fatty acids induce both effector and regulatory T cells by suppression of histone deacetylases and regulation of the mTOR-S6K pathway. Mucosal Immunol 8(1):80–93. https://doi.org/10.1038/mi.2014.44

Poupeau A, Garde C, Sulek K, Citirikkaya K, Treebak JT, Arumugam M, Simar D, Olofsson LE, Bäckhed F, Barrès R (2019) Genes controlling the activation of natural killer lymphocytes are epigenetically remodeled in intestinal cells from germ-free mice. FASEB J 33(2):2719–2731. https://doi.org/10.1096/fj.201800787R

Qin Y, Roberts JD, Grimm SA, Lih FB, Deterding LJ, Li R, Chrysovergis K, Wade PA (2018) An obesity-associated gut microbiome reprograms the intestinal epigenome and leads to altered colonic gene expression. Genome Biol 19(1):7. https://doi.org/10.1186/s13059-018-1389-1

Qu S, Yu Z, Zhou Y, Wang S, Jia M, Chen T, Zhang X (2024) Gut microbiota modulates neurotransmitter and gut-brain signaling. Microbiol Res 287:127858. https://doi.org/10.1016/j.micres.2024.127858

Ramos C, Magistro D, Walton GE, Whitham A, Camp N, Poveda C, Gibson GR, Hough J, Kinnear W, Hunter K (2025) Assessing the gut microbiota composition in older adults: connections to physical activity and healthy ageing. Geroscience. https://doi.org/10.1007/s11357-025-01605-w

Ramos-Molina B, Sánchez-Alcoholado L, Cabrera-Mulero A, Lopez-Dominguez R, Carmona-Saez P, Garcia-Fuentes E, Moreno-Indias I, Tinahones FJ (2019) Gut microbiota composition is associated with the global DNA methylation pattern in obesity. Front Genet 10:613. https://doi.org/10.3389/fgene.2019.00613

Remely M, Aumueller E, Merold C, Dworzak S, Hippe B, Zanner J, Pointner A, Brath H, Haslberger AG (2014) Effects of short chain fatty acid producing bacteria on epigenetic regulation of FFAR3 in type 2 diabetes and obesity. Gene 537(1):85–92. https://doi.org/10.1016/j.gene.2013.11.081

Ren J, Zhou L, Li S, Zhang Q, Xiao X (2024) The roles of the gut microbiota, metabolites, and epigenetics in the effects of maternal exercise on offspring metabolism. Am J Physiol Endocrinol Metab 327(6):E760–e772. https://doi.org/10.1152/ajpendo.00200.2024

Roostaee A, Benoit YD, Boudjadi S, Beaulieu JF (2016) Epigenetics in intestinal epithelial cell renewal. J Cell Physiol 231(11):2361–2367. https://doi.org/10.1002/jcp.25401

Sanchez HN, Moroney JB, Gan H, Shen T, Im JL, Li T, Taylor JR, Zan H, Casali P (2020) B cell-intrinsic epigenetic modulation of antibody responses by dietary fiber-derived short-chain fatty acids. Nat Commun 11(1):60. https://doi.org/10.1038/s41467-019-13603-6

Santarossa S, Sitarik AR, Johnson CC, Li J, Lynch SV, Ownby DR, Ramirez A, Yong GL, Cassidy-Bushrow AE (2021) Associations of physical activity with gut microbiota in pre-adolescent children. Phys Act Nutr 25(4):24–37. https://doi.org/10.20463/pan.2021.0023

Santarossa S, Sitarik AR, Cassidy-Bushrow AE, Comstock SS (2023) Prenatal physical activity and the gut microbiota of pregnant women: results from a preliminary investigation. Phys Act Nutr 27(2):1–7. https://doi.org/10.20463/pan.2023.0011

Schulthess J, Pandey S, Capitani M, Rue-Albrecht KC, Arnold I, Franchini F, Chomka A, Ilott NE, Johnston DGW, Pires E, McCullagh J, Sansom SN, Arancibia-Cárcamo CV, Uhlig HH, Powrie F (2019) The short chain fatty acid butyrate imprints an antimicrobial program in macrophages. Immunity 50(2):432–445.e7. https://doi.org/10.1016/j.immuni.2018.12.018

Si J, Vázquez-Castellanos JF, Gregory AC, Decommer L, Rymenans L, Proost S, Centelles Lodeiro J, Weger M, Notdurfter M, Leitner C, Santer P, Rungger G, Willeit J, Willeit P, Pechlaner R, Grabherr F, Kiechl S, Tilg H, Raes J (2022) Long-term life history predicts current gut microbiome in a population-based cohort study. Nat Aging 2(10):885–895. https://doi.org/10.1038/s43587-022-00286-w

Smith PM, Howitt MR, Panikov N, Michaud M, Gallini CA, Bohlooly YM, Glickman JN, Garrett WS (2013) The microbial metabolites, short-chain fatty acids, regulate colonic Treg cell homeostasis. Science 341(6145):569–573. https://doi.org/10.1126/science.1241165

Sun S, Xu X, Liang L, Wang X, Bai X, Zhu L, He Q, Liang H, Xin X, Wang L, Lou C, Cao X, Chen X, Li B, Wang B, Zhao J (2021) Lactic acid-producing probiotic Saccharomyces cerevisiae attenuates ulcerative colitis via suppressing macrophage pyroptosis and modulating gut microbiota. Front Immunol 12:777665. https://doi.org/10.3389/fimmu.2021.777665

Tahara T, Hirata I, Nakano N, Tahara S, Horiguchi N, Kawamura T, Okubo M, Ishizuka T, Yamada H, Yoshida D, Ohmori T, Maeda K, Komura N, Ikuno H, Jodai Y, Kamano T, Nagasaka M, Nakagawa Y, Tuskamoto T, Urano M, Shibata T, Kuroda M, Ohmiya N (2017) Potential link between Fusobacterium enrichment and DNA methylation accumulation in the inflammatory colonic mucosa in ulcerative colitis. Oncotarget 8(37):61917–61926. https://doi.org/10.18632/oncotarget.18716

Tang R, Liu R, Zha H, Cheng Y, Ling Z, Li L (2024) Gut microbiota induced epigenetic modifications in the non-alcoholic fatty liver disease pathogenesis. Eng Life Sci 24(5):2300016. https://doi.org/10.1002/elsc.202300016

Tarallo S, Ferrero G, De Filippis F, Francavilla A, Pasolli E, Panero V, Cordero F, Segata N, Grioni S, Pensa RG, Pardini B, Ercolini D, Naccarati A (2022) Stool microRNA profiles reflect different dietary and gut microbiome patterns in healthy individuals. Gut 71(7):1302–1314. https://doi.org/10.1136/gutjnl-2021-325168

Urban S, Chmura O, Wątor J, Panek P, Zapała B (2024) The intensive physical activity causes changes in the composition of gut and oral microbiota. Sci Rep 14(1):20858. https://doi.org/10.1038/s41598-024-71684-w

Virtue AT, McCright SJ, Wright JM, Jimenez MT, Mowel WK, Kotzin JJ, Joannas L, Basavappa MG, Spencer SP, Clark ML, Eisennagel SH, Williams A, Levy M, Manne S, Henrickson SE, Wherry EJ, Thaiss CA, Elinav E, Henao-Mejia J (2019) The gut microbiota regulates white adipose tissue inflammation and obesity via a family of microRNAs. Sci Transl Med 11(496). https://doi.org/10.1126/scitranslmed.aav1892

de Vos WM, Tilg H, Van Hul M, Cani PD (2022) Gut microbiome and health: mechanistic insights. Gut 71(5):1020–1032. https://doi.org/10.1136/gutjnl-2021-326789

Wang J, Shi ZH, Yang J, Wei Y, Wang XY, Zhao YY (2020) Gut microbiota dysbiosis in preeclampsia patients in the second and third trimesters. Chin Med J 133(9):1057–1065. https://doi.org/10.1097/cm9.0000000000000734

Wang T, Ye Z, Li Z, Jing DS, Fan GX, Liu MQ, Zhuo QF, Ji SR, Yu XJ, Xu XW, Qin Y (2023) Lactate-induced protein lactylation: a bridge between epigenetics and metabolic reprogramming in cancer. Cell Prolif 56(10):e13478. https://doi.org/10.1111/cpr.13478

Wang M, Ma G, Li Y, Li J, Xie J, He J, He C, He Y, Jia K, Feng X, Tian T, Li H, Liao X, Liu X (2025a) Potential modulatory roles of gut microbiota and metabolites in the associations of macronutrient-to-physical activity ratios with dyslipidemia. J Am Heart Assoc:e040042. https://doi.org/10.1161/jaha.124.040042

Wang Y, Bai S, Yang T, Guo J, Zhu X, Dong Y (2025b) Impact of exercise-induced alterations on gut microbiota diversity and composition: comparing effects of different training modalities. Cell Regen 14(1):28. https://doi.org/10.1186/s13619-025-00244-y

Warbeck C, Dowd AJ, Kronlund L, Parmar C, Daun JT, Wytsma-Fisher K, Millet GY, Schick A, Reimer RA, Fung T, Culos-Reed SN (2021) Feasibility and effects on the gut microbiota of a 12-week high-intensity interval training plus lifestyle education intervention on inactive adults with celiac disease. Appl Physiol Nutr Metab 46(4):325–336. https://doi.org/10.1139/apnm-2020-0459

Whitt J, Woo V, Lee P, Moncivaiz J, Haberman Y, Denson L, Tso P, Alenghat T (2018) Disruption of epithelial HDAC3 in intestine prevents diet-induced obesity in mice. Gastroenterology 155(2):501–513. https://doi.org/10.1053/j.gastro.2018.04.017

Wieërs G, Belkhir L, Enaud R, Leclercq S, Philippart de Foy JM, Dequenne I, de Timary P, Cani PD (2019) How probiotics affect the microbiota. Front Cell Infect Microbiol 9:454. https://doi.org/10.3389/fcimb.2019.00454

Woo V, Alenghat T (2022) Epigenetic regulation by gut microbiota. Gut Microbes 14(1):2022407. https://doi.org/10.1080/19490976.2021.2022407

Wu SE, Hashimoto-Hill S, Woo V, Eshleman EM, Whitt J, Engleman L, Karns R, Denson LA, Haslam DB, Alenghat T (2020) Microbiota-derived metabolite promotes HDAC3 activity in the gut. Nature 586(7827):108–112. https://doi.org/10.1038/s41586-020-2604-2

Wu Y, Wang CZ, Wan JY, Yao H, Yuan CS (2021) Dissecting the interplay mechanism between epigenetics and gut microbiota: health maintenance and disease prevention. Int J Mol Sci 22(13). https://doi.org/10.3390/ijms22136933

Xie Y, Tobin LA, Camps J, Wangsa D, Yang J, Rao M, Witasp E, Awad KS, Yoo N, Ried T, Kwong KF (2013) MicroRNA-24 regulates XIAP to reduce the apoptosis threshold in cancer cells. Oncogene 32(19):2442–2451. https://doi.org/10.1038/onc.2012.258

Xie A, Ensink E, Li P, Gordevičius J, Marshall LL, George S, Pospisilik JA, Aho VTE, Houser MC, Pereira PAB, Rudi K, Paulin L, Tansey MG, Auvinen P, Brundin P, Brundin L, Labrie V, Scheperjans F (2022) Bacterial butyrate in Parkinson's disease is linked to epigenetic changes and depressive symptoms. Mov Disord 37(8):1644–1653. https://doi.org/10.1002/mds.29128

Yoon EJ, Lee SR, Ortutu BF, Kim J-O, Jaiswal V, Baek S, Yoon S-I, Lee SK, Yoon JH, Lee H-J (2024) Effect of endurance exercise training on gut microbiota and ER stress. Int J Mol Sci 25(19):10742

Yu DH, Gadkari M, Zhou Q, Yu S, Gao N, Guan Y, Schady D, Roshan TN, Chen MH, Laritsky E, Ge Z, Wang H, Chen R, Westwater C, Bry L, Waterland RA, Moriarty C, Hwang C, Swennes AG, Moore SR, Shen L (2015) Postnatal epigenetic regulation of intestinal stem cells requires DNA methylation and is guided by the microbiome. Genome Biol 16:211. https://doi.org/10.1186/s13059-015-0763-5

Zahn LM (2017) Archaeal diversity and evolution. Science 357(6351):560–562

Zeng X, Sunkara LT, Jiang W, Bible M, Carter S, Ma X, Qiao S, Zhang G (2013) Induction of porcine host defense peptide gene expression by short-chain fatty acids and their analogs. PLoS One 8(8):e72922. https://doi.org/10.1371/journal.pone.0072922

Zhang K, Guo H, Zhang X, Yang H, Yuan G, Zhu Z, Lu X, Zhang J, Du J, Shi H, Jin G, Ren J, Hao J, Sun Y, Su P, Zhang Z (2024) Effects of aerobic exercise or Tai Chi Chuan interventions on problematic mobile phone use and the potential role of intestinal flora: a multi-arm randomized controlled trial. J Psychiatr Res 170:394–407. https://doi.org/10.1016/j.jpsychires.2024.01.012

Zhang Y, Wei Y, Han X, Shi L, Yu H, Ji X, Gao Y, Gao Q, Zhang L, Duan Y, Li W, Yuan Y, Shi J, Cheng L, Li Y (2025) Faecalibacterium prausnitzii prevents age-related heart failure by suppressing ferroptosis in cardiomyocytes through butyrate-mediated LCN2 regulation. Gut Microbes 17(1):2505119. https://doi.org/10.1080/19490976.2025.2505119

Zheng L, Kelly CJ, Battista KD, Schaefer R, Lanis JM, Alexeev EE, Wang RX, Onyiah JC, Kominsky DJ, Colgan SP (2017) Microbial-derived butyrate promotes epithelial barrier function through IL-10 receptor-dependent repression of claudin-2. J Immunol 199(8):2976–2984. https://doi.org/10.4049/jimmunol.1700105

Zhou Z, Zhang C-J, Liu P-F, Fu L, Laso-Pérez R, Yang L, Bai L-P, Li J, Yang M, Lin J-Z (2022) Non-syntrophic methanogenic hydrocarbon degradation by an archaeal species. Nature 601(7892):257–262

Chapter 12
Extracellular Vesicles in Exercise-Induced Health Benefits

Xinxin Cui, Xiaotong Ding, Tian Hao, Guoping Li, Lijun Wang, and Junjie Xiao

Abstract Extracellular vesicles (EVs) have garnered considerable interest owing to their essential functions in intercellular communication and their potential to influence various physiological responses. Exercise has been shown to modulate the release and uptake of EVs, with effects varying according to the intensity of physical activity and the specific cell types involved. The involvement of EVs in exercise-induced adaptations is multifaceted, influencing numerous physiological systems. They are considered key mediators of the health benefits associated with exercise, facilitating cellular signaling and contributing to tissue protection mechanisms. In this chapter, we provide a concise overview of the classification, origin, isolation, and characterization of EVs. We specifically focus on current knowledge regarding EV release and regulation in the context of systemic adaptations to exercise training. Additionally, we summarize the effects of EVs in mediating the health-promoting effects of exercise, particularly in the prevention and treatment of diseases such as cardiovascular disorders, diabetes mellitus, and cancer. Finally, we discuss the involvement of EVs in inter-tissue communication during exercise, outline future research directions, and highlight key challenges in the field.

Keywords Extracellular vesicles · Exercise · Cardioprotection · Metabolic regulation · Tissue crosstalk

X. Cui · X. Ding · L. Wang (✉) · J. Xiao (✉)
Cardiac Regeneration and Ageing Lab, Institute of Geriatrics (Shanghai University), Affiliated Nantong Hospital of Shanghai University (The Sixth People's Hospital of Nantong) and School of Life Science, Shanghai University, Nantong, China

Institute of Cardiovascular Sciences, Shanghai Engineering Research Center of Organ Repair, Joint International Research Laboratory of Biomaterials and Biotechnology in Organ Repair (Ministry of Education), School of Life Science, Shanghai University, Shanghai, China
e-mail: lijunwang@shu.edu.cn; junjiexiao@shu.edu.cn

T. Hao · G. Li
Department of Anesthesia, Critical Care and Pain Medicine, Massachusetts General Hospital, Harvard Medical School, Boston, MA, USA

F. Gao, X. Zhang (eds.), *Exercise, Epigenetics and Human Health*, Epigenetics and Human Health 15, https://doi.org/10.1007/978-3-032-08045-5_12

Abbreviations

AMPK	AMP-activated protein kinase
ANGPT2	Angiopoietin-2
CAV1	Caveolin-1
eNAMPT	extracellular nicotinamide phosphoribosyltransferase
ERK1/2	Extracellular signal-regulated kinases ½
EVs	Extracellular vesicles
FNDC5	Fibronectin type III domain-containing protein 5
HSP27	Heat shock protein 27
IL-10	Interleukin-10
IL-6	Interleukin-6
lncRNA	Long non-coding RNA
MAPK	Mitogen-activated protein kinase
miRNA	MicroRNA
MVBs	Multivesicular bodies
NF-κB	Nuclear factor kappa-light-chain-enhancer of activated B cells
NRF2	Nuclear factor erythroid 2–related factor 2
TNF-α	Tumor necrosis factor alpha
VO_2peak	Peak oxygen uptake

12.1 Introduction

Extracellular vesicles (EVs) have garnered significant attention in recent years due to their crucial role in intercellular communication and their potential to mediate diverse physiological responses. EVs-including exosomes, microvesicles, and apoptotic bodies, are typically classified by their size, cellular origin, and mechanisms of release. Their biogenesis is a complex process involving endosomal formation and plasma membrane budding, ultimately leading to their secretion into the extracellular environment. Owing to their capacity to transport bioactive molecules such as proteins, lipids, and RNAs between cells, the study of EV isolation, characterization, and regulatory mechanisms has become a central focus in biomedical research (McIlvenna and Whitham 2023).

Exercise has been shown to modulate both the release and uptake of EVs, with the effects varying according to the intensity, modality and duration of physical activity, as well as the specific cell types involved. Various studies have reported that exercise induces an increase in circulating small vesicles and exosomes, suggesting that EVs may play a central role in mediating systemic adaptations to exercise (Lisi et al. 2023; Maggio et al. 2023). The composition of these EVs is influenced by their tissue of origin, which may include skeletal muscle, adipose tissue, and the cardiovascular system. These exercise-induced EVs contribute to a wide range of biological effects, promoting inter-tissue communication and potentially supporting overall physiological health (Conkright et al. 2024).

The role of EVs in exercise-induced adaptations is multifaceted, influencing a wide range of physiological processes. EVs are integral to the cardioprotective effects of exercise, mediating cellular signaling and contributing to tissue repair and protection. They are also increasingly recognized for their roles in improving metabolic health, particularly in the context of diseases such as diabetes mellitus, and for their emerging therapeutic potential in cancer. Beyond metabolic regulation, EVs have been implicated in the anti-cancer effects of exercise, supporting immune modulation and maintaining tissue homeostasis. Moreover, EVs facilitate inter-organ communication, enabling tissue crosstalk that enhances the body's coordinated response to physical stress. This chapter summarizes current knowledge on the role of EVs in systemic adaptations to exercise training, emphasizing their contributions to the prevention and treatment of human diseases, including heart diseases, diabetes mellitus, and cancers. The mechanisms underlying EV-mediated tissue crosstalk during exercise are also explored.

12.2 Extracellular Vesicles

EVs are membrane-bound particles secreted by cells into the extracellular space. They are abundantly present in various body fluids and play crucial roles in intercellular communication, immune regulation, and progression of numerous disease (Wu et al. 2023).

12.2.1 Classification and Origin

EVs are a heterogeneous group of nanosized lipid bilayer-enclosed vesicles released by various cell types. Based on their biogenesis, EVs are broadly categorized into exosomes and ectosomes (Li et al. 2023). Within multivesicular bodies (MVBs), inward budding of the endosomal membrane generates intraluminal vesicles that encapsulate various intracellular components. Upon fusion of MVBs with the plasma membrane, these vesicles are released into the extracellular space as exosomes (Zhang et al. 2019; Mas-Bargues and Alique 2023). Exosomes, typically 40–160 nm in diameter, are enriched with proteins, RNAs, lipids, and other bioactive molecules (Théry et al. 2018; Li et al. 2023). In contrast, ectosomes are directly shed from the plasma membrane of diverse cell types and include microvesicles, migrasomes, exophers, apoptotic bodies, and large oncosomes, ranging in size from approximately 50 nm to 10 μm (Zhang et al. 2021; Buzas 2023; Li et al. 2023). In recent years, several types of non-vesicular extracellular particles have also been identified to carry distinct protein or RNA cargo and participate in intercellular communication (Mittelbrunn and Sánchez-Madrid 2012; Zhang et al. 2021). They

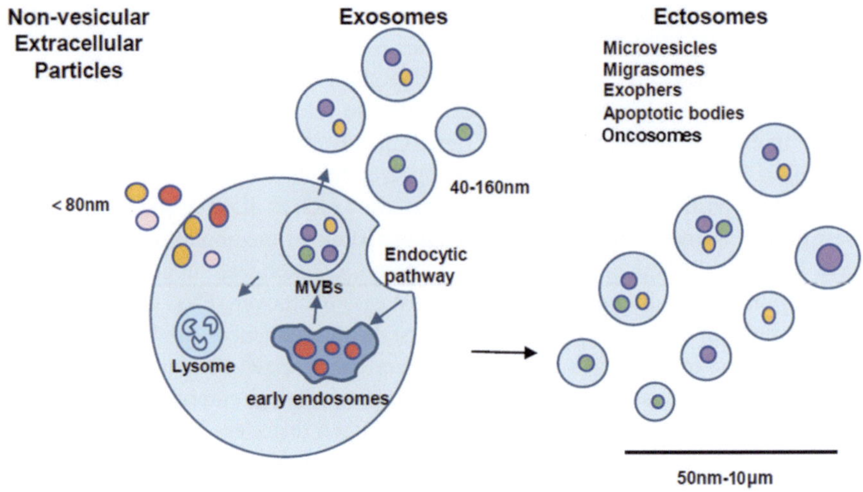

Fig. 12.1 Overview of extracellular vesicle (EV) biogenesis and classification. EVs are divided into exosomes, originating from the endosomal pathway, and ectosomes, which are directly shed from the plasma membrane. Both types mediate intercellular communication by transferring diverse bioactive cargos, contributing to physiological and pathological processes. Non-vesicular extracellular particles also participate in cell-to-cell communication through the delivery of functional molecules

mediate intercellular communication by interacting with receptors on recipient cells or through membrane fusion, thereby modulating cellular functions and contributing to diverse physiological and pathological processes (Nowak et al. 2023) (Fig. 12.1).

In contrast, microvesicles are generated through the direct outward budding of the plasma membrane and are generally larger in size, typically ranging from 100 to 1000 nm in diameter (Gregory and Rimmer 2023). Under the regulation of both intracellular and extracellular signals, the plasma membrane undergoes localized expansion and outward budding, resulting in vesicles that enclose cellular contents. Microvesicles encapsulate various cellular components, including proteins, lipids, RNAs. They can interact with target cells by direct membrane fusion, playing essential roles in signal transduction, intercellular communication, and immune regulation (Lange 2023).

Finally, apoptotic bodies are larger particles generated by cells undergoing apoptosis, typically ranging from 1 to 5 μm in diameter (Wu et al. 2023). As cells undergo apoptosis, the cell membrane ruptures, enclosing the cytoplasm and nucleus in vesicles to form apoptotic bodies. These apoptotic bodies are usually recognized and cleared by phagocytic cells, such as macrophages, and are involved in the regulation of immune responses (Du et al. 2023; Nowak et al. 2023).

In summary, EVs are classified into three main types—exosomes, microvesicles, and apoptotic bodies based on their biogenesis and functional characteristics. Exosomes are derived from the endocytic pathway, microvesicles are formed by the

Table 12.1 Classification of extracellular vesicles (EVs)

Type of EVs	Size	Formation mechanism	Subtypes	Function	References
Exosomes	40–160 nm	Generated by the endocytic pathway; released upon fusion of endosomal MVBs with plasma membrane	Exosomes	Transfer bioactive molecules; mediate paracrine/systemic signaling	Théry et al. (2018), Li et al. (2023)
Ectosome	50 nm–10 μm	Directly shed from the plasma membrane of diverse cell types	Microvesicles, migrasomes, exophers, apoptotic bodies, large oncosomes	Transfer bioactive molecules; mediate intercellular communication	Zhang et al. (2021), Buzas (2023), and Li et al. (2023)
Non-vesicular extracellular particles	<80 nm	Non-vesicular particles carrying distinct protein or RNA cargo	Lipoproteins, exomeres, supermeres, chromatimeres	Carry protein/RNA cargo; mediate intercellular communication	Mittelbrunn and Sánchez-Madrid (2012), Zhang et al. (2021)

outward budding of the plasma membrane, and apoptotic bodies originate from apoptotic cells (Théry et al. 2018). Each type of EV plays crucial roles in mediating the transfer of biological materials and regulating cellular signaling, highlighting their importance not only in intercellular communication but also in disease diagnostics and therapeutic applications (Table 12.1).

12.2.2 Isolation and Characterization

The isolation and characterization of EVs are critical for understanding their biological functions. As EVs are typically present at low concentrations in body fluids and display considerable heterogeneity, the development of efficient and reliable methodologies is essential. EV isolation techniques primarily rely on size, density, and surface markers. Frequently employed isolation technologies include ultracentrifugation, density gradient centrifugation, immunocapture, size-exclusion chromatography, and ultrafiltration (Mazahir and Yadav 2024).

Ultracentrifugation is the most widely used traditional method for isolating EVs, relying on high-speed centrifugation to precipitate vesicles based on their size and density. While this method is straightforward and efficient, it is less sensitive to low-density EVs and can be time consuming (Dilsiz 2024). Density gradient centrifugation separates EVs based on density differences and can purify exosomes more efficiently, although the procedure is more technically demanding (Gao et al. 2022). Immunocapture techniques use antibodies targeting specific EV surface markers, allowing for the selective isolation of defined EV subpopulations. Size-exclusion chromatography separates EVs based on size, proving a mild and efficient method suitable for large-scale sample processing (Foers et al. 2018; Gao et al. 2022). Ultrafiltration utilizes membrane separation with defined pore sizes for straight forward separation, but may overlook smaller EVs due to size limitations. Precipitation methods, such as those using polyethylene glycol (PEG), induce EV aggregation and sedimentation. These are fast, simple, and scalable, but often result in lower purity due to co-precipitation of non-EV components. (Li et al. 2021; Park et al. 2021). To enhance purity, precipitation is frequently combined with other techniques like ultracentrifugation or immunocapture (Martínez-Greene et al. 2021). In recent years, microfluidic platforms have emerged as a promising tool for EV isolation. These systems enable precise control of fluid dynamics and separation processes (Leong et al. 2022; Chen et al. 2023). The advantages of this approach include high throughput, low loss, and high resolution, although there may be some limitations in handling large sample volumes (Priedols et al. 2023). Despite the availability of various EV isolation techniques, the choice of isolation technique depends on the sample type, experimental objectives, and conditions (Cable et al. 2023). Continued development and optimization of EV isolation strategies remain critical for achieving high yield, purity, and functional integrity.

Characterization techniques for exosomes are essential for confirming their purity, morphology, size distribution, and molecular composition, ensuring that the isolated EVs meet research requirements. Nanoparticle tracking analysis (NTA), monitors particle movement in real-time and provides size and concentration distributions (Vrablova et al. 2023). Transmission electron microscopy (TEM) offers ultra-high-resolution imaging, allowing direct visualization of exosome morphology and bilayer membrane structure. Dynamic light scattering is a rapid method for assessing particle size distribution, although it lacks sensitivity for distinguishing between vesicle subtypes with similar sizes (Shami-Shah et al. 2023). Western blotting is widely used to detect exosome-specific markers (such as CD9, CD63, Alix, TSG101) to confirm exosome purity and identification (Qazi et al. 2023; Vrablova et al. 2023). Flow cytometry allows for high-throughput analysis of surface markers, making it especially useful for identifying and quantifying specific exosome subpopulations. Mass spectrometry can be used to identify exosome proteomes, lipidomes, and RNAs, providing further insight into their composition and biological functions (Qazi et al. 2023). Additionally, PCR or RNA-Seq are commonly used to analyze RNA molecules, including miRNAs and mRNAs, to explore their roles in intercellular communication and disease progression (Dilsiz 2024). Typically, these techniques are used in combination to achieve a comprehensive characterization of EVs, enhancing the reliability and reproducibility of experimental outcomes.

12.3 Regulation of EVs During and After Exercise

Among the various types of EVs, exosomes have garnered the most attention in the context of exercise research. Due to their nanoscale size, endosomal origin, and ability to mediate intercellular communication by transferring bioactive molecules, exosomes play a crucial role in physiological adaptations to exercise. Compared to microvesicles and apoptotic bodies, exosomes are more extensively studied, particularly for their potential roles as biomarkers and mediators of exercise-induced benefits. Accordingly, the following sections will focus primarily on exosomes, exploring their functional properties and relevance in exercise-related adaptations (Fig. 12.2).

12.3.1 Release and Regulation of Exosomes During Exercise

Exercise significantly promotes the release of EVs, which transport a variety of bioactive molecules—including proteins, lipids, and nucleic acids (such as miRNAs), that regulate the functions of multiple organs (Fruhbeis et al. 2015; Liu et al. 2022) (Table 12.2). Studies have shown that exhaustive exercise, such as cycling or running to exhaustion, leads to a marked increase in circulating exosomes in healthy individuals, with levels typically returning to baseline within 90 minutes post-exercise (Fruhbeis et al. 2015). This rapid release of EVs is closely associated with exercise-induced physiological adaptations and plays a pivotal role in metabolic regulation during physical activity. The concentration of exosomes in the

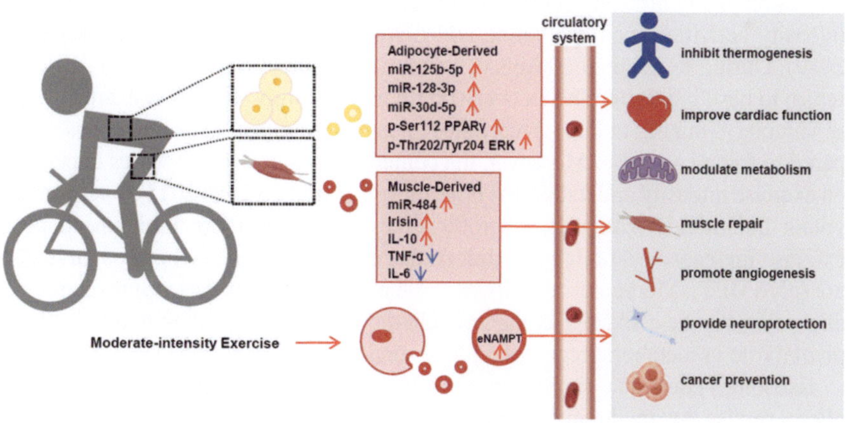

Fig. 12.2 Release and regulation of exosomes during exercise. Exercise promotes the release of exosomes, which are exosomes carrying proteins, lipids, and nucleic acids (such as miRNAs) that facilitate inter-organ signaling and regulate physiological functions. Exercise-induced exosomes play a crucial role in muscle repair, metabolic regulation, inflammatory responses, and neuroprotection

Table 12.2 Exercise-induced exosome release and its physiological functions

Physiological effect/ function	Exosome changes	Key components	References
Exercise adaptation and metabolic regulation	Significant increase in exosome count in circulation post-exercise, returning to baseline within 90 minutes	Flot1, Hsp/Hsc70, Tsg101 and αIIb	Fruhbeis et al. (2015), Chong et al. (2022)
Energy metabolism adaptation	Sustained exosome release during exercise, gradually returning to baseline during recovery	Altered exosome composition, such as eNAMPT	Chong et al. (2022)
Muscle repair and growth	Increased muscle-derived exosomes post-exercise	FDNC5, PGC-1α, IL-6	Huang et al. (2024a)
Neuroprotection and synaptic plasticity	Exosomes enriched with neurotrophic factors (BDNF, FNDC5/Irisin) post-exercise	BDNF, FNDC5/Irisin	Huang et al. (2024a)
Inflammation regulation and immune response	Exosomes regulate inflammatory cytokines post-exercise	IL-10, TNF-α, IL-6 , AMPKα, Nrf2 , HMOX1	Tao et al. (2024)
Mitochondrial bioenergetics and ferroptosis inhibition	Exosomes released from muscle activate the AMPK-Nrf2 pathway	Irisin, Cav1, AMPKα, HMOX1, SLC7A11, ACSL4	Tao et al. (2024), Huang et al. (2024b)
Maternal-fetal adaptation	Pregnant women release higher levels of exosomes post-exercise compared to non-pregnant controls	No significant effect on trophoblast cell proliferation or angiogenesis	Mohammad et al. (2023)

bloodstream significantly increases following exercise, particularly after high-intensity activities such as cycling and running (Whitham et al. 2018; Bei et al. 2019). During exercise, the levels of exosome temporarily rise and then gradually return to baseline during the recovery phase. Interestingly, voluntary exercise, such as voluntary running induces a more sustained modulation of exosome release compared to forced exercise. Moreover, the composition of exosomes varies depending on exercise intensity and individual fitness levels (Fruhbeis et al. 2015). For instance, young individuals with higher aerobic fitness exhibit a more pronounced post-exercise increase in circulating extracellular nicotinamide phosphoribosyltransferase (eNAMPT). Moderate-intensity exercise not only promotes exosome release but also enhances the eNAMPT content within the vesicles, suggesting that EVs play a pivotal role in metabolic regulation post-exercise.

Exosomes contribute significantly to various physiological processes during and after exercise by mediating intercellular communication, particularly in muscle repair, immune responses, and metabolic adaptations (Chong et al. 2022). During muscle repair, exercise-induced exosomes are enriched with myokines and microRNAs that promote the activation and proliferation of muscle stem cells (satellite cells) and promote muscle fiber regeneration (Huang et al. 2024a). Neuronal innervation also plays an essential role in regulating the secretion and composition of muscle-derived exosomes. A neuromuscular co-culture model demonstrated that exercise-induced exosomes contain neurotrophic factors, such as brain-derived neurotrophic factor (BDNF) and FNDC5/Irisin, and enhance the secretion of neurotrophic microRNAs upon glutamate stimulation, ultimately promoting neuronal survival and synaptic plasticity (Huang et al. 2024a; Tao et al. 2024). In terms of immune modulation, exercise-induced exosome release downregulates pro-inflammatory cytokines (e.g., TNF-α and IL-6) while upregulating anti-inflammatory factors (e.g., IL-10), thereby contributing to the homeostasis of the muscle microenvironment (Tao et al. 2024). Furthermore, skeletal muscle-derived exosomes transport irisin can activate the AMPK-Nrf2 signaling pathway, enhance mitochondrial bioenergetics, and inhibit programmed cell death pathways, including ferroptosis, which is particularly significant for neuroprotection following ischemic injury (Huang et al. 2024b). Also, compared to non-pregnant controls, pregnant women release higher levels of exosomes into the circulation following exercise. These exosomes interact with trophoblast cells but do not significantly alter trophoblast proliferation, angiogenic marker gene expression, or β-hCG secretion (Mohammad et al. 2023). This suggests that exosomes may act as conveyors of exercise-induced benefits across the maternal–fetal interface. In summary, the release of exosomes post-exercise not only exhibits dynamic variations across genders and exercise intensities, but also participates in physiological processes such as muscle repair, inflammatory responses, and metabolic adaptations by regulating the release of metabolic-related factors, highlighting the importance of exosomes in mediating the systemic benefits of exercise.

12.3.2 Role of Adipocyte-Derived Exosomes

Exercise effectively regulates lipid metabolism, reduces body weight, and promotes the browning of white adipose tissue. Beyond the classical cytokine-mediated signaling pathways, recent evidence suggests that exercise also facilitates inter-tissue communication via the release of EVs, particularly exosomes. Circulating exosomes have been shown to promote the browning of white adipose tissue in mice, thereby enhancing lipid energy metabolism and thermogenesis, ultimately contributing to weight loss (Shi et al. 2024). Furthermore, exercise mediates the transport of adipose tissue-derived miRNAs through exosomes, which regulate insulin signaling pathways and improve metabolic function. Notably, miRNAs such as miR-125b-5p, miR-128-3p, and miR-30d-5p are enriched in adipose-derived exosomes following exercise and are implicated in cardioprotection (Zhao et al. 2022). These findings highlight the potential of exosomes as key modulators of metabolic adaptation, cardiac protection, and systemic inflammatory regulation. Other studies have identified that miRNAs released from adipose-derived exosomes, particularly miR-23b from white adipose tissue-derived exosomes, can inhibit thermogenesis by targeting Elf4 and modulating the transcription of GLP-1R (Wang et al. 2024). The modulation of thermogenesis and insulin signaling pathways contribute to the progression of obesity and improves metabolic function post-exercise (Wang et al. 2024). Exercise training also alters the component of adipose-derived stem cell-released exosomes to suppress adipocyte differentiation (Osawa et al. 2024). Mechanically, exercise training-driven exosomal miR-323–5p increased phosph-Ser112 on PPARγ and phosph-Thr202/Tyr204 on ERK levels while reducing DUSP3 expression (Osawa et al. 2024). Additionally, adipose tissue-derived exosomes released under physical stress play an essential role in mediating the protective effects of exercise by facilitating tissue cross communication (Wang et al. 2022). For instance, exercise can promote the brown adipose tissue-derived exosomes delivery to the heart to suppress MAPK activation, thereby subsequently reducing myocardial ischemia/reperfusion-induced cardiac injury and improving cardiac function (Doncheva et al. 2022). Collectively, these findings highlight the important roles of adipocyte-derived exosomes in maintaining tissue homeostasis, mediating exercise-induced benefits, and contributing to the prevention or resolution of metabolic complications.

12.3.3 Role of Skeletal Muscle-Derived Exosomes

Regular exercise plays a crucial role in maintaining body metabolic profiles, particularly through the regulation of exosome secretion by skeletal muscle, which is integral to various physiological processes. A 12-week swimming training program has been shown to significantly improve glucose tolerance in both obese wild-type and ApoE knockout mice, reduce visceral fat accumulation, alleviate liver damage, and inhibit the progression of atherosclerosis (Wang et al. 2023). Exercise can partially reshape body metabolism, leading to health beneficial effects through skeletal muscle-derived exosomes. Exercise-induced skeletal muscle-derived exosomes have been shown to promote muscle regeneration by transporting proteins and RNAs involved in muscle cell proliferation and differentiation (Anderson 2022; Watanabe et al. 2022). For example, irisin—an exercise-induced myokine—has been identified as a key cargo of muscle-derived exosomes (Watanabe et al. 2022; Tao et al. 2024). Exosomal irisin can activate AMPK signaling and promote bone formation, offering mechanistic insight into the therapeutic potential of exercise in treating osteoporosis (Watanabe et al. 2022; Tao et al. 2024). Moreover, skeletal muscle-derived exosomes have been demonstrated to promote angiogenesis by enhancing endothelial cell proliferation, migration, and tube formation via VEGF-independent pathways (Nie et al. 2019). This alternative mechanism suggests a novel pathway for improving capillary rarefaction in conditions like diabetes and aging. In the nervous system, exercise-induced exosomal miR-484 protects neurons from ischemic injury by reducing lipid peroxidation and inhibiting ferroptosis, providing neuroprotection against stroke (Nie et al. 2019; Huang et al. 2024b). In the context of cancer, exercise has been associated with reduced tumor progression, in part through muscle-derived exosomes that modulate immune responses and influence tumor microenvironment dynamics (Papadopetraki et al. 2022; Vitucci et al. 2023). In conclusion, skeletal muscle-derived exosomes serve as critical carriers of exercise-induced beneficial factors for the prevention and treatment of many diseases; by regulating its secretion and content composition, exosomes provide an effective mechanism for tissue crosstalk during exercise.

12.4 EVs in Exercise-Induced Adaptations

Exercise promotes the release of EVs from various tissues, including adipose tissue, the heart, and skeletal muscles. Following endurance training, the number of EVs in the bloodstream increases significantly. These vesicles carry a diverse array of bioactive molecules such as miRNAs, metabolites, cytokines, myokines, which play a crucial role in exercise's health benefits (Fig. 12.3, Table 12.3).

12.4.1 The Effects of EVs in Exercise-Induced Cardioprotective Benefits

The role of circulating EVs in mediating the cardioprotective benefits has received increasing attention. Studies have demonstrated that EVs derived from plasma after exercise enhance antioxidant enzyme activity (such as glutathione reductase and

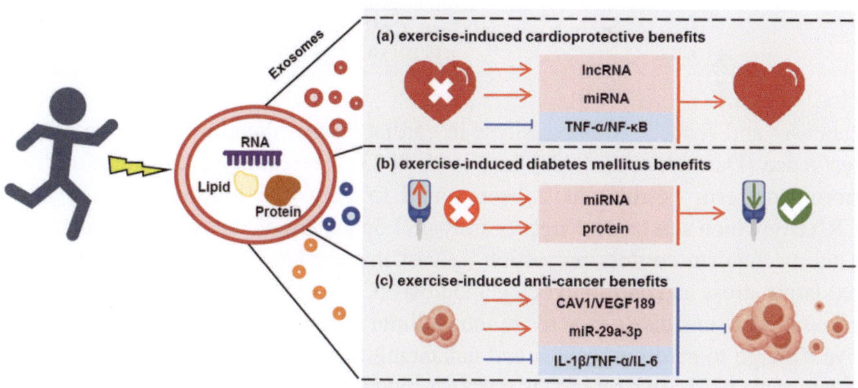

Fig. 12.3 Exercise-induced extracellular vesicles (EVs) contribute to systemic health by facilitating inter-organ communication and delivering bioactive molecules. (**a**) EVs mediate cardioprotective benefits by reducing oxidative stress, enhancing antioxidant enzyme activity, and modulating TNF-α/NF-κB and other stress-responsive mechanisms, thereby mitigating pathological remodeling and improving myocardial function. (**b**) In diabetes mellitus, EVs enhance insulin sensitivity, regulate glucose metabolism, and reduce adipose tissue inflammation, promoting metabolic homeostasis. (**c**) EVs exert anti-cancer effects by modulating the tumor microenvironment, suppressing tumor progression, and enhancing immune surveillance

Table 12.3 The effects of extracellular vesicles (EVs) in exercise-induced health benefits

Health benefit	Mechanism	Key findings	References
Cardioprotection	EVs from BAT and skeletal muscle carry bioactive molecules (e.g., miRNAs) that protect the heart.	BAT-derived EVs (miR-125b-5p, miR-128-3p, miR-30d-5p) inhibit MAPK signaling and reduce oxidative stress. EVs enhance antioxidant enzyme activity (e.g., GR, CAT). EVs suppress TNF-α/NF-κB signaling, reducing myocardial hypertrophy. Exercise elevated serum EVs and conferred anti-apoptotic effects via ERK1/2-HSP27 signaling.	Zhao et al. (2022), Bei et al. (2017), Hou et al. (2019), and Fandl et al. (2023)
Diabetes management	EVs improve insulin sensitivity, glucose metabolism, lipid metabolism, and wound healing.	EVs (miR-133b) enhance insulin sensitivity and glucose metabolism. EVs promote angiogenesis and accelerate wound healing. EVs regulate lipid metabolism and reduce insulin resistance.	Wang et al. (2024), Musi et al. (2001), Castaño et al. (2022), and Castaño et al. (2022)
Anti-cancer effects	EVs modulate the tumor microenvironment and immune response, inhibiting tumor progression.	Muscle-derived EVs (miR-29a-3p) inhibit ECM remodeling and enhance T-cell infiltration. EVs downregulate pro-inflammatory cytokines (e.g., IL-1β, TNF-α, IL-6). Exercise may counteract tumor-promoting EV effects (e.g., CAV1-positive EVs, VEGF189-positive EVs).	Sadovska et al. (2021), Luo et al. (2024), Nilsson et al. (2019), Campos et al. (2018), and Xie et al. (2020)

catalase) and reduce oxidative stress in cardiac tissue, thereby improving cardiac resilience (Darragh et al. 2021; Lisi et al. 2023). For example, exosomes derived from long-term exercise have been found to carry the long non-coding RNA CRNDE, which acts as a sponge for miR-489-3p and upregulates NRF2 expression. This mechanism confers protection against myocardial infarction by attenuating oxidative stress and cardiomyocyte apoptosis (Chen et al. 2024). Additionally, EVs released after exercise can activate intracellular defense pathways, reducing oxidative damage to cardiomyocytes and enhancing the heart's intrinsic repair capacity (Zhao et al. 2022). In a rat model, exercise-induced exosomes were shown to protect against myocardial ischemia/reperfusion (I/R) injury. (Hou et al. 2019). One key molecule, exosomal miR-342-5p, exerted cardioprotective effects by inhibiting endothelial apoptosis through targeting Caspase 9 and JNK2 (Hou et al. 2019). In addition to changing the exosome components, exercise increased serum EVs by ~1.85-fold in mice and provided an anti-apoptotic effect by ERK1/2 and HSP27 signaling activation (Bei et al. 2017). After exercise, EVs can modulate gene expression in the heart, inhibiting detrimental signaling pathways such as TNF-α/NF-κB to suppress pathological cardiac remodeling and contribute to the long-term cardiac function maintenance (Fandl et al. 2023). Notably, exercise-induced organ crosstalk is also mediated by exosomes. A recent study reports that exercise induces the release of brown adipose tissue-derived exosomes to mediate communication with remote organ heart and exert cardioprotective effects via delivering cardioprotective miRNAs such as miR-125b-5p, miR-128-3p, and miR-30d-5p (Zhao et al. 2022). These miRNAs act by inhibiting MAPK signaling, reducing oxidative stress, and preventing myocardial ischemia/reperfusion injury (Zhao et al. 2022). Collectively, these findings highlight that exercise-induced EVs, through delivering beneficial cargos, represent a potent mechanism by which exercise exerts cardioprotection effects. Their therapeutic potential represents a promising avenue for the prevention and treatment of cardiac diseases.

12.4.2 The Effects of EVs in Exercise-Induced Diabetes Mellitus Benefits

Diabetes has emerged as a significant global health challenge, affecting millions of individuals worldwide. While pharmacological therapies remain essential for symptom management, exercise serves as a non-pharmacological therapeutic approach that has demonstrated effective impact on diabetes management. Among its various benefits, exercise promotes the release of EVs, which serve as biomarkers and mediators of intercellular communication by conveying physiological signals generated during physical activity. EVs play a critical role in enhancing exercise-induced insulin sensitivity. It has demonstrated that exercise-induced EVs can regulate hepatic and muscle metabolic processes by carrying specific miRNAs, such as miR-133b, thereby improving insulin sensitivity and glucose metabolism (Musi et al. 2001; Castaño et al. 2022). By applying exercise-induced EVs to obese mice, investigators observed significant improvements in insulin sensitivity, insulin secretion, and glucose tolerance (Castaño et al. 2022). Beyond metabolic regulation, exercise-induced EVs also contribute to tissue repair and regeneration. EVs exhibit unique therapeutic potential in diabetic wound healing and exercise-induced EVs promote angiogenesis through the delivery of angiogenic factors, thus accelerating the healing of diabetic wounds (Egger et al. 2013). Exercise not only alters the levels of miRNAs and various proteins in exosomes, but also reduces endothelial-derived EVs in prediabetic adults while improving cardiovascular endurance (VO_{2peak}) (Eichner et al. 2020; Aas et al. 2023). Moreover, combining exercise-derived EVs with decellularized dermal matrix hydrogels has been shown to enhance skin regeneration in diabetic models, promoting faster wound healing and increased vascularization (Liu et al. 2022). These findings highlight the therapeutic potential of exercise-induced EVs in improving diabetic wound healing and vascular regeneration.

Exercise improves overall metabolic health in individuals with diabetes by regulating lipid metabolism. The beneficial role of EVs in this process is largely mediated through their cargo of microRNAs (miRNAs) and proteins. Specifically, exercise-induced EVs can reduce insulin resistance by regulating hepatic lipid metabolism and muscle fatty acid oxidation (Dubé et al. 2008). Furthermore, exercise has been shown to restore mitochondrial function, enhance intramyocellular lipid storage, and improve insulin sensitivity in type 2 diabetes (Meex et al. 2010). A 12-week training program significantly increased mitochondrial function, insulin-mediated glucose disposal, and metabolic flexibility, despite a slight increase in intramyocellular lipid content (Meex et al. 2010). These improvement in lipid metabolism are linked to bioactive components carried by EVs, including metabolic regulators such as miR-27a and miR-30d-5p (Kacerovsky-Bielesz et al. 2012; Wang et al. 2022; Zheng et al. 2024). As carriers of biological signals generated in response to exercise, EVs demonstrate broad potential in the treatment of diabetes. Exercise-induced EVs not only improve insulin sensitivity, regulate lipid metabolism, and enhance wound healing but may also serve as a novel therapeutic tool for diabetes treatment in the future, particularly in personalized therapy and regenerative medicine.

12.4.3 The Effects of EVs in Exercise-Induced Anti-cancer Benefits

Exercise has been increasingly recognized for its beneficial effects in cancer prevention and treatment. It has been indicated that exercise promotes the secretion of EVs, which can inhibit tumor growth and metastasis (Sadovska et al. 2021). A recent study utilizing the F344 rat prostate cancer model revealed that exercise-induced EVs modulate tumor progression by altering their RNA content (including mRNA, miRNA, and rRNA), which significantly reduces tumor volume and potentially inhibits metastasis (Sadovska et al. 2021). Lifelong aerobic exercise has also been associated with reduced cancer incidence and extended health span, likely due to its anti-inflammatory effects, including decreased levels of pro-inflammatory cytokine such as IL-1β, TNF-α, and IL-6 (Nilsson et al. 2019). Beyond systemic effects, exercise-induced EVs influence the tumor microenvironment by delivering bioactive molecules that modulate immune responses and tissue remodeling. For instance, muscle-derived EVs enriched in miR-29a-3p target Col1a1, inhibiting the tumor extracellular matrix, promoting immune cell infiltration, and enhancing the efficacy of immunotherapy (Luo et al. 2024). Clinically, the levels of miR-29a-3p are associated with reduced extracellular matrix, increased T cell infiltration, and improved responses to immunotherapy (Luo et al. 2024). In contrast, cancer-derived EVs contribute to tumor malignancy and angiogenesis by carrying specific proteins. For example, CAV1-positive EVs from breast cancer cells enhance cellular migration and invasion by transferring cell adhesion-related proteins, while hepatocellular carcinoma-derived exosomes induce angiogenesis via ANGPT2 (Campos et al. 2018)·(Xie et al. 2020). These studies highlight the dual role of EVs within the tumor microenvironment, which may promote cancer progression while also suppressing tumor growth through exercise-induced mechanisms. Future research should aim to elucidate the specific molecular mechanisms underlying exercise-induced EV functions and explore their potential clinical applications, thereby providing more effective strategies for cancer prevention and treatment.

12.5 Further Research Direction and Challenges

In summary, exercise induces physiological stress that perturbs tissue homeostasis and facilitates inter-organ communication. A key mediator of this systemic cross-talk is the population of EVs released in response to physical activity. Exercise-induced EVs play pivotal roles in coordinating cellular and tissue responses across various systems. In the circulatory system, these EVs significantly enhance angiogenic functions by promoting endothelial cell proliferation, migration, and tube formation, thereby accelerating wound healing in diabetic conditions (Liu et al. 2022). Exercise also significantly alters the protein composition of EVs, increasing levels of immune-regulatory proteins such as ANXA1, LYZ, and S100A8, which rapidly enter circulation post-exercise and modulate systemic immune responses (Chong et al. 2024). During exercise, EVs carry a broad array of bioactive molecules that regulate both metabolic and immune functions (Whitham et al. 2018; Chong et al. 2024). Acute exercise significantly increases exosome release, especially in young and healthy individuals, enriching them with immune-related proteins like ADAMTS13, ANXA6, and ITGB2 (Chong et al. 2024). Furthermore, exercise-induced EVs regulate the metabolic state in obese and type 2 diabetes patients by transporting long-chain fatty acid transport proteins such as CD36 and FAT (Nielsen et al. 2019). These effects suggest that exercise-induced EVs not only modulate metabolism but also enhance the body's adaptability by regulating immune responses. While their role in intercellular communication is increasingly recognized, the specific mechanisms and clinical applications of exercise-derived EVs remain to be fully elucidated. Future research could focus on the roles of specific miRNAs and proteins carried by EVs in exercise-induced metabolism, immunity, and tissue repair. In addition, developing EV-based therapeutic strategies such as utilizing exercise-induced EVs to enhance tissue repair or suppress inflammation-may provide new insights for treating various diseases. In conclusion, the role of exercise-induced EVs in mediating organ communication offers a novel perspective on understanding the extensive benefits of exercise for health and holds significant clinical application potential.

Despite a growing body of evidence highlighting the role of EVs in exercise-induced health benefits, several critical questions remain unresolved. First, the precise molecular mechanisms governing EV biogenesis, cargo loading, and release in response to different exercise modalities require further elucidation. Understanding these mechanisms will be critical for designing exercise regimens that optimize EV-mediated benefits. Second, the heterogeneity of EVs poses a challenge for standardizing their isolation, characterization, and functional analysis. Developing more precise and high-throughput analytical techniques to distinguish EV subtypes and their specific bioactive cargos is urgently needed. Third, while preclinical and in vitro studies have demonstrated the potential of EVs in mediating exercise-induced benefits, their clinical translation remains limited. Future research should prioritize validating these findings in human clinical trials and assessing the safety and efficacy of EV-based interventions. Additionally, the role of EVs in mediating systemic adaptations through inter-tissue communication warrants further investigation. Such insights could reveal how exercise-induced EVs contribute to tissue-specific and whole-body physiological improvements. Addressing these challenges will not only expand our understanding of EV biology in the context of exercise but also facilitate the development of innovative EV-based therapeutic strategies for a wide range of diseases.

Compliance with Ethical Standards

Funding This work was supported by the grants from National Natural Science Foundation of China (82020108002 and 82225005 to J.X., 82470369 and 82270291 to L.W.), the grant from Science and Technology Commission of Shanghai Municipality (23410750100, 20DZ2255400 and 21XD1421300 to J.X.), and the "Dawn" Program of Shanghai Education Commission (19SG34 to J.X.). Shanghai Rising-Star Program (24QA2703000 to L.W.). Natural Science Foundation of Shanghai (23ZR1423000 to L.W.). G.L. is supported by the American Heart Association (23CDA1045944).

Conflict of Interest All authors declare no conflict of interest.

Ethical Approval This chapter is a review of previously published accounts, as such, no animal or human studies were performed.

References

Aas V, Øvstebø R, Brusletto BS, Aspelin T, Trøseid AS, Qureshi S, Eid DSO, Olstad OK, Nyman TA, Haug KBF (2023) Distinct microRNA and protein profiles of extracellular vesicles secreted from myotubes from morbidly obese donors with type 2 diabetes in response to electrical pulse stimulation. Front Physiol 14:1143966. https://doi.org/10.3389/fphys.2023.1143966

Anderson JE (2022) Key concepts in muscle regeneration: muscle "cellular ecology" integrates a gestalt of cellular cross-talk, motility, and activity to remodel structure and restore function. Eur J Appl Physiol 122(2):273–300. https://doi.org/10.1007/s00421-021-04865-4

Bei Y, Xu T, Lv D, Yu P, Xu J, Che L, Das A, Tigges J, Toxavidis V, Ghiran I, Shah R, Li Y, Zhang Y, Das S, Xiao J (2017) Exercise-induced circulating extracellular vesicles protect against cardiac ischemia-reperfusion injury. Basic Res Cardiol 112(4):38. https://doi.org/10.1007/s00395-017-0628-z

Bei Y, Xu T, Lv D, Yu P, Xu J, Che L, Das A, Tigges J, Toxavidis V, Ghiran I, Shah R, Li Y, Zhang Y, Das S, Xiao J (2019) Correction to: exercise-induced circulating extracellular vesicles protect against cardiac ischemia-reperfusion injury. Basic Res Cardiol 114(6):44. https://doi.org/10.1007/s00395-019-0751-0

Buzas EI (2023) The roles of extracellular vesicles in the immune system. Nat Rev Immunol 23(4):236–250. https://doi.org/10.1038/s41577-022-00763-8

Cable J, Witwer KW, Coffey RJ, Milosavljevic A, von Lersner AK, Jimenez L, Pucci F, Barr MM, Dekker N, Barman B, Humphrys D, Williams J, de Palma M, Guo W, Bastos N, Hill AF, Levy E, Hantak MP, Crewe C, Aikawa E, Adamczyk AM, Zanotto TM, Ostrowski M, Arab T, Rabe DC, Sheikh A, da Silva DR, Jones JC, Okeoma C, Gaborski T, Zhang Q, Gololobova O (2023) Exosomes, microvesicles, and other extracellular vesicles-a keystone symposia report. Ann N Y Acad Sci 1523(1):24–37. https://doi.org/10.1111/nyas.14974

Campos A, Salomon C, Bustos R, Díaz J, Martínez S, Silva V, Reyes C, Díaz-Valdivia N, Varas-Godoy M, Lobos-González L, Quest AF (2018) Caveolin-1-containing extracellular vesicles transport adhesion proteins and promote malignancy in breast cancer cell lines. Nanomedicine (Lond) 13(20):2597–2609. https://doi.org/10.2217/nnm-2018-0094

Castaño C, Meza-Ramos A, Batlle M, Guasch E, Novials A, Párrizas M (2022) Treatment with EV-miRNAs alleviates obesity-associated metabolic dysfunction in mice. Int J Mol Sci 23(23). https://doi.org/10.3390/ijms232314920

Chen M, Lin S, Zhou C, Cui D, Haick H, Tang N (2023) From conventional to microfluidic: progress in extracellular vesicle separation and individual characterization. Adv Healthc Mater 12(8):e2202437. https://doi.org/10.1002/adhm.202202437

Chen W, Ye Q, Dong Y (2024) Long term exercise-derived exosomal LncRNA CRNDE mitigates myocardial infarction injury through miR-489-3p/Nrf2 signaling axis. Nanomedicine 55:102717. https://doi.org/10.1016/j.nano.2023.102717

Chong MC, Silva A, James PF, Wu SSX, Howitt J (2022) Exercise increases the release of NAMPT in extracellular vesicles and alters NAD(+) activity in recipient cells. Aging Cell 21(7):e13647. https://doi.org/10.1111/acel.13647. Available at: https://www.ncbi.nlm.nih.gov/pubmed/35661560, https://pmc.ncbi.nlm.nih.gov/articles/PMC9282849/pdf/ACEL-21-e13647.pdf

Chong MC, Shah AD, Schittenhelm RB, Silva A, James PF, Wu SSX, Howitt J (2024) Acute exercise-induced release of innate immune proteins via small extracellular vesicles changes with aerobic fitness and age. Acta Physiol (Oxf) 240(3):e14095. https://doi.org/10.1111/apha.14095

Conkright WR, Kargl CK, Hubal MJ, Tiede DR, Beckner ME, Sterczala AJ, Krajewski KT, Martin BJ, Flanagan SD, Greeves JP, O'Leary TJ, Wardle SL, Sahu A, Ambrosio F, Nindl BC (2024) Acute resistance exercise modifies extracellular vesicle miRNAs targeting anabolic gene pathways: a prospective cohort study. Med Sci Sports Exerc 56(7):1225–1232. https://doi.org/10.1249/MSS.0000000000003408. Available at: https://www.ncbi.nlm.nih.gov/pubmed/38377006

Darragh IAJ, O'Driscoll L, Egan B (2021) Exercise training and circulating small extracellular vesicles: appraisal of methodological approaches and current knowledge. Front Physiol 12:738333. https://doi.org/10.3389/fphys.2021.738333. Available at: https://www.ncbi.nlm.nih.gov/pubmed/34777006, https://pmc.ncbi.nlm.nih.gov/articles/PMC8581208/pdf/fphys-12-738333.pdf

Dilsiz N (2024) A comprehensive review on recent advances in exosome isolation and characterization: toward clinical applications. Transl Oncol 50:102121. https://doi.org/10.1016/j.tranon.2024.102121

Doncheva AI, Romero S, Ramirez-Garrastacho M, Lee S, Kolnes KJ, Tangen DS, Olsen T, Drevon CA, Llorente A, Dalen KT, Hjorth M (2022) Extracellular vesicles and microRNAs are altered in response to exercise, insulin sensitivity and overweight. Acta Physiol (Oxf) 236(4):e13862. https://doi.org/10.1111/apha.13862. Available at: https://www.ncbi.nlm.nih.gov/pubmed/36377504, https://pmc.ncbi.nlm.nih.gov/articles/PMC9788120/pdf/APHA-236-e13862.pdf

Du Y, Wu L, Wang L, Reiter RJ, Lip GYH, Ren J (2023) Extracellular vesicles in cardiovascular diseases: from pathophysiology to diagnosis and therapy. Cytokine Growth Factor Rev 74:40–55. https://doi.org/10.1016/j.cytogfr.2023.09.006

Dubé JJ, Amati F, Stefanovic-Racic M, Toledo FG, Sauers SE, Goodpaster BH (2008) Exercise-induced alterations in intramyocellular lipids and insulin resistance: the athlete's paradox revisited. Am J Physiol Endocrinol Metab 294(5):E882–E888. https://doi.org/10.1152/ajpendo.00769.2007

Egger A, Kreis R, Allemann S, Stettler C, Diem P, Buehler T, Boesch C, Christ ER (2013) The effect of aerobic exercise on intrahepatocellular and intramyocellular lipids in healthy subjects. PLoS One 8(8):e70865. https://doi.org/10.1371/journal.pone.0070865

Eichner NZM, Gilbertson NM, Heiston EM, Musante L, Salvia SLA, Weltman A, Erdbrugger U, Malin SK (2020) Interval exercise lowers circulating CD105 extracellular vesicles in prediabetes. Med Sci Sports Exerc 52(3):729–735. https://doi.org/10.1249/mss.0000000000002185

Fandl HK, Garcia VP, Treuth JW, Brewster LM, Greiner JJ, Davy KP, Stauffer BL, Desouza CA (2023) Endothelial-derived extracellular vesicles from obese/hypertensive adults increase factors associated with hypertrophy and fibrosis in cardiomyocytes. Am J Physiol Heart Circ Physiol 324(5):H675–h685. https://doi.org/10.1152/ajpheart.00035.2023

Foers AD, Chatfield S, Dagley LF, Scicluna BJ, Webb AI, Cheng L, Hill AF, Wicks IP, Pang KC (2018) Enrichment of extracellular vesicles from human synovial fluid using size exclusion chromatography. J Extracell Vesicles 7(1):1490145. https://doi.org/10.1080/20013078.2018.1490145

Fruhbeis C, Helmig S, Tug S, Simon P, Kramer-Albers EM (2015) Physical exercise induces rapid release of small extracellular vesicles into the circulation. J Extracell Vesicles 4:28239. https://doi.org/10.3402/jev.v4.28239. Available at: https://www.ncbi.nlm.nih.gov/pubmed/26142461, https://www.ncbi.nlm.nih.gov/pmc/articles/PMC4491306/pdf/JEV-4-28239.pdf

Gao J, Li A, Hu J, Feng L, Liu L, Shen Z (2022) Recent developments in isolating methods for exosomes. Front Bioeng Biotechnol 10:1100892. https://doi.org/10.3389/fbioe.2022.1100892

Gregory CD, Rimmer MP (2023) Extracellular vesicles arising from apoptosis: forms, functions, and applications. J Pathol 260(5):592–608. https://doi.org/10.1002/path.6138

Hou Z, Qin X, Hu Y, Zhang X, Li G, Wu J, Li J, Sha J, Chen J, Xia J, Wang L, Gao F (2019) Longterm exercise-derived exosomal miR-342-5p: a novel exerkine for cardioprotection. Circ Res 124(9):1386–1400. https://doi.org/10.1161/circresaha.118.314635

Huang KY, Upadhyay G, Ahn Y, Sakakura M, Pagan-Diaz GJ, Cho Y, Weiss AC, Huang C, Mitchell JW, Li J, Tan Y, Deng YH, Ellis-Mohr A, Dou Z, Zhang X, Kang S, Chen Q, Sweedler JV, Im SG, Bashir R, Chung HJ, Popescu G, Gillette MU, Gazzola M, Kong H (2024a) Neuronal innervation regulates the secretion of neurotrophic myokines and exosomes from skeletal muscle. Proc Natl Acad Sci USA 121(19):e2313590121. https://doi.org/10.1073/pnas.2313590121

Huang M, Cheng S, Li Z, Chen J, Wang C, Li J, Zheng H (2024b) Preconditioning exercise inhibits neuron ferroptosis and ameliorates brain ischemia damage by skeletal muscle-derived exosomes via regulating miR-484/ACSL4 axis. Antioxid Redox Signal 41(13–15):769–792. https://doi.org/10.1089/ars.2023.0492. Available at: https://www.ncbi.nlm.nih.gov/pubmed/38545792

Kacerovsky-Bielesz G, Kacerovsky M, Chmelik M, Farukuoye M, Ling C, Pokan R, Tschan H, Szendroedi J, Schmid AI, Gruber S, Herder C, Wolzt M, Moser E, Pacini G, Smekal G, Groop L, Roden M (2012) A single nucleotide polymorphism associates with the response of muscle ATP synthesis to long-term exercise training in relatives of type 2 diabetic humans. Diabetes Care 35(2):350–357. https://doi.org/10.2337/dc11-1426

Lange S (2023) Extracellular vesicles in phylogeny. Int J Mol Sci 24(13). https://doi.org/10.3390/ijms241310466

Leong SY, Ong HB, Tay HM, Kong F, Upadya M, Gong L, Dao M, Dalan R, Hou HW (2022) Microfluidic size exclusion chromatography (μSEC) for extracellular vesicles and plasma protein separation. Small 18(6):e2104470. https://doi.org/10.1002/smll.202104470

Li M, Lou D, Chen J, Shi K, Wang Y, Zhu Q, Liu F, Zhang Y (2021) Deep dive on the proteome of salivary extracellular vesicles: comparison between ultracentrifugation and polymer-based precipitation isolation. Anal Bioanal Chem 413(2):365–375. https://doi.org/10.1007/s00216-020-03004-w

Li G, Chen T, Dahlman J, Eniola-Adefeso L, Ghiran IC, Kurre P, Lam WA, Lang JK, Marban E, Martin P, Momma S, Moos M, Nelson DJ, Raffai RL, Ren X, Sluijter JPG, Stott SL, Vunjak-Novakovic G, Walker ND, Wang Z, Witwer KW, Yang PC, Lundberg MS, Ochocinska MJ, Wong R, Zhou G, Chan SY, Das S, Sundd P (2023) Current challenges and future directions for engineering extracellular vesicles for heart, lung, blood and sleep diseases. J Extracell Vesicles 12(2):e12305. https://doi.org/10.1002/jev2.12305. Available at: https://www.ncbi.nlm.nih.gov/pubmed/36775986, https://pmc.ncbi.nlm.nih.gov/articles/PMC9923045/pdf/JEV2-12-12305.pdf

Lisi V, Senesi G, Bertola N, Pecoraro M, Bolis S, Gualerzi A, Picciolini S, Raimondi A, Fantini C, Moretti E, Parisi A, Sgro P, Di Luigi L, Geiger R, Ravera S, Vassalli G, Caporossi D, Balbi C (2023) Plasma-derived extracellular vesicles released after endurance exercise exert cardioprotective activity through the activation of antioxidant pathways. Redox Biol 63:102737. https://doi.org/10.1016/j.redox.2023.102737. Available at: https://www.ncbi.nlm.nih.gov/pubmed/37236143, https://www.ncbi.nlm.nih.gov/pmc/articles/PMC10220283/pdf/main.pdf

Liu H, Wu B, Shi X, Cao Y, Zhao X, Liang D, Qin Q, Liang X, Lu W, Wang D, Liu J (2022) Aerobic exercise-induced circulating extracellular vesicle combined decellularized dermal matrix hydrogel facilitates diabetic wound healing by promoting angiogenesis. Front Bioeng Biotechnol 10:903779. https://doi.org/10.3389/fbioe.2022.903779

Luo Z, Mei J, Wang X, Wang R, He Z, Geffen Y, Sun X, Zhang X, Xu J, Wan R, Feng X, Jiao C, Su X, Sun J, Chen S, Chen J, Mao W, Yang Y, Sun Y (2024) Voluntary exercise sensitizes cancer immunotherapy via the collagen inhibition-orchestrated inflammatory tumor immune microenvironment. Cell Rep 43(9):114697. https://doi.org/10.1016/j.celrep.2024.114697

Maggio S, Canonico B, Ceccaroli P, Polidori E, Cioccoloni A, Giacomelli L, Ferri Marini C, Annibalini G, Gervasi M, Benelli P, Fabbri F, Del Coco L, Fanizzi FP, Giudetti AM, Lucertini F, Guescini M (2023) Modulation of the circulating extracellular vesicles in response to different exercise regimens and study of their inflammatory effects. Int J Mol Sci 24(3). https://doi.org/10.3390/ijms24033039

Martínez-Greene JA, Hernández-Ortega K, Quiroz-Baez R, Resendis-Antonio O, Pichardo-Casas I, Sinclair DA, Budnik B, Hidalgo-Miranda A, Uribe-Querol E, Ramos-Godínez MDP, Martínez-Martínez E (2021) Quantitative proteomic analysis of extracellular vesicle subgroups isolated by an optimized method combining polymer-based precipitation and size exclusion chromatography. J Extracell Vesicles 10(6):e12087. https://doi.org/10.1002/jev2.12087

Mas-Bargues C, Alique M (2023) Extracellular vesicles as "very important particles" (VIPs) in aging. Int J Mol Sci 24(4). https://doi.org/10.3390/ijms24044250

Mazahir F, Yadav AK (2024) Recent progress in engineered extracellular vesicles and their biomedical applications. Life Sci 350:122747. https://doi.org/10.1016/j.lfs.2024.122747

McIlvenna LC, Whitham M (2023) Exercise, healthy ageing, and the potential role of small extracellular vesicles. J Physiol 601(22):4937–4951. https://doi.org/10.1113/JP282468. Available at: https://www.ncbi.nlm.nih.gov/pubmed/35388915, https://pmc.ncbi.nlm.nih.gov/articles/PMC10952297/pdf/TJP-601-4937.pdf

Meex RC, Schrauwen-Hinderling VB, Moonen-Kornips E, Schaart G, Mensink M, Phielix E, van de Weijer T, Sels JP, Schrauwen P, Hesselink MK (2010) Restoration of muscle mitochondrial function and metabolic flexibility in type 2 diabetes by exercise training is paralleled by increased myocellular fat storage and improved insulin sensitivity. Diabetes 59(3):572–579. https://doi.org/10.2337/db09-1322

Mittelbrunn M, Sánchez-Madrid F (2012) Intercellular communication: diverse structures for exchange of genetic information. Nat Rev Mol Cell Biol 13(5):328–335. https://doi.org/10.1038/nrm3335

Mohammad S, Bhattacharjee J, Tzaneva V, Hutchinson KA, Shaikh M, Fernandes da Silva D, Burger D, Adamo KB (2023) The influence of exercise-associated small extracellular vesicles on trophoblasts in vitro. Biomedicine 11(3). https://doi.org/10.3390/biomedicines11030857.

Available at: https://www.ncbi.nlm.nih.gov/pubmed/36979835, https://mdpi-res.com/d_attachment/biomedicines/biomedicines-11-00857/article_deploy/biomedicines-11-00857.pdf?version=1678526945
Musi N, Fujii N, Hirshman MF, Ekberg I, Fröberg S, Ljungqvist O, Thorell A, Goodyear LJ (2001) AMP-activated protein kinase (AMPK) is activated in muscle of subjects with type 2 diabetes during exercise. Diabetes 50(5):921–927. https://doi.org/10.2337/diabetes.50.5.921
Nie Y, Sato Y, Garner RT, Kargl C, Wang C, Kuang S, Gilpin CJ, Gavin TP (2019) Skeletal muscle-derived exosomes regulate endothelial cell functions via reactive oxygen species-activated nuclear factor-kappaB signalling. Exp Physiol 104(8):1262–1273. https://doi.org/10.1113/EP087396. Available at: https://www.ncbi.nlm.nih.gov/pubmed/31115069
Nielsen MH, Sabaratnam R, Pedersen AJT, Højlund K, Handberg A (2019) Acute exercise increases plasma levels of muscle-derived microvesicles carrying fatty acid transport proteins. J Clin Endocrinol Metab 104(10):4804–4814. https://doi.org/10.1210/jc.2018-02547
Nilsson MI, Bourgeois JM, Nederveen JP, Leite MR, Hettinga BP, Bujak AL, May L, Lin E, Crozier M, Rusiecki DR, Moffatt C, Azzopardi P, Young J, Yang Y, Nguyen J, Adler E, Lan L, Tarnopolsky MA (2019) Lifelong aerobic exercise protects against inflammaging and cancer. PLoS One 14(1):e0210863. https://doi.org/10.1371/journal.pone.0210863
Nowak M, Górczyńska J, Kołodzińska K, Rubin J, Choromańska A (2023) Extracellular vesicles as drug transporters. Int J Mol Sci 24(12). https://doi.org/10.3390/ijms241210267
Osawa S, Kato H, Kemmoku D, Yamaguchi S, Jiang L, Tsuchiya Y, Takakura H, Izawa T (2024) Exercise training-driven exosomal miRNA-323-5p activity suppresses adipogenic conversion of 3T3-L1 cells via the DUSP3/ERK pathway. Biochem Biophys Res Commun 734:150447. https://doi.org/10.1016/j.bbrc.2024.150447
Papadopetraki A, Maridaki M, Zagouri F, Dimopoulos MA, Koutsilieris M, Philippou A (2022) Physical exercise restrains cancer progression through muscle-derived factors. Cancers (Basel) 14(8). https://doi.org/10.3390/cancers14081892
Park J, Go EB, Oh JS, Lee JK, Lee SY (2021) Multiple-cycle polymeric extracellular vesicle precipitation and its evaluation by targeted mass spectrometry. Int J Mol Sci 22(9). https://doi.org/10.3390/ijms22094311
Priedols M, Paidere G, Santos CB, Miscenko A, Bergmanis RG, Spule A, Bekere B, Mozolevskis G, Abols A, Rimsa R (2023) Bifurcated asymmetric field flow fractionation of nanoparticles in PDMS-free microfluidic devices for applications in label-free extracellular vesicle separation. Polymers (Basel) 15(4). https://doi.org/10.3390/polym15040789
Qazi REM, Sajid Z, Zhao C, Hussain I, Iftikhar F, Jameel M, Rehman FU, Mian AA (2023) Lyophilization based isolation of exosomes. Int J Mol Sci 24(13). https://doi.org/10.3390/ijms241310477
Sadovska L, Auders J, Keisa L, Romanchikova N, Silamikele L, Kreismane M, Zayakin P, Takahashi S, Kalnina Z, Line A (2021) Exercise-induced extracellular vesicles delay the progression of prostate cancer. Front Mol Biosci 8:784080. https://doi.org/10.3389/fmolb.2021.784080. Available at: https://www.ncbi.nlm.nih.gov/pubmed/35087866, https://pmc.ncbi.nlm.nih.gov/articles/PMC8787363/pdf/fmolb-08-784080.pdf
Shami-Shah A, Travis BG, Walt DR (2023) Advances in extracellular vesicle isolation methods: a path towards cell-type specific EV isolation. Extracell Vesicles Circ Nucl Acids 4(3):447–460. https://doi.org/10.20517/evcna.2023.14
Shi H, Hao X, Sun Y, Zhao Y, Wang Y, Cao X, Gong Z, Ji S, Lu J, Yan Y, Yu X, Luo X, Wang J, Wang H (2024) Exercise-inducible circulating extracellular vesicle irisin promotes browning and the thermogenic program in white adipose tissue. Acta Physiol (Oxf) 240(3):e14103. https://doi.org/10.1111/apha.14103. Available at: https://www.ncbi.nlm.nih.gov/pubmed/38288566
Tao L, Wang J, Wang K, Liu Q, Li H, Xu S, Gu C, Zhu Y (2024) Exerkine FNDC5/irisin-enriched exosomes promote proliferation and inhibit ferroptosis of osteoblasts through interaction with Caveolin-1. Aging Cell 23(8):e14181. https://doi.org/10.1111/acel.14181. Available at: https://www.ncbi.nlm.nih.gov/pubmed/38689463, https://pmc.ncbi.nlm.nih.gov/articles/PMC11320359/pdf/ACEL-23-e14181.pdf

Théry C, Witwer KW, Aikawa E, Alcaraz MJ, Anderson JD, Andriantsitohaina R, Antoniou A, Arab T, Archer F, Atkin-Smith GK, Ayre DC, Bach JM, Bachurski D, Baharvand H, Balaj L, Baldacchino S, Bauer NN, Baxter AA, Bebawy M, Beckham C, Bedina Zavec A, Benmoussa A, Berardi AC, Bergese P, Bielska E, Blenkiron C, Bobis-Wozowicz S, Boilard E, Boireau W, Bongiovanni A, Borràs FE, Bosch S, Boulanger CM, Breakefield X, Breglio AM, Brennan M, Brigstock DR, Brisson A, Broekman ML, Bromberg JF, Bryl-Górecka P, Buch S, Buck AH, Burger D, Busatto S, Buschmann D, Bussolati B, Buzás EI, Byrd JB, Camussi G, Carter DR, Caruso S, Chamley LW, Chang YT, Chen C, Chen S, Cheng L, Chin AR, Clayton A, Clerici SP, Cocks A, Cocucci E, Coffey RJ, Cordeiro-da-Silva A, Couch Y, Coumans FA, Coyle B, Crescitelli R, Criado MF, D'Souza-Schorey C, Das S, Datta Chaudhuri A, de Candia P, De Santana EF, De Wever O, Del Portillo HA, Demaret T, Deville S, Devitt A, Dhondt B, Di Vizio D, Dieterich LC, Dolo V, Dominguez Rubio AP, Dominici M, Dourado MR, Driedonks TA, Duarte FV, Duncan HM, Eichenberger RM, Ekström K, El Andaloussi S, Elie-Caille C, Erdbrügger U, Falcón-Pérez JM, Fatima F, Fish JE, Flores-Bellver M, Försönits A, Frelet-Barrand A, Fricke F, Fuhrmann G, Gabrielsson S, Gámez-Valero A, Gardiner C, Gärtner K, Gaudin R, Gho YS, Giebel B, Gilbert C, Gimona M, Giusti I, Goberdhan DC, Görgens A, Gorski SM, Greening DW, Gross JC, Gualerzi A, Gupta GN, Gustafson D, Handberg A, Haraszti RA, Harrison P, Hegyesi H, Hendrix A, Hill AF, Hochberg FH, Hoffmann KF, Holder B, Holthofer H, Hosseinkhani B, Hu G, Huang Y, Huber V, Hunt S, Ibrahim AG, Ikezu T, Inal JM, Isin M, Ivanova A, Jackson HK, Jacobsen S, Jay SM, Jayachandran M, Jenster G, Jiang L, Johnson SM, Jones JC, Jong A, Jovanovic-Talisman T, Jung S, Kalluri R, Kano SI, Kaur S, Kawamura Y, Keller ET, Khamari D, Khomyakova E, Khvorova A, Kierulf P, Kim KP, Kislinger T, Klingeborn M, Klinke DJ, Kornek M 2nd, Kosanović MM, Kovács ÁF, Krämer-Albers EM, Krasemann S, Krause M, Kurochkin IV, Kusuma GD, Kuypers S, Laitinen S, Langevin SM, Languino LR, Lannigan J, Lässer C, Laurent LC, Lavieu G, Lázaro-Ibáñez E, Le Lay S, Lee MS, Lee YXF, Lemos DS, Lenassi M, Leszczynska A, Li IT, Liao K, Libregts SF, Ligeti E, Lim R, Lim SK, Linē A, Linnemannstöns K, Llorente A, Lombard CA, Lorenowicz MJ, Lörincz ÁM, Lötvall J, Lovett J, Lowry MC, Loyer X, Lu Q, Lukomska B, Lunavat TR, Maas SL, Malhi H, Marcilla A, Mariani J, Mariscal J, Martens-Uzunova ES, Martin-Jaular L, Martinez MC, Martins VR, Mathieu M, Mathivanan S, Maugeri M, McGinnis LK, McVey MJ, Meckes DG, Meehan JKL, Mertens I, Minciacchi VR, Möller A, Møller Jørgensen M, Morales-Kastresana A, Morhayim J, Mullier F, Muraca M, Musante L, Mussack V, Muth DC, Myburgh KH, Najrana T, Nawaz M, Nazarenko I, Nejsum P, Neri C, Neri T, Nieuwland R, Nimrichter L, Nolan JP, Nolte-'t Hoen EN, Noren Hooten N, O'Driscoll L, O'Grady T, O'Loghlen A, Ochiya T, Olivier M, Ortiz A, Ortiz LA, Osteikoetxea X, Østergaard O, Ostrowski M, Park J, Pegtel DM, Peinado H, Perut F, Pfaffl MW, Phinney DG, Pieters BC, Pink RC, Pisetsky DS, Pogge von Strandmann E, Polakovicova I, Poon IK, Powell BH, Prada I, Pulliam L, Quesenberry P, Radeghieri A, Raffai RL, Raimondo S, Rak J, Ramirez MI, Raposo G, Rayyan MS, Regev-Rudzki N, Ricklefs FL, Robbins PD, Roberts DD, Rodrigues SC, Rohde E, Rome S, Rouschop KM, Rughetti A, Russell AE, Saá P, Sahoo S, Salas-Huenuleo E, Sánchez C, Saugstad JA, Saul MJ, Schiffelers RM, Schneider R, Schøyen TH, Scott A, Shahaj E, Sharma S, Shatnyeva O, Shekari F, Shelke GV, Shetty AK, Shiba K, Siljander PR, Silva AM, Skowronek A, Snyder OL 2nd, Soares RP, Sódar BW, Soekmadji C, Sotillo J, Stahl PD, Stoorvogel W, Stott SL, Strasser EF, Swift S, Tahara H, Tewari M, Timms K, Tiwari S, Tixeira R, Tkach M, Toh WS, Tomasini R, Torrecilhas AC, Tosar JP, Toxavidis V, Urbanelli L, Vader P, van Balkom BW, van der Grein SG, Van Deun J, van Herwijnen MJ, Van Keuren-Jensen K, van Niel G, van Royen ME, van Wijnen AJ, Vasconcelos MH, Vechetti IJJ, Veit TD, Vella LJ, Velot É, Verweij FJ, Vestad B, Viñas JL, Visnovitz T, Vukman KV, Wahlgren J, Watson DC, Wauben MH, Weaver A, Webber JP, Weber V, Wehman AM, Weiss DJ, Welsh JA, Wendt S, Wheelock AM, Wiener Z, Witte L, Wolfram J, Xagorari A, Xander P, Xu J, Yan X, Yáñez-Mó M, Yin H, Yuana Y, Zappulli V, Zarubova J, Žėkas V, Zhang JY, Zhao Z, Zheng L, Zheutlin AR, Zickler AM, Zimmermann P, Zivkovic AM, Zocco D, Zuba-Surma EK (2018) Minimal information for studies of extracellular vesicles 2018 (MISEV2018): a position statement of the International Society for Extracellular Vesicles and update of the MISEV2014 guidelines. J Extracell Vesicles 7(1):1535750. https://doi.org/10.1080/20013078.2018.1535750

Vitucci D, Martone D, Alfieri A, Buono P (2023) Muscle-derived exosomes and exercise in cancer prevention. Front Mol Med 3:1202190. https://doi.org/10.3389/fmmed.2023.1202190

Vrablova V, Kosutova N, Blsakova A, Bertokova A, Kasak P, Bertok T, Tkac J (2023) Glycosylation in extracellular vesicles: isolation, characterization, composition, analysis and clinical applications. Biotechnol Adv 67:108196. https://doi.org/10.1016/j.biotechadv.2023.108196

Wang D, Zhang X, Li Y, Jia L, Zhai L, Wei W, Zhang L, Jiang H, Bai Y (2022) Exercise-induced browning of white adipose tissue and improving skeletal muscle insulin sensitivity in obese/non-obese growing mice: do not neglect exosomal miR-27a. Front Nutr 9:940673. https://doi.org/10.3389/fnut.2022.940673

Wang Y, Liu Y, Zhang S, Li N, Xing C, Wang C, Wang J, Wei M, Yang G, Yuan L (2023) Exercise improves metabolism and alleviates atherosclerosis via muscle-derived extracellular vesicles. Aging Dis 14(3):952–965. https://doi.org/10.14336/AD.2022.1131. Available at: https://www.ncbi.nlm.nih.gov/pubmed/37191422, https://pmc.ncbi.nlm.nih.gov/articles/PMC10187707/pdf/ad-14-3-952.pdf

Wang Q, Du J, Ma R (2024) White adipocyte-derived exosomal miR-23b inhibits thermogenesis by targeting Elf4 to regulate GLP-1R transcription. Naunyn Schmiedebergs Arch Pharmacol 397(8):5847–5860. https://doi.org/10.1007/s00210-024-02984-1

Watanabe S, Sudo Y, Makino T, Kimura S, Tomita K, Noguchi M, Sakurai H, Shimizu M, Takahashi Y, Sato R, Yamauchi Y (2022) Skeletal muscle releases extracellular vesicles with distinct protein and microRNA signatures that function in the muscle microenvironment. PNAS Nexus 1(4):pgac173. https://doi.org/10.1093/pnasnexus/pgac173

Whitham M, Parker BL, Friedrichsen M, Hingst JR, Hjorth M, Hughes WE, Egan CL, Cron L, Watt KI, Kuchel RP, Jayasooriah N, Estevez E, Petzold T, Suter CM, Gregorevic P, Kiens B, Richter EA, James DE, Wojtaszewski JFP, Febbraio MA (2018) Extracellular vesicles provide a means for tissue crosstalk during exercise. Cell Metab 27(1):237–251.e4. https://doi.org/10.1016/j.cmet.2017.12.001

Wu J, Ma Y, Chen Y (2023) Extracellular vesicles and COPD: foe or friend? J Nanobiotechnol 21(1):147. https://doi.org/10.1186/s12951-023-01911-5

Xie JY, Wei JX, Lv LH, Han QF, Yang WB, Li GL, Wang PX, Wu SB, Duan JX, Zhuo WF, Liu PQ, Min J (2020) Angiopoietin-2 induces angiogenesis via exosomes in human hepatocellular carcinoma. Cell Commun Signal 18(1):46. https://doi.org/10.1186/s12964-020-00535-8

Zhang Q, Higginbotham JN, Jeppesen DK, Yang YP, Li W, McKinley ET, Graves-Deal R, Ping J, Britain CM, Dorsett KA, Hartman CL, Ford DA, Allen RM, Vickers KC, Liu Q, Franklin JL, Bellis SL, Coffey RJ (2019) Transfer of functional cargo in exomeres. Cell Rep 27(3):940–954.e6. https://doi.org/10.1016/j.celrep.2019.01.009

Zhang Q, Jeppesen DK, Higginbotham JN, Graves-Deal R, Trinh VQ, Ramirez MA, Sohn Y, Neininger AC, Taneja N, McKinley ET, Niitsu H, Cao Z, Evans R, Glass SE, Ray KC, Fissell WH, Hill S, Rose KL, Huh WJ, Washington MK, Ayers GD, Burnette DT, Sharma S, Rome LH, Franklin JL, Lee YA, Liu Q, Coffey RJ (2021) Supermeres are functional extracellular nanoparticles replete with disease biomarkers and therapeutic targets. Nat Cell Biol 23(12):1240–1254. https://doi.org/10.1038/s41556-021-00805-8

Zhao H, Chen X, Hu G, Li C, Guo L, Zhang L, Sun F, Xia Y, Yan W, Cui Z, Guo Y, Guo X, Huang C, Fan M, Wang S, Zhang F, Tao L (2022) Small extracellular vesicles from brown adipose tissue mediate exercise cardioprotection. Circ Res 130(10):1490–1506. https://doi.org/10.1161/CIRCRESAHA.121.320458. Available at: https://www.ncbi.nlm.nih.gov/pubmed/35387487, https://www.ingentaconnect.com/content/wk/res/2022/00000130/00000010/art00003;jsessionid=b26dcf1q92q32.x-ic-live-03

Zheng L, Rao Z, Wu J, Ma X, Jiang Z, Xiao W (2024) Resistance exercise improves glycolipid metabolism and mitochondrial biogenesis in skeletal muscle of T2DM mice via miR-30d-5p/SIRT1/PGC-1α Axis. Int J Mol Sci 25(22). https://doi.org/10.3390/ijms252212416

Chapter 13
The Epigenetic Regulation of Exerkines

Yihao Hou, Shujuan Wang, Yin Jia, and Hang Zhao

Abstract Exerkines are exercise-induced signaling molecules secreted by skeletal muscle, adipose tissue, and other organs, playing crucial roles in metabolic, cardiovascular, and cognitive regulation. Recent studies have uncovered that epigenetic mechanisms—including DNA methylation, histone modifications, and non-coding RNAs—serve as pivotal regulators of exerkine expression and secretion. This review highlights how DNA methylation dynamically alters the expression of key exerkines such as Irisin, IL-6, BDNF, and FGF21; how histone acetylation, methylation, and lactylation modulate genes like Myostatin and IL-6; and how various non-coding RNAs (miRNAs, lncRNAs, and circRNAs) influence the post-transcriptional regulation and extracellular vesicle-mediated transport of exerkines. These epigenetic modifications act as molecular bridges connecting physical exercise to long-term physiological adaptation and disease prevention. Understanding these regulatory networks provides valuable insight into the mechanisms by which exercise exerts health-promoting effects and opens new avenues for therapeutic strategies in metabolic and neurological disorders.

Keywords Exerkines · DNA methylation · Histone modificationsm · Non-coding RNAs

Y. Hou
Department of Cardiology, Xijing Hospital, Air Force Medical University (Fourth Military Medical University), Xi'an, Shaanxi, China

S. Wang · H. Zhao (✉)
Department of Health Service, Air Force Medical University (Fourth Military Medical University), Xi'an, Shaanxi, China

Y. Jia
Student of the Basic Medical College, Air Force Medical University (Fourth Military Medical University), Xi'an, Shaanxi, China

F. Gao, X. Zhang (eds.), *Exercise, Epigenetics and Human Health*, Epigenetics and Human Health 15, https://doi.org/10.1007/978-3-032-08045-5_13

Abbreviations

BDNF	Brain-derived neurotrophic factor
DNMT1	DNA methyltransferase 1
FGF21	Fibroblast growth factor 21
FNDC5	Fibronectin type III domain-containing protein 5
FSTL1	Follistatin-like protein 1
HATs	Histone acetyltransferases
HCC	Hepatocellular carcinoma
HDACs	Histone deacetylases
IL-6	Interleukin-6
MSTN	Myostatin
PPARα	Peroxisome proliferator-activated receptor alpha

13.1 Introduction

Exerkines are signaling molecules secreted by various tissues, including skeletal muscle, adipose tissue and liver, in response to physical exercise. They mediate numerous beneficial systemic effects, such as improved metabolism, enhanced cardiovascular function, and cognitive health. Recent studies highlight the critical role of epigenetic mechanisms in regulating the expression and secretion of exerkines, providing insights into how physical activity translates into long-term physiological adaptations. The primary epigenetic regulations involved in modulating exerkine production include DNA methylation, histone modifications (acetylation, methylation, phosphorylation, and lactylation), and non-coding RNAs (miRNAs, lncRNAs, and circRNAs) (Zhang et al. 2015; Zhao et al. 2019; Orange et al. 2023). These mechanisms dynamically respond to exercise stimuli, altering gene expression patterns to support adaptive biological responses, thus offering potential targets for therapeutic intervention in metabolic, cardiovascular, and neurological diseases. In this review, we will summarize recent advances in understanding how DNA methylation, histone modifications, and non-coding RNAs regulate exerkine secretion, providing valuable insights and inspiration for future research direction (Fig. 13.1).

13.2 DNA Methylation and Its Impact on Exerkine Secretion

DNA methylation involves the addition of methyl groups to the cytosine residues of DNA, typically at CpG sites, which can lead to gene silencing. In the context of exerkines, DNA methylation plays a significant role in regulating the expression of genes involved in exercise-induced signaling. Changes in DNA methylation patterns can modulate the secretion of various exerkines that mediate the beneficial effects of physical activity, such as metabolic regulation, immune function and cognitive improvement.

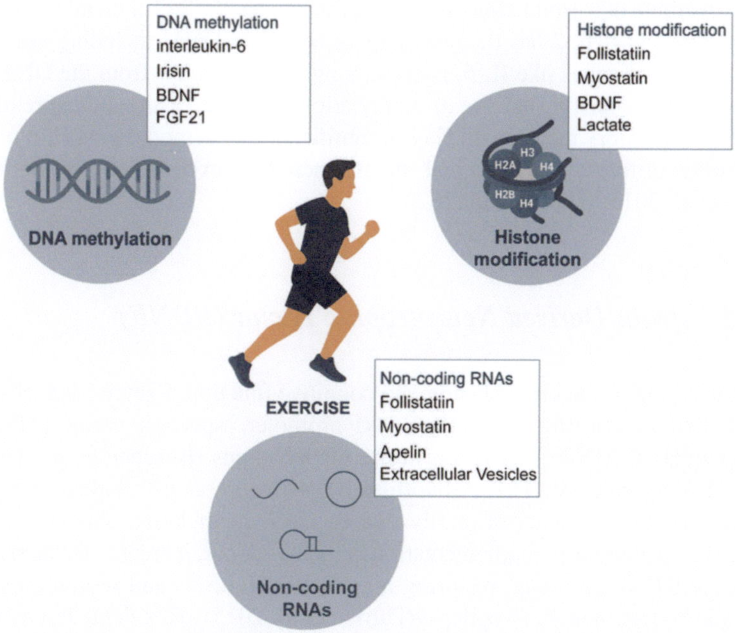

Fig. 13.1 Exercise induced different epigenetic regulation of virous exerkines

13.3 Impact on Exerkines

13.3.1 Irisin

Irisin is produced through the cleavage of fibronectin type III domain-containing protein 5 (FNDC5), and its expression can be regulated by DNA methylation. Methylation of this CpG island leads to reduced expression of FNDC5 and consequently lower levels of Irisin. Methylation-specific PCR analyses and ChIP assays showed that treatments with demethylating agents like 5-Aza and HDAC inhibitors (e.g., NaB) can increase FNDC5 mRNA levels, indicating that demethylation enhances Irisin expression (Kim et al. 2017).

13.3.2 Interleukin-6 (IL-6)

IL-6 is a cytokine produced and released by skeletal muscles during exercise, where it plays an important role in inflammation and metabolism. The expression of the IL6 gene is regulated by DNA methylation. Specifically, inhibition of DNMT1 (DNA methyltransferase 1) protein leads to a decrease in IL-6 expression, particularly at specific CpG sites within the IL-6 gene's promoter region (Bai et al. 2023).

One example of a protein that influences this process is Tet2, a member of the Tet family of proteins. Tet2 has the ability to demethylate DNA, including at the promoter regions of genes like IL-6. By removing methyl groups from the DNA, Tet2 can reverse the silencing effects of methylation, thus increasing the expression of IL-6. This regulatory mechanism plays a significant role in modulating IL-6 production during different physiological states, including exercise and inflammation (Zhang et al. 2015; Orange et al. 2023).

13.3.3 Brain-Derived Neurotrophic Factor (BDNF)

BDNF is crucial for neuronal health and cognitive function. Exercise induces DNA demethylation at specific sites in the BDNF promoter, especially promoter IV, leading to increased BDNF expression in the hippocampus (Boschen et al. 2017). In temporal lobe epilepsy (TLE), abnormal BDNF expression is linked to seizure-induced cognitive deficits, primarily due to DNA methylation changes (Cuomo et al. 2021). Methionine supplementation, a methyl donor, increases DNA methylation at the BDNF promoter, restoring normal BDNF levels and rescuing memory deficits in rodent models of epilepsy(Parrish et al. 2015). This demethylation process enhances synaptic plasticity and improves cognitive performance.

13.3.4 Fibroblast Growth Factor 21 (FGF21)

FGF21 plays a crucial role in regulating glucose and lipid metabolism, particularly in the liver. Its expression can be epigenetically regulated by DNA methylation. Specifically, reduced DNA methylation at certain CpG sites in the FGF21 promoter region has been linked to increased expression of FGF21. This increase in FGF21 expression may contribute to improved metabolic outcomes, such as enhanced lipid metabolism and better insulin sensitivity (Geissler et al. 2022).

Additionally, factors such as exercise and diet can influence the activity of the peroxisome proliferator-activated receptor alpha (PPARα), a key regulator of lipid metabolism. PPARα activation can modulate the DNA methylation status of the FGF21 gene, leading to changes in its expression. This regulation of FGF21 by PPARα may further impact its secretion and activity, highlighting the interplay between epigenetic modifications and metabolic regulation (Shinozaki et al. 2024).

In summary, DNA methylation serves as a dynamic regulatory mechanism for the expression and secretion of key exercise-induced factors like Irisin, IL-6, BDNF, and FGF21. Physical activity can modulate the methylation status of genes encoding these exerkines, thereby influencing their levels and contributing to the beneficial effects of exercise on metabolic and cognitive health.

13.4 Histone Modifications and Its Impact on Exerkine Secretion

Histone modifications, including acetylation, methylation, and phosphorylation, play a pivotal role in regulating the expression and secretion of exercise-induced factors, known as exerkines. These epigenetic alterations influence chromatin structure, thereby modulating gene accessibility and transcriptional activity during physical activity.

Brain-Derived Neurotrophic Factor (BDNF), Myostatin, Decorin, Osteonectin (SPARC), and Interleukin-6 (IL-6). These epigenetic alterations influence chromatin structure, thereby modulating gene transcription in response to physical exercise. Below is a detailed overview of how specific histone modifications affect the expression of these exerkines, along with pertinent references:.

13.4.1 Brain-Derived Neurotrophic Factor (BDNF)

Exercise induces histone acetylation at the BDNF gene promoter, particularly increasing acetylation at histone H3 lysine 9 (H3K9ac) and histone H4 lysine 12 (H4K12ac). H3K9ac, regulated by histone acetyltransferases (HATs) such as p300/CBP, is linked to increased transcription of the BDNF gene. This modification relaxes the chromatin structure, making the gene more accessible to transcription factors. As a result, H3K9ac promotes enhanced neuronal plasticity and cognitive functions such as learning and memory (Li et al. 2024b).

Similarly, H4K12ac (acetylation of histone H4 at lysine 12) also enhances BDNF transcription by improving chromatin accessibility, thus allowing better binding of transcription factors. This acetylation is regulated by HATs like p300 and plays a crucial role in neuronal activity-induced BDNF expression, supporting synaptic plasticity and cognitive performance. Together, H3K9ac and H4K12ac are essential for the epigenetic regulation of BDNF, promoting neuronal health and cognitive function (Intlekofer et al. 2013). .

13.4.2 Myostatin

Exercise and maternal diet influence myostatin (MSTN) gene expression through histone modifications, especially H3K4me3. Enrichment of H3K4me3 at the MSTN promoter is linked to increased MSTN expression, contributing to muscle growth regulation and potentially muscle atrophy. H3K9Ac also enhances MSTN transcription by loosening chromatin structure, particularly under conditions like exercise or a low-protein diet (Jia et al. 2016).In contrast, H3K27me3 is a well-established repressive marker of gene expression showed no significant changes at the MSTN promoter in these studies. While transcription factors like FoxO3 and Glucocorticoid Receptor (GR) also enhance MSTN gene expression, their binding to the MSTN promoter, influenced by factors like maternal diet, highlights the complex regulation of muscle growth (Li et al. 2024a).

13.4.3 Interleukin-6 (IL-6)

Exercise regulates IL-6 expression through various histone modifications. H3K4me3, a histone mark associated with gene activation, increases at the IL-6 promoter during exercise, facilitating its transcription. This modification is regulated by histone methyltransferases, which enhances H3K4me3 levels, promoting IL-6 gene expression (Villamil-Parra and Moscoso-Loaiza 2024). Additionally, H3K27me3, typically a repressive mark, is involved in modulating IL-6 expression. EZH2, a methyltransferase, mediates H3K27me3 enrichment, which may suppress IL-6 production under certain conditions (Villamil-Parra and Moscoso-Loaiza 2024).

Histone deacetylases (HDACs) also play a role in this regulation. The decreased activity of HDAC2 during strenuous exercise enhances H3K9ac levels, promoting IL-6 gene transcription. Furthermore, KDM2B, a histone demethylase, has been shown to regulate IL-6 expression by facilitating chromatin remodeling at the IL-6 promoter. KDM2B interacts with Brg1, a component of the SWI/SNF chromatin remodeling complex, to increase chromatin accessibility and recruit RNA polymerase II, thus promoting IL-6 transcription (Zhang et al. 2015; Zhou et al. 2020). .

13.4.4 Lactate

Histone lactylation, a novel post-translational modification where lactate molecules attach to histone lysine residues, plays a significant role in regulating gene expression in response to metabolic changes induced by exercise. Lactate also act as a key substrate in histone lactylation. During intense exercise, lactate is produced as a byproduct of anaerobic glycolysis. This increase in lactate not only serves as an energy source but also functions as a substrate for histone lactylation. IL-6 production is influenced by histone modifications such as H3K18 lactylation (Chu et al. 2021). The presence of lactate in muscle and other tissues during and after exercise thus acts as a key signal, inducing lactylation of histones that control the expression of genes involved in both metabolic adaptation and immune regulation. In essence, lactate helps convert the transient effects of exercise into long-term physiological adaptations by promoting gene expression through histone lactylation (Yan et al. 2024).

13.4.5 Regulation of Immune Exerkines

Beyond metabolism, histone modification also modulates immune responses. During exercise, lactate-induced histone lactylation is involved in regulating inflammatory pathways by increasing the expression of anti-inflammatory exerkines, such as IL-10 and IL-1Ra (Chu et al. 2021). This helps mitigate the inflammatory response triggered by exercise and supports quicker recovery from exercise-induced muscle damage.

In summary, histone modifications serve as key epigenetic regulators of exerkine expression, mediating the beneficial effects of exercise on muscle growth, neuroplasticity, and immune function. Understanding these mechanisms provides insight into how physical activity influences gene expression to promote health and adaptation.

13.5 Non-coding RNAs and Their Impact on Exerkine Secretion

Non-coding RNAs (ncRNAs), including microRNAs (miRNAs), long non-coding RNAs (lncRNAs), and circular RNAs (circRNAs), play a crucial role in regulating gene expression at the post-transcriptional level. These regulatory RNAs do not encode proteins but modulate gene expression by influencing mRNA stability, translation, and splicing. Exercise-induced alterations in ncRNA levels significantly affect the secretion of exerkines, which are essential for the beneficial effects of physical activity. Below, we break down the impact of different types of ncRNAs on exerkine secretion.

13.5.1 Myostatin

Myostatin is a negative regulator of muscle growth. Certain miRNAs, such as miR-27a and miR-27b, have been shown to target myostatin mRNA, leading to its degradation and thus promoting muscle hypertrophy (Archacka et al. 2021).

MiR-181b-5p has been identified as a key regulator of myostatin in tilapia, directly targeting its 3′ untranslated region (UTR). The binding of miR-181b-5p to this region reduces the expression of MSTN, thereby promoting muscle growth by modulating myostatin levels. Additionally, miR-27a and miR-206 have been shown to regulate MSTN expression by targeting the 3′ UTR of MSTN in both tilapia and mammals (Zhao et al. 2019). Additionally, lncRNAs like lnc-mg can act as molecular sponges for these miRNAs, thereby modulating myostatin expression indirectly.

13.5.2 Apelin

Apelin plays a role in cardiovascular function and metabolism. Specific miRNAs, such as miR-424, have been identified to target apelin mRNA, affecting its expression levels. Exercise-induced changes in ncRNA profiles may thus influence apelin-mediated pathways, contributing to cardiovascular adaptations (Tian et al. 2021). non-coding RNAs, particularly long non-coding RNA NEAT1, influence the expression of myostatin through their effects on mRNA stability. NEAT1 modulates the expression of Apelin by recruiting UPF1, an RNA-binding protein involved in mRNA decay. This interaction destabilizes Apelin mRNA, thereby influencing downstream signaling pathways that may indirectly regulate myostatin expression (Chen et al. 2024).

Long non-coding RNA BACE1-AS has been found to play an oncogenic role in hepatocellular carcinoma (HCC) cells through its regulation of the miR-214-3p/Apelin axis. BACE1-AS, a long non-coding RNA, acts as a molecular sponge for miR-214-3p, a microRNA known for its tumor-suppressive effects. By sequestering miR-214-3p, BACE1-AS indirectly upregulates Apelin (Tian et al. 2021).

13.5.3 Follistatin

Non-coding RNAs, particularly microRNAs and long non-coding RNAs (lncRNAs), have been shown to regulate the expression of follistatin-like protein 1 (FSTL1). In asthma, miR-200b-3p targets FSTL1, and its expression is downregulated in the airway epithelium of asthmatic patients, which correlates with FSTL1 upregulation. miR-200b-3p regulates airway remodeling and inflammation by inhibiting FSTL1 through a competitive endogenous RNA (ceRNA) mechanism, where the lncRNA PCAT19 also plays a role by sponging miR-200b-3p. This interplay between miR-200b-3p and lncRNA PCAT19 highlights a complex regulatory network impacting FSTL1 expression in the pathogenesis of asthma (Lorenzo-Almoros et al. 2019; Liu et al. 2022).

13.5.4 RNAs in Extracellular Vesicles

Extracellular vesicles (EVs), including exosomes, are small membrane-bound vesicles released by cells into the extracellular space. These vesicles play an essential role in intercellular communication by transferring bioactive molecules, such as proteins, lipids, and nucleic acids, between cells. RNA molecules, particularly miRNAs, long non-coding RNAs (lncRNAs), and other small RNAs, are frequently carried by EVs and contribute significantly to the regulation of exerkines in response to exercise. EVs mediate the transfer of exercise-induced regulatory RNAs from muscle cells to other tissues, influencing systemic adaptations to physical activity.

13.6 Implications for Health and Disease

The epigenetic regulation of exerkines is particularly significant for metabolic health. Regular physical activity-induced epigenetic changes can lead to sustained improvements in glucose metabolism, lipid profiles, and overall energy balance, which are crucial for preventing and managing metabolic disorders such as obesity, type 2 diabetes, and non-alcoholic fatty liver disease (NAFLD).

Additionally, understanding these mechanisms opens avenues for developing novel therapeutic strategies that mimic the beneficial effects of exercise. For instance, pharmacological agents that modulate epigenetic marks could potentially enhance the production of beneficial exerkines, providing new approaches to treat metabolic diseases and improve overall health outcomes.

13.7 Conclusion

The epigenetic regulation of exerkines represents a vital link between physical activity and long-term health benefits. By elucidating the mechanisms through which exercise influences gene expression, researchers can develop targeted interventions to harness these benefits, offering new hope for combating metabolic diseases and enhancing quality of life.

Compliance with Ethical Standards

Conflict of Interest All authors declare no conflict of interest.

Ethical Approval This chapter is a review of previously published accounts; as such, no animal or human studies were performed.

References

Archacka K, Ciemerych MA, Florkowska A, Romanczuk K (2021) Non-coding RNAs as regulators of myogenesis and postexercise muscle regeneration. Int J Mol Sci 22(21). https://doi.org/10.3390/ijms222111568. Available at: https://www.ncbi.nlm.nih.gov/pubmed/34768999

Bai Z, Gao T, Zhang R, Lu Y, Tian J, Wang T, Zhao K, Wang H (2023) Inhibition of IL-6 methylation by Saikosaponin C regulates neuroinflammation to alleviate depression. Int Immunopharmacol 118:110043. https://doi.org/10.1016/j.intimp.2023.110043

Boschen KE, McKeown SE, Roth TL, Klintsova AY (2017) Impact of exercise and a complex environment on hippocampal dendritic morphology, Bdnf gene expression, and DNA methylation in male rat pups neonatally exposed to alcohol. Dev Neurobiol 77(6):708–725. https://doi.org/10.1002/dneu.22448. Available at: https://www.ncbi.nlm.nih.gov/pubmed/27597545

Chen K, Ou B, Huang Q, Deng D, Xiang Y, Hu F (2024) LncRNA NEAT1 aggravates human microvascular endothelial cell injury by inhibiting the Apelin/Nrf2/HO-1 signalling pathway in type 2 diabetes mellitus with obstructive sleep apnoea. Epigenetics 19(1):2293409. https://doi.org/10.1080/15592294.2023.2293409. Available at: https://www.ncbi.nlm.nih.gov/pubmed/38232183

Chu X, Di C, Chang P, Li L, Feng Z, Xiao S, Yan X, Xu X, Li H, Qi R, Gong H, Zhao Y, Xiao F, Chang Z (2021) Lactylated histone H3K18 as a potential biomarker for the diagnosis and predicting the severity of septic shock. Front Immunol 12:786666. https://doi.org/10.3389/fimmu.2021.786666. Available at: https://www.ncbi.nlm.nih.gov/pubmed/35069560

Cuomo M, Borrelli L, Della Monica R, Coretti L, De Riso G, Lancellotti D'A, di Durazzo L, Fioretti A, Lembo F, Dinan TG, Cryan JF, Cocozza S, Chiariotti L (2021) DNA methylation profiles of Tph1A and BDNF in gut and brain of L. Rhamnosus-Treated Zebrafish. Biomolecules 11(2). https://doi.org/10.3390/biom11020142. Available at: https://www.ncbi.nlm.nih.gov/pubmed/33499115

Geissler C, Krause C, Neumann AM, Britsemmer JH, Taege N, Grohs M, Kaehler M, Cascorbi I, Lewis AG, Seeley RJ, Oster H, Kirchner H (2022) Dietary induction of obesity and insulin resistance is associated with changes in Fgf21 DNA methylation in liver of mice. J Nutr Biochem 100:108907. https://doi.org/10.1016/j.jnutbio.2021.108907. Available at: https://www.ncbi.nlm.nih.gov/pubmed/34801693

Intlekofer KA, Berchtold NC, Malvaez M, Carlos AJ, McQuown SC, Cunningham MJ, Wood MA, Cotman CW (2013) Exercise and sodium butyrate transform a subthreshold learning event into long-term memory via a brain-derived neurotrophic factor-dependent mechanism. Neuropsychopharmacology 38(10):2027–2034. https://doi.org/10.1038/npp.2013.104. Available at: https://www.ncbi.nlm.nih.gov/pubmed/23615664

Jia Y, Gao G, Song H, Cai D, Yang X, Zhao R (2016) Low-protein diet fed to crossbred sows during pregnancy and lactation enhances myostatin gene expression through epigenetic regulation in skeletal muscle of weaning piglets. Eur J Nutr 55(3):1307–1314. https://doi.org/10.1007/s00394-015-0949-3. Available at: https://www.ncbi.nlm.nih.gov/pubmed/26066356

Kim HK, Jeong YJ, Song IS, Noh YH, Seo KW, Kim M, Han J (2017) Glucocorticoid receptor positively regulates transcription of FNDC5 in the liver. Sci Rep 7:43296. https://doi.org/10.1038/srep43296. Available at: https://www.ncbi.nlm.nih.gov/pubmed/28240298

Li J, Zhang S, Li C, Zhang X, Shan Y, Zhang Z, Bo H, Zhang Y (2024a) Endurance exercise-induced histone methylation modification involved in skeletal muscle fiber type transition and mitochondrial biogenesis. Sci Rep 14(1):21154. https://doi.org/10.1038/s41598-024-72088-6. Available at: https://www.ncbi.nlm.nih.gov/pubmed/39256490

Li Y, Liu A, Chen K, Li L, Zhang X, Zou F, Zhang X, Meng X (2024b) Sodium butyrate alleviates lead-induced neuroinflammation and improves cognitive and memory impairment through the ACSS2/H3K9ac/BDNF pathway. Environ Int 184:108479. https://doi.org/10.1016/j.envint.2024.108479. Available at: https://www.ncbi.nlm.nih.gov/pubmed/38340407

Liu F, Zhang J, Zhang D, Qi Q, Cui W, Pan Y, Liu X, Xu J, Qiao X, Wang Z, Dong L (2022) Follistatin-related protein 1 in asthma: miR-200b-3p interactions affect airway remodeling and inflammation phenotype. Int Immunopharmacol 109:108793. https://doi.org/10.1016/j.intimp.2022.108793. Available at: https://www.ncbi.nlm.nih.gov/pubmed/35483234

Lorenzo-Almoros A, Hang T, Peiro C, Soriano-Guillen L, Egido J, Tunon J, Lorenzo O (2019) Predictive and diagnostic biomarkers for gestational diabetes and its associated metabolic and cardiovascular diseases. Cardiovasc Diabetol 18(1):140. https://doi.org/10.1186/s12933-019-0935-9. Available at: https://www.ncbi.nlm.nih.gov/pubmed/31666083

Orange ST, Leslie J, Ross M, Mann DA, Wackerhage H (2023) The exercise IL-6 enigma in cancer. Trends Endocrinol Metab 34(11):749–763. https://doi.org/10.1016/j.tem.2023.08.001. Available at: https://www.ncbi.nlm.nih.gov/pubmed/37633799

Parrish RR, Buckingham SC, Mascia KL, Johnson JJ, Matyjasik MM, Lockhart RM, Lubin FD (2015) Methionine increases BDNF DNA methylation and improves memory in epilepsy. Ann Clin Transl Neurol 2(4):401–416. https://doi.org/10.1002/acn3.183. Available at: https://www.ncbi.nlm.nih.gov/pubmed/25909085

Shinozaki H, Kawai S, Gamo-Kawasaki M, Takei A, Tsujikado K, Fukuda K, Yamauchi M, Hara K, Tsuchiya T, Takebayashi K, Hashimoto K (2024) Analysis of serum levels and DNA methylation of fibroblast growth factor 21 using peripheral blood-derived genomes in patients with obesity. Endocr J 71(9):907–924. https://doi.org/10.1507/endocrj.EJ23-0570. Available at: https://www.ncbi.nlm.nih.gov/pubmed/38910123

Tian Q, Yan X, Yang L, Liu Z, Yuan Z, Zhang Y (2021) Long non-coding RNA BACE1-AS plays an oncogenic role in hepatocellular carcinoma cells through miR-214-3p/APLN axis. Acta Biochim Biophys Sin Shanghai 53(11):1538–1546. https://doi.org/10.1093/abbs/gmab134. Available at: https://www.ncbi.nlm.nih.gov/pubmed/34636395

Villamil-Parra W, Moscoso-Loaiza L (2024) Effects of physical exercise on irisin and BDNF concentrations, and their relationship with cardiometabolic and mental health of individuals with metabolic syndrome: a systematic review. Exp Gerontol 198:112640. https://doi.org/10.1016/j.exger.2024.112640. Available at: https://www.ncbi.nlm.nih.gov/pubmed/39579805

Yan L, Wang Y, Hu H, Yang D, Wang W, Luo Z, Wang Y, Yang F, So KF, Zhang L (2024) Physical exercise mediates cortical synaptic protein lactylation to improve stress resilience. Cell Metab 36(9):2104–2117.e4. https://doi.org/10.1016/j.cmet.2024.07.018. Available at: https://www.ncbi.nlm.nih.gov/pubmed/39163863

Zhang Q, Zhao K, Shen Q, Han Y, Gu Y, Li X, Zhao D, Liu Y, Wang C, Zhang X, Su X, Liu J, Ge W, Levine RL, Li N, Cao X (2015) Tet2 is required to resolve inflammation by recruiting Hdac2 to specifically repress IL-6. Nature 525(7569):389–393. https://doi.org/10.1038/nature15252. Available at: https://www.ncbi.nlm.nih.gov/pubmed/26287468

Zhao Z, Yu X, Jia J, Yang G, Sun C, Li W (2019) miR-181b-5p may regulate muscle growth in tilapia by targeting myostatin b. Front Endocrinol (Lausanne) 10:812. https://doi.org/10.3389/fendo.2019.00812. Available at: https://www.ncbi.nlm.nih.gov/pubmed/31849840

Zhou Q, Zhang Y, Wang B, Zhou W, Bi Y, Huai W, Chen X, Chen Y, Liu Z, Liu X, Zhan Z (2020) KDM2B promotes IL-6 production and inflammatory responses through Brg1-mediated chromatin remodeling. Cell Mol Immunol 17(8):834–842. https://doi.org/10.1038/s41423-019-0251-z. Available at: https://www.ncbi.nlm.nih.gov/pubmed/31197256

Chapter 14
Exercise, Epigenetics, and Cell Death

Xinxin Xu and Guohua Song

Abstract Exercise is widely recognized for its health benefits, including its effects on the prevention and treatment of various diseases. Many studies have elucidated the molecular mechanisms underlying these benefits, including epigenetics. The role of epigenetics has received a lot of attention recently. Epigenetics refers to heritable changes in gene expression that do not involve alterations to the underlying DNA sequence. Exercise has been shown to induce changes in DNA methylation, histone modifications, non-coding RNAs and mitochondrial function, which are crucial components of the epigenetic machinery. These changes are capable of modifying the expression of genes involved in energy metabolism, oxidative stress response, inflammation and other aspects. The modulation of gene expression that occurs through exercise can have profound effects on cellular homeostasis and the balance between cell survival and death. As a basic pathological process, cell death plays an important role in a variety of diseases. The intervention of cell death can play a role in the treatment or prevention of diseases. The research on the effect of exercise on epigenetic changes related to cell death is constantly being updated, and the specific mechanisms by which epigenetics influence various cell death modalities are being elucidated in more detail. Understanding the interplay among aerobic exercise, epigenetic regulation, and cell death provides a promising avenue for the development of personalized exercise regimens and therapeutic interventions that can enhance human health.

Keywords Exercise · Epigenetics · Gene expression · Cell death · Therapeutic interventions

---

X. Xu · G. Song (✉)
School of Clinical and Basic Medical Sciences, Shandong First Medical University & Shandong Academy of Medical Science, Jinan, Shandong, China
e-mail: ghsong@sdfmu.edu.cn

F. Gao, X. Zhang (eds.), *Exercise, Epigenetics and Human Health*, Epigenetics and Human Health 15, https://doi.org/10.1007/978-3-032-08045-5_14

Abbreviations

ALKBH5	AlkBhomologue5
BDNF	Brain-derived neurotrophic factor
DNMT	DNA methyltransferase
Fh1	Fumarate hydratase 1
FTO	Fat mass and obesity-associated protein
I/R	Ischemia-reperfusion
Klf10	Kruppcl-likc factor 10
LncRNAs	Long non-coding RNAs
METTL14	methyltransferase like 14
METTL3	methyltransferase like 3
miRNAs	microRNAs
mtDNA	mitochondrial DNA
NADPH	Nicotinamide adenine dinucleotide phosphate
NAFLD	Non-alcoholic fatty liver disease
ncRNAs	non-coding RNAs
ROS	Reactive oxygen species
T2DM	Type 2 diabetes mellitus
Tfam	Transcription factor A
WTAP	Wilms tumor 1 associated protein

14.1 Introduction

The complex interplay between physical exercise, epigenetics, and cell death provides a compelling framework within the fields of modern medicine and biology. The beneficial effects of exercise on health have long been recognized as a potent modulator of epigenetic mechanisms, which, in turn, play a crucial role in the regulation of cell death pathways (McGee and Hargreaves 2019; Tetti et al. 2022).

Epigenetics, the study of changes in gene function that do not involve alterations to the underlying DNA sequence, encompasses a variety of modifications, including DNA methylation, histone modifications, and the regulation of non-coding RNAs (miRNAs, LncRNAs, circRNAs, etc.) (Ling and Ronn 2019). These modifications are influenced by a range of biological and environmental factors and can also be modulated through interventions such as physical exercise (Gomez-Pinilla and Thapak 2024). The epigenome is highly dynamic and can adapt to the physiological demands imposed on the body during exercise, inducing lasting changes in gene expression that contribute to the health benefits linked to a healthy lifestyle (Alegria-Torres et al. 2011).

One of the most extensively studied aspects of epigenetics is its role in regulating cell death. As a fundamental biological process, cell death is central to the pathogenesis of numerous diseases, including cancer, neurodegenerative disorders, cardiovascular diseases, and autoimmune conditions (Kayagaki et al. 2024). For instance, the death of cardiomyocytes can exacerbate myocardial infarction and heart failure; neuronal cell death may contribute to cognitive dysfunction, and the demise of cancer cells can facilitate tumor regression.

Currently, numerous studies investigate cell death, including apoptosis, necrosis, pyroptosis, ferroptosis, cuproptosis, and disulfidptosis (Newton et al. 2024). Since the concept of apoptosis was first introduced in 1972, there has been a clearer understanding of cell death mechanisms, and necrosis was subsequently recognized as another mode of cell death. In 2012, the journal *Cell* introduced a distinct form of iron-dependent cell death, ferroptosis, which differs from apoptosis. During ferroptosis, an imbalance between the generation and degradation of intracellular lipid reactive oxygen species (ROS) results in the accumulation of lipid peroxides, ultimately leading to cell death. Recent studies over the past 2 years have identified two novel forms of cell death: cuproptosis and disulfidptosis. These forms of cell death are regulated by a variety of mechanisms, in which epigenetics plays a critical role. The epigenome influences cell death by modulating the expression of key genes, thereby impacting the associated diseases (Li et al. 2023).

The modulation of epigenetic marks by exercise is a rapidly advancing area of research. For example, it is well established that exercise can activate "writers" that add epigenetic marks to DNA or histones, "readers" that recognize these marks, and "erasers" that remove them, thereby influencing the cellular decision between survival and cell death (McGee 2007; Horsburgh et al. 2015). This delicate balance is essential for maintaining tissue homeostasis and has significant implications for the prevention and treatment of various diseases.

Understanding the mechanisms through which exercise influences the epigenome and cell death pathways could facilitate the development of personalized exercise regimens and therapeutic interventions. These interventions have the potential to enhance human health by reducing disease risk, promoting recovery, and improving quality of life. As our understanding of the epigenome expands, so too does the potential for targeted interventions that harness the benefits of exercise to optimize health outcomes.

In this chapter, we explore the molecular mechanisms underlying the effects of exercise on the epigenome, with a particular focus on how these changes modulate cell death pathways. We review the current literature, discuss its implications for human health, and examine potential future research directions in this dynamic and rapidly evolving field.

14.2 Epigenetics

Epigenetics is the study of gene expression regulation and focuses on how heritable changes in gene expression or cell phenotype can occur without alterations to the DNA sequence. This field encompasses mechanisms such as DNA and RNA methylation, histone modifications, and non-coding RNAs (ncRNAs). Recent studies have demonstrated that epigenetics plays a critical role in regulating cell survival and death, and is also implicated in the onset and progression of various diseases.

14.2.1 DNA Methylation

DNA methylation is a fundamental epigenetic modification that plays a crucial role in the regulation of gene expression. This process involves the covalent addition of a methyl group to the 5-carbon position of the cytosine base within a CpG dinucleotide, a reaction primarily catalyzed by DNA methyltransferase (DNMT) enzymes (Nishiyama and Nakanishi 2021). DNA methylation is essential for several biological processes, including embryonic development, regulation of gene expression, maintenance of chromosomal stability, and silencing of transposable elements.

The presence or absence of methylation at specific genomic sites can determine whether a gene is expressed or silenced. Abnormal DNA methylation patterns are implicated in numerous diseases, particularly cancer, where hypermethylation of tumor suppressor genes and hypomethylation of oncogenes can drive uncontrolled cell growth and tumorigenesis (Klutstein et al. 2016).

Furthermore, DNA methylation is a dynamic modification that can be regulated during cell differentiation and in response to environmental stimuli. The dynamic nature of DNA methylation adds an additional layer of regulation, enabling cells to adapt to changing conditions. This suggests that DNA methylation can be regulated and influenced by the external environment to facilitate cellular adaptation.

14.2.2 m^{6}A RNA Methylation

m^{6}A RNA methylation is a key epigenetic modification that plays a critical role in several biological processes, including the regulation of RNA splicing, translation, stability, and translocation. This modification entails the addition of a methyl group to the nitrogen-6 position of adenosine residues in RNA molecules, primarily catalyzed by a methyltransferase complex that includes proteins such as methyltransferase like 3 (METTL3), methyltransferase like 14 (METTL14), and Wilms tumor 1 associated protein (WTAP). Additionally, these marks can be removed by "eraser" demethylases, including fat mass and obesity-associated protein (FTO) and AlkBhomologue5 (ALKBH5).

Regarding disease associations, m^{6}A RNA methylation has been implicated in several diseases. For instance, in acute myeloid leukemia, m^{6}A methylation can enhance the self-renewal capacity of leukemic stem cells (Vu et al. 2017). In contrast, inhibition of m^{6}A demethylation may stabilize the proliferation of glioma stem cells (Zhang et al. 2017). With advancing research, it has been demonstrated that dysregulation of m^{6}A RNA methylation occurs in various diseases, including non-alcoholic fatty liver disease (NAFLD), azoospermia, and heart failure (Jiang et al. 2021). These findings underscore the potential of targeting the m^{6}A RNA methylation pathway for therapeutic intervention in a range of diseases, including cancer.

14.2.3 Histone Modification

The core histones (H2A, H2B, H3, and H4) form the structural core around which DNA is wrapped, while H1 and its variants bind to the entry and exit points of the DNA, facilitating higher-order chromatin compaction (Bannister and Kouzarides 2011). Histone modifications represent a complex and essential layer of epigenetic regulation that significantly influence gene expression without altering the underlying DNA sequence. These modifications entail chemical changes to the histone proteins around which DNA is wrapped, contributing to the formation of the chromatin structure within the cell nucleus (Allfrey et al. 1964).

Histone modifications constitute a broad class of protein post-translational modifications that occur covalently, including lysine (K) and arginine (R) methylation, serine (S) and threonine (T) phosphorylation, as well as lysine acetylation (Ac), butyrylation, and ubiquitination. Among these, histone methylation and acetylation primarily occur in the N-terminal tails and can modulate gene transcription. Given their early discovery, substantial research has been conducted on these modifications (Bannister and Kouzarides 2011). In recent years, there have been significant advancements in the understanding of histone modification patterns, including palmitoylation, propionylation, crotonylation, and lactylation, which have also generated a renewed wave of research interest.

In conclusion, histone modifications represent a dynamic and complex system that plays a pivotal role in the regulation of gene expression. Understanding the mechanisms by which these modifications are established, interpreted, and erased is essential for deciphering the intricate language of the epigenome and holds significant promise for the development of novel therapeutic strategies for a wide range of diseases.

14.2.4 Non-coding RNAs

NcRNAs and epigenetics constitute two intricate and interconnected fields that have considerably broadened our understanding of gene regulation, extending beyond the traditional paradigm of DNA encoding proteins. Unlike mRNAs, which translate genetic information into proteins, ncRNAs function without translation and play essential roles in regulating gene expression at both the transcriptional and post-transcriptional levels.

A considerable body of research underscores the interplay between ncRNAs and epigenetic mechanisms in diverse biological processes and diseases. For example, microRNAs (miRNAs), as a well-studied class of small ncRNAs, can direct the silencing of target mRNAs by promoting their degradation or inhibiting their translation, thereby modulating gene expression (Diener et al. 2022). Moreover, long non-coding RNAs (LncRNAs) have been implicated in modulating chromatin structure and function, as well as acting as scaffolds for chromatin-modifying complexes that reshape the epigenetic landscape of specific genomic regions (Herman et al. 2022).

The convergence of ncRNA research and epigenetic studies not only enhances our understanding of gene regulatory networks but also opens new avenues for therapeutic interventions. Targeting ncRNAs or epigenetic modifications presents promising strategies for treating a wide range of diseases, including cancer, cardiovascular disorders, and neurological conditions. As our understanding of these intricate regulatory systems advances, it is expected that novel biomarkers and precision medicine approaches will emerge, propelling the next generation of medical advancements.

14.3 Exercise, Epigenetics, and Cell Death

Various forms of cell death have been proposed, and this section will elucidate the specific molecular mechanisms underlying exercise-mediated epigenetic effects on cell death.

14.3.1 Exercise, Epigenetics, and Apoptosis

Cell death is an inevitable and essential process in the life cycle of all cells. Among the various forms of cell death, apoptosis is distinguished as a highly regulated and controlled mechanism that plays a critical role in maintaining homeostasis within an organism. Apoptosis, derived from the Greek word "apo," meaning "away from," and "ptosis," meaning "falling off," was first described by Kerr et al. in 1972 as a form of cell death distinct from necrosis, characterized by a series of morphological and biochemical changes within the cell (Kerr et al. 1972).

Apoptosis is a tightly regulated and energy-dependent process, characterized by distinct morphological changes, including cell shrinkage, chromatin condensation, and DNA fragmentation. Apoptosis is a complex process involving multiple signaling pathways and molecular mechanisms, ensuring that cells are eliminated in a manner that is both efficient and free of inflammation (Bertheloot et al. 2021). It is a crucial component of development, tissue remodeling, and the immune system's response to infection. Moreover, dysregulation of apoptosis has been implicated in a range of diseases, including cancer, cardiovascular disease, autoimmune disorders, reproductive system disorders, and neurodegenerative diseases (Newton et al. 2024).

As a key mechanism in mediating cell apoptosis, epigenetics plays a crucial role in the process, with epigenetic modifications altering the biological functions of cells and potentially being passed on to subsequent generations. The intersection of apoptosis and epigenetics resides in the regulation of genes and pathways that govern cell fate (Karlic et al. 2014). Epigenetic modifications can regulate the expression of genes involved in the apoptotic process, thereby influencing the cell's fate between survival and death. For example, alterations in histone modifications or DNA methylation patterns can either promote or suppress the expression of pro-apoptotic and anti-apoptotic genes, respectively (Karlic et al. 2014; Patra et al. 2019). As a well-established beneficial exogenous intervention, exercise has been shown to exert positive effects on various diseases; however, the specific molecular mechanisms underlying its impact on epigenetics and subsequent apoptosis remain unclear. In the following, we will explore the specific effects of exercise on apoptosis through different epigenetic mechanisms.

Recent studies have emphasized the significant role of exercise in reducing cellular apoptosis and modulating gene expression via m^6A methylation pathways, particularly in the context of myocardial ischemia-reperfusion (I/R) injury and heart failure. For example, exercise training has been demonstrated to reduce WTAP levels, thereby downregulating the m^6A modification of FOXO3a mRNA and mitigating I/R-induced apoptosis in cardiomyocytes (Wang et al. 2023a). Furthermore, exercise regulates the expression of another m^6A methyltransferase METTL14, which subsequently modulates the m^6A modification of Phlpp2 mRNA. This regulation activates Akt-S473, thereby promoting cardiomyocyte growth and inhibiting apoptosis during myocardial I/R injury (Wang et al. 2022). In addition, exercise training can mitigate the progression of heart failure with preserved ejection fraction (HFpEF) by modulating the m^6A methylation levels of various genes, particularly those involved in apoptosis-related signaling pathways, such as caspase 12, JUN, AGTR1, and Fas (Liu et al. 2022). Meanwhile, a moderate-intensity exercise regimen significantly downregulated the expression of the demethylase FTO in the hearts of HFpEF mice. Overexpression of FTO in the heart leads to increased cardiomyocyte apoptosis and fibrosis, thereby counteracting the protective effects of exercise on heart failure (Liu et al. 2022). In addition to its benefits for cardiovascular health, exercise training has been shown to reduce m^6A methylation levels in the testicular tissue of obese males, thereby decreasing apoptosis and potentially ameliorating obesity-associated reproductive dysfunction (Xu et al. 2022). In summary, these findings collectively highlight the therapeutic potential of exercise in modulating m^6A methylation and reducing apoptosis.

Exercise can regulate DNA methylation and modulate apoptosis. Several studies have highlighted the significant impact of exercise on DNA methylation and its subsequent effects on cellular functions and apoptotic pathways. For example, the ACADVL gene undergoes significant hypermethylation following high-intensity interval training (HIIT), which activates downstream caspase-mediated actin depolymerization and cell death pathways. The hypermethylation of the ACADVL gene, which plays a critical role in mitochondrial fatty acid oxidation, may result in mitochondrial dysfunction, thereby triggering cell apoptosis. These findings suggest that exercise could enhance cardiac function in heart failure patients by modulating DNA methylation and apoptotic pathways (Hsu et al. 2023). In the context of osteoporosis, Nrf2 expression in the femur is significantly reduced, accompanied by elevated expression of DNMT and hypermethylation of the Nrf2 promoter. The suppression of Nrf2 results in increased oxidative stress and apoptosis. However, 1 hour of daily aerobic exercise significantly mitigates these epigenetic alterations, upregulating Nrf2 expression, enhancing femoral bone mass, and reducing muscle loss (Chen et al. 2021). Moreover, exercise has been shown to enhance DNA methylation in prostate cancer cells, downregulating the expression of DNMT genes. Previous studies conducted by the same research group revealed that trained rats exhibited slower prostate tumor progression and increased cancer cell apoptosis. These findings suggest that physical exercise may promote cancer control by modulating the epigenetic regulation of prostate tumors (Gueritat et al. 2014). In summary, these studies collectively underscore the beneficial effects of exercise on DNA methylation and its role in regulating apoptosis and cellular functions. Whether improving cardiac function in heart failure patients, enhancing bone health in osteoporosis, or regulating prostate cancer growth, exercise-induced epigenetic modifications play a crucial role in mediating these health benefits.

The aforementioned research has elucidated the significant role of exercise in modulating epigenetic mechanisms and improving various health conditions. Moreover, exercise can influence cell death processes by mediating histone modifications. For instance, kruppel-like factor 10 (Klf10) has been identified as a downstream effector of exercise, facilitating the expression of fumarate hydratase 1 (Fh1) and reducing fumarate accumulation in hepatocytes. The reduction in fumarate levels lowers H3K4me3 levels on the promoters of lipogenic genes, thereby mitigating hepatic lipogenesis. Consequently, exercise helps mitigate fatty acid-induced hepatocyte steatosis, apoptosis, and insulin resistance (Luo et al. 2024). Moreover, brain-derived neurotrophic factor (BDNF) plays a critical role in the central nervous system, supporting neuronal survival, structural remodeling, and plasticity. Exercise stimulates the production of β-hydroxybutyrate, which inhibits HDAC2/HDAC3 activity and promotes histone H3 acetylation, thereby enhancing BDNF expression. The upregulation of BDNF alleviates neuronal apoptosis in neurodegenerative diseases, underscoring its therapeutic potential (Sleiman et al. 2016). Furthermore, several studies have shown that exercise influences histone modifications and contributes to cell apoptosis by impacting mitochondrial DNA. Endurance training induced more pronounced beneficial mitochondrial adaptive changes in aged animals, including increased relative mitochondrial DNA (mtDNA) content, enhanced mitochondrial biosynthesis, and improved antioxidant capacity. It also reduced the level of phosphorylated histone H2AX and mitigated DNA damage, thereby alleviating pathological processes such as apoptosis in aged testes (Joseph et al. 2014).

These findings collectively underscore the profound effects of exercise on epigenetic regulation and its beneficial impact on cardiac function, liver metabolism, and neuronal health. By modulating HDAC activity, histone modifications, and transcription factors, exercise emerges as a powerful regulator of cellular processes, presenting promising avenues for the treatment of heart failure, NAFLD, and neurodegenerative disorders.

NcRNAs play a central role in mediating the beneficial effects of exercise on various physiological processes, particularly in mitigating aging, neurotoxicity, and cardiac dysfunction. Recent research has elucidated how exercise-induced modulation of specific non-coding RNAs can confer substantial health benefits. The study demonstrated that exercise downregulates miR-196b-5p, thereby enhancing the expression of N-Ras in aging skeletal muscle. N-Ras, as an anti-apoptotic protein, is essential for cell survival, thereby mitigating apoptosis during muscle aging (Liang et al. 2023). This mechanism contributes to slowing the loss of muscle mass and age-related degeneration. In the context of neuroprotection, fluoride exposure is known to impair hippocampal mitochondrial function and disrupt myelin structure, leading to alterations in metabolite and miRNA levels. Exercise has been shown to upregulate miR-206-3p, which effectively reduces fluoride-induced cell death by approximately 75%. The underlying molecular mechanism involves miR-206-3p-mediated downregulation of JunD and HDAC4 expression, underscoring its protective role against neurotoxicity (Chai et al. 2024). Moreover, running exercise enhances cognitive function in mice with cerebral ischemia-reperfusion injury by upregulating the expression of LncRNA MALAT1. This upregulation mitigates neuronal apoptosis in the hippocampus, thereby improving cognitive performance (Shang et al. 2018). Exercise can also influence the onset and progression of diseases by modulating the expression of circular RNAs (circRNAs). A 4-week swimming regimen significantly enhances the expression of circUtrn in the mouse myocardium, promoting physiological cardiac hypertrophy. Overexpression of circUtrn mitigates myocardial ischemia-reperfusion injury, reduces cardiac fibrosis and myocardial cell apoptosis, and improves cardiac function (Wang et al. 2023b). In summary, these findings underscore the crucial role of ncRNAs in mediating the health benefits of exercise. By modulating specific miRNAs and LncRNAs, exercise can provide protective effects against muscle aging, neurotoxicity, and cardiac dysfunction, presenting promising therapeutic strategies for these conditions.

Apoptosis is a highly regulated, energy-dependent process that involves distinct morphological changes and signaling pathways. Dysregulation of apoptosis is associated with diseases such as cancer and cardiovascular disorders. Epigenetics plays a critical role in regulating apoptosis by modulating gene expression. Exercise has been shown to alter epigenetic markers, including m^6A methylation, DNA methylation, and histone modifications, thereby mitigating apoptosis in various tissues. NcRNAs play a crucial role in regulating cellular processes, with exercise modulating their expression to alleviate aging, neurotoxicity, and cardiac dysfunction. These findings underscore the therapeutic potential of exercise in managing various diseases.

14.3.2 Exercise, Epigenetics and Necrosis

Necrosis is a form of irreversible cell death that results from severe cellular damage. Unlike apoptosis, necrosis is typically induced by external factors such as trauma, infection, toxins, or hypoxia, resulting in substantial structural and functional disruption within the cell. During necrosis, the loss of cell membrane integrity leads to the release of intracellular enzymes into the surrounding tissue, subsequently triggering an inflammatory response (Guo et al. 2023a). This unregulated, non-programmed form of cell death often results in greater harm to the organism. Necrosis is strongly associated with a variety of diseases. For example, in myocardial infarction, interrupted blood flow induces hypoxia in cardiac cells, leading to necrosis and subsequent impairment of heart function. Hepatitis virus infections can trigger necrosis in hepatocytes, impairing liver function. In summary, necrosis plays a pivotal role in the initiation and progression of both acute and chronic diseases.

The relationship between necrosis and epigenetics has attracted growing attention in recent years. Studies have demonstrated that epigenetic modifications play a critical role in regulating cellular responses to damage and the progression of necrosis. For instance, under hypoxic conditions, alterations in histone modifications and DNA methylation patterns regulate cellular responses to hypoxia, thereby modulating cell survival and apoptosis (Zhang et al. 2024). Moreover, specific ncRNAs, including miRNAs and LncRNAs, play critical roles in the necrosis process by modulating the expression of key genes, thereby influencing cell fate (Su et al. 2021).

Exercise not only promotes overall health but also modulates the process of necrosis through its effects on epigenetic modifications. Research indicates that regular physical activity can influence DNA methylation and histone modifications, thereby regulating the expression of genes linked to inflammation and oxidative stress, ultimately enhancing cellular resilience to damage (McGee and Hargreaves 2020). Exercise can decrease the methylation of genes associated with the inflammatory response, thereby reducing the expression of inflammatory mediators and attenuating the inflammatory response induced by necrosis (Swiatowy et al. 2021). Moreover, exercise can enhance the expression of antioxidant genes, mitigating oxidative stress-induced cellular damage and reducing the incidence of necrosis (Gomez-Cabrera et al. 2008). Necrosis is closely associated with epigenetics, and exercise, as a critical lifestyle intervention, can modulate the process of cell necrosis through epigenetic mechanisms, thereby playing a beneficial role in disease prevention and treatment.

Epigenetics plays a critical role in mediating the effects of various forms of exercise at the cellular and molecular levels. For example, aerobic exercise has been shown to mitigate hepatic steatosis and patchy necrosis in mice by downregulating the expression of LncRNA SRA (Wu et al. 2022). Moreover, long-term exercise training exerts anti-inflammatory and endothelial-protective effects by modulating the expression levels of various LncRNAs in the aortic endothelium, thereby reducing endothelial cell necrosis (Liu et al. 2018). In addition to ncRNAs, exercise can also influence histone modifications. Resistance training, as a form of exercise, upregulates the expression of the muscle-specific histone methyltransferase Smyd1, thereby influencing histone methylation. This regulatory mechanism plays a crucial role in enhancing cardiac function following myocardial infarction by alleviating necrosis and cell death in the infarcted myocardium (Liang et al. 2020). Collectively, these findings underscore the profound impact of exercise-induced epigenetic modifications in enhancing cellular health and mitigating disease pathogenesis.

Exercise, epigenetics, and necrosis are intricately interconnected. Exercise modulates epigenetic modifications that regulate gene expression associated with inflammation and oxidative stress. These modifications can mitigate necrosis by enhancing cellular resilience and improving cellular functions. NcRNAs modulated by exercise also play crucial roles in regulating necrosis. Therefore, exercise-induced epigenetic alterations significantly contribute to alleviating cellular necrosis and promoting overall health.

14.3.3 Exercise, Epigenetics, and Pyroptosis

Pyroptosis is a form of programmed cell death distinct from both apoptosis and necrosis, characterized by the activation of inflammatory responses. It is mediated by gasdermin proteins, which oligomerize to form pores in the cell membrane, resulting in cell lysis and the subsequent release of pro-inflammatory cytokines, such as IL-1β and IL-18 (Broz et al. 2020). Pyroptosis plays a critical role in immune defense against infections by eliminating pathogen-infected cells and activating the immune system. However, dysregulated pyroptosis is implicated in several diseases, including sepsis, atherosclerosis, and neurodegenerative disorders, where it contributes to pathological inflammation and tissue damage (Vasudevan et al. 2023).

The relationship between pyroptosis and epigenetics is an emerging area of research. Epigenetic modifications, including DNA methylation, histone modifications, and ncRNAs, regulate the expression of genes involved in pyroptosis (Zhang et al. 2023). These modifications modulate the sensitivity of cells to pyroptotic signals and the production of pro-inflammatory cytokines. Moreover, lifestyle factors, including physical exercise, have been shown to influence the epigenetic landscape. Exercise can induce epigenetic modifications that regulate the expression of genes associated with inflammation and cell death, thereby potentially influencing the onset and progression of pyroptosis (Yang et al. 2023). Physical exercise is one of the key modulators of epigenetic changes, exerting profound effects on a variety of physiological processes. Understanding the epigenetic regulation of pyroptosis paves the way for novel therapeutic interventions targeting the modulation of inflammatory responses in various diseases.

Recent studies have elucidated the role of exercise in modulating bone health and metabolic diseases through epigenetic mechanisms. Exercise promotes skeletal development and has been demonstrated to mitigate bone loss and the deterioration of mechanical strength in Type 2 Diabetes Mellitus (T2DM) mouse models. A 2022 study confirmed that exercise promotes bone formation and mitigates bone loss through epigenetic pathways. Specifically, exercise downregulates the expression of miR-150 and upregulates the levels of FNDC5/Irisin in skeletal muscle (Behera et al. 2022). This regulation inhibits pyroptosis in skeletal muscle cells, thereby mitigating diabetes-induced bone loss and promoting bone formation. Furthermore, exercise modulates the progression of atherosclerosis by regulating LncRNA NEAT1 through m^6A modifications (Yang et al. 2023). Exercise modulates m^6A modifications through the regulation of METTL14, resulting in the downregulation of NEAT1 expression. This epigenetic mechanism reduces endothelial cell pyroptosis and mitigates the progression of atherosclerosis.

In summary, physical exercise exerts significant epigenetic effects that regulate pyroptosis and contribute to the prevention of diseases such as T2DM-induced bone loss and atherosclerosis. These findings underscore the significance of incorporating regular physical activity into health management strategies to leverage its beneficial epigenetic effects.

14.3.4 Exercise, Epigenetics, and Ferroptosis

Ferroptosis is a recently identified form of regulated cell death, characterized by the accumulation of lipid peroxides and iron-dependent ROS within cells. Unlike apoptosis and necrosis, ferroptosis is driven by iron-dependent catalysis, resulting in oxidative damage to cellular membranes. The term *ferroptosis* was first introduced in 2012 by researchers who discovered that this distinct form of cell death could be inhibited by iron chelators and lipophilic antioxidants (Dixon et al. 2012). Ferroptosis has been implicated in a range of diseases, including neurodegenerative disorders, cancer, and ischemia-reperfusion injury, underscoring its relevance in both physiological and pathological contexts. The relationship between ferroptosis and epigenetics represents an emerging area of research. Epigenetic modifications, such as DNA methylation, histone modifications, and ncRNAs, are pivotal in regulating the expression of genes associated with ferroptosis (Wu et al. 2020).

Physical exercise has been shown to modulate epigenetic mechanisms, which, in turn, can influence ferroptosis. Exercise-induced epigenetic modifications can regulate the expression of key genes involved in iron homeostasis and antioxidant defense. By modulating these pathways, exercise has the potential to reduce the risk of diseases associated with ferroptosis. For example, exercise can modulate the expression of LncRNAs and miRNAs, thereby enhancing antioxidant capacity and improving cellular resistance to oxidative stress (Thirupathi et al. 2021). Understanding these mechanisms provides new opportunities for therapeutic interventions aimed at preventing or treating ferroptosis-related conditions.

Several studies have emphasized the substantial impact of physical exercise on epigenetic modifications, particularly in relation to ferroptosis. Research has shown that aerobic exercise reduces iron levels and inhibits excessive ROS production in mitochondria (Wang et al. 2024). This inhibition prevents mitochondrial DNA damage and ferroptosis, thereby mitigating azithromycin-induced cardiotoxicity.

Meanwhile, aerobic exercise increases the expression of p-AMPK/AMPKα2 in cardiac tissue, significantly reducing myocardial damage, including necrosis, apoptosis, and ferroptosis. The upregulation of p-AMPK/AMPKα2 expression activates PGC-1α, thereby maintaining antioxidant enzyme activity and preventing mitochondrial dysfunction. Furthermore, acute exercise can induce the translocation of the nuclear protein p53 to the mitochondria, where it forms complexes with mitochondrial transcription factor A (TFAM) and mtDNA (Saleem and Hood 2013). This interaction positively regulates the transcription of mtDNA, contributing to mitochondrial dysfunction and triggering cell death pathways, including apoptosis and ferroptosis. Preconditioning exercise also exhibits neuroprotective effects by upregulating the levels of miR-484 in exosomes (Huang et al. 2024). MiR-484 downregulates the protein expression of ACSL4 by inhibiting its translation, thereby suppressing ferroptosis in neuronal cells under ischemic and hypoxic conditions. This mechanism provides neuroprotection against cerebral ischemia.

In summary, physical exercise exerts significant epigenetic effects that regulate ferroptosis and other forms of cell death. These findings underscore the potential of exercise-induced epigenetic modifications in preventing or alleviating ferroptosis-related diseases and emphasize the importance of incorporating regular physical activity into therapeutic strategies.

14.3.5 Exercise, Epigenetics, and Cuproptosis

Cuproptosis is a recently discovered cell death pathway characterized by the accumulation of copper. This form of cell death is distinct from apoptosis, ferroptosis, and necrosis, representing a novel type of regulated cell death. The discovery of cuproptosis stems from a 2022 study published in journal *Science*, which demonstrated that copper ions can induce cell death, even when traditional cell death pathways are inhibited (Tsvetkov et al. 2022). The hallmarks of cuproptosis include mitochondrial contraction, cell membrane rupture, endoplasmic reticulum disruption, and chromatin fragmentation. Although these characteristics resemble apoptosis, the underlying mechanisms differ from those of other established cell death pathways.

The mechanism of cuproptosis involves the direct binding of copper ions to lipoylated proteins within the mitochondrial TCA cycle, leading to the abnormal aggregation of these proteins and the loss of Fe-S cluster proteins in the respiratory chain complex. This triggers a protein toxicity stress response, ultimately resulting in cell death (Tsvetkov et al. 2022). Research on cuproptosis is crucial for understanding diseases associated with copper metabolism imbalances, such as cardiovascular diseases and cancer. The discovery of cuproptosis offers new insights and potential therapeutic targets for treating these diseases. For example, in cancer therapy, copper ion carriers can be utilized to selectively induce cuproptosis in cancer cells, thereby inhibiting tumor growth (Guo et al. 2023b). Furthermore, research on cuproptosis may contribute to the development of novel treatment strategies, such as the modulation of copper metabolism to prevent and manage related diseases.

As noted earlier, epigenetics, defined as heritable changes in gene expression that do not alter the DNA sequence, plays a critical role in regulating various cell death pathways, including cuproptosis. Epigenetic modifications, including DNA methylation, histone modifications, and regulation of ncRNAs, can modulate the expression of genes involved in copper homeostasis and the cellular response to copper-induced stress. Cuproptosis is frequently observed in cancer. The concentration of copper ions is significantly elevated in gastric cancer. Further screening revealed that METTL16 facilitates the m^6A methylation of the key gene FDX1, thereby triggering the onset of cuproptosis (Sun et al. 2023a). Additionally, LncRNAs, as epigenetic regulators, have been identified as potential predictive markers for cancer, strongly associated with cuproptosis (Sun et al. 2023b; Zhang et al. 2022).

Exercise has long been acknowledged for its numerous health benefits; recent research has begun to elucidate its specific effects on cellular processes, such as copper-dependent cell death, also referred to as cuproptosis. A 2022 study identified IL-15 as a potential molecular biomarker for predicting prognosis, immune response, and cuproptosis in patients with pan-cancer (Luo et al. 2022). This study further demonstrated that exercise, as a beneficial lifestyle intervention, induces alterations in IL-15 levels, which are linked to improved prognosis and tumor suppression across various cancer types. Exercise, as a lifestyle intervention, has been shown to modulate oxidative phosphorylation. This modulation may contribute to the suppression of cuproptosis by alleviating protein-induced stress responses (Kordi et al. 2024). However, the precise molecular mechanisms underlying this protective effect remain to be fully elucidated. The correlation between exercise and cuproptosis highlights the potential of physical activity not only in general health promotion but also in specific cancer therapies.

Although a correlation between exercise and cuproptosis has been suggested, and a link between epigenetics and copper-induced cell death has been independently identified, these findings tentatively highlight the significant role of cuproptosis in various diseases. However, given the recent emergence of the concept of cuproptosis, research on the potential intervention of aerobic exercise in cuproptosis through the regulation of epigenetic mechanisms is still limited, with no conclusive studies published to date. Additional research is required to elucidate the complex molecular pathways by which exercise modulates epigenetic mechanisms to regulate cuproptosis, with the ultimate goal of leveraging this knowledge to enhance cancer treatment strategies.

14.3.6 Exercise, Epigenetics, and Disulfidptosis

Disulfidptosis is a newly identified form of regulated cell death, first proposed by Professor Gan Boyi and his team at the MD Anderson Cancer Center in 2023 (Liu et al. 2023). This form of cell death is associated with the aberrant formation of disulfide bonds within the cell, particularly under conditions of glucose deprivation. Inhibition of the pentose phosphate pathway (PPP) results in a deficiency of nicotinamide adenine dinucleotide phosphate (NADPH), leading to the accumulation of cysteine within the cell. This accumulation, coupled with the deficiency of NADPH, results in heightened disulfide bond stress within the cell, leading to the formation of multiple disulfide bonds in F-actin. This induces the contraction and detachment of F-actin from the cell membrane, ultimately leading to cell death.

The cellular features of disulfidptosis include disrupted migration of actin cytoskeletal proteins, contraction and detachment of F-actin from the plasma membrane, as well as the abnormal formation and accumulation of disulfide bond molecules within the cell (Liu et al. 2023). At the molecular level, studies have shown that in cells with elevated expression of SLC7A11, glucose deprivation triggers NADPH depletion and disulfide stress, leading to the formation of multiple intermolecular disulfide bonds in actin cytoskeletal proteins. The formation of these abnormal disulfide bonds destabilizes the actin cytoskeleton, ultimately leading to cell death. Disulfidptosis holds significant potential in the context of cancer metabolic therapy (Koppula et al. 2021). Studies have demonstrated that inducing disulfidptosis can enhance sensitivity to anticancer agents, thereby offering potential therapeutic targets for clinical intervention. For instance, glucose transporter inhibitors can induce disulfidptosis in cancer cells exhibiting elevated SLC7A11 expression, thereby effectively suppressing tumor growth (Zheng et al. 2023a). Additionally, disulfidptosis may play a role in the prevention, diagnosis, and treatment of other diseases, especially those related to metabolic reprogramming. Furthermore, disulfidptosis may contribute to the prevention, diagnosis, and treatment of various diseases, particularly those associated with metabolic reprogramming (Zheng et al. 2023b).

The epigenetic landscape, encompassing modifications such as m5C and m7G methylation, plays a critical role in regulating diverse cell death mechanisms, including disulfidptosis (Zhou et al. 2024). These modifications regulate gene expression and protein synthesis, thereby modulating cellular responses to stress and therapeutic interventions. The relationship between disulfidptosis and epigenetics is especially significant in the context of cancer. Epigenetic modifications can influence the susceptibility of cancer cells to disulfidptosis by regulating the expression of key genes involved in disulfide stress responses. For example, the overexpression of the cystine transporter SLC7A11, which is epigenetically regulated, has been shown to increase the sensitivity of tumor cells to disulfidptosis induced by GLUT inhibitors (Zheng et al. 2023b; Liu et al. 2024).

Although a preliminary correlation between disulfidptosis and epigenetics has been suggested, these studies remain in the early stages of development. Research on exercise as an intervention capable of modulating epigenetic modifications has advanced. However, the hypothesis that exercise may influence disulfidptosis-related mortality through epigenetic modifications has yet to be proposed. In the future, further investigation into the impact of exercise on disulfidptosis through epigenetic mechanisms remains a promising area of research.

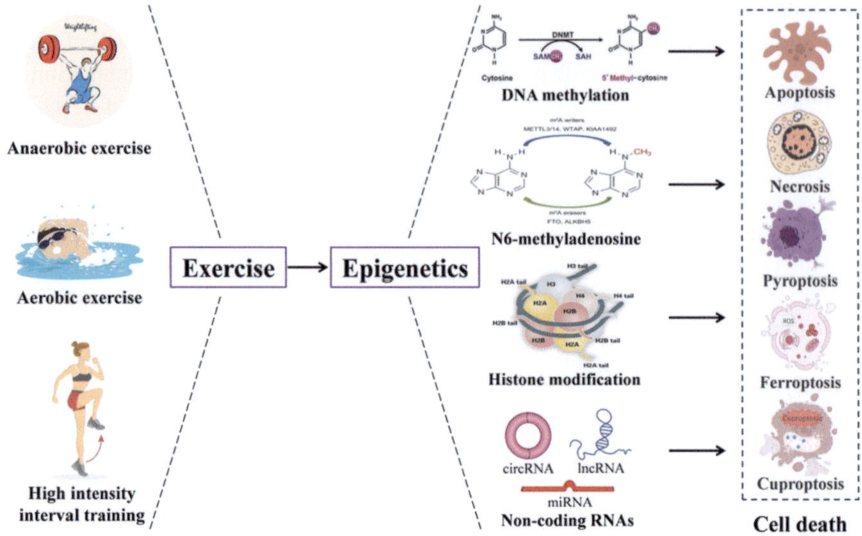

Fig. 14.1 Exercise, epigenetics and cell death

14.4 Conclusions

The summary above underscores the significant impact of exercise on cellular survival via epigenetic mechanisms. Exercise-induced epigenetic modifications are critical in modulating various forms of cell death, such as apoptosis, necroptosis, and autophagy. By synthesizing the findings of existing studies, it is evident that physical activity induces changes in DNA methylation, histone modifications, and ncRNA expression, which, in turn, influence gene expression patterns associated with cell survival and death pathways (Fig. 14.1). By elucidating the relationship between exercise and epigenetic regulation, we underscore the potential of exercise as a non-pharmacological intervention to modulate cellular functions and enhance health. These insights deepen our understanding of how lifestyle factors, such as exercise, influence cellular aging and the progression of disease. Future research should investigate the specific epigenetic markers influenced by various types and intensities of exercise, and how these alterations can be leveraged for therapeutic applications. Overall, significant progress has been made in research, and these findings provide compelling evidence that exercise is a potent modulator of the epigenetic landscape, influencing multiple cell death pathways and, consequently, contributing to enhanced cellular homeostasis and resilience.

Compliance with Ethical Standards

Conflict of Interest All authors declare no conflict of interest.

Ethical Approval This chapter is a review of previously published accounts; as such, no animal or human studies were performed.

Funding Funding for this research was provided by the National Natural Science Foundation of China (No.82370446).

References

Alegria-Torres JA, Baccarelli A, Bollati V (2011) Epigenetics and lifestyle. Epigenomics 3:267–277. https://doi.org/10.2217/epi.11.22

Allfrey VG, Faulkner R, Mirsky AE (1964) Acetylation and methylation of histones and their possible role in the regulation of Rna synthesis. Proc Natl Acad Sci USA 51:786–794. https://doi.org/10.1073/pnas.51.5.786

Bannister AJ, Kouzarides T (2011) Regulation of chromatin by histone modifications. Cell Res 21:381–395. https://doi.org/10.1038/cr.2011.22

Behera J, Ison J, Voor MJ, Tyagi N (2022) Exercise-linked skeletal irisin ameliorates diabetes-associated osteoporosis by inhibiting the oxidative damage-dependent miR-150-FNDC5/pyroptosis axis. Diabetes 71:2777–2792. https://doi.org/10.2337/db21-0573

Bertheloot D, Latz E, Franklin BS (2021) Necroptosis, pyroptosis and apoptosis: an intricate game of cell death. Cell Mol Immunol 18:1106–1121. https://doi.org/10.1038/s41423-020-00630-3

Broz P, Pelegrin P, Shao F (2020) The gasdermins, a protein family executing cell death and inflammation. Nat Rev Immunol 20:143–157. https://doi.org/10.1038/s41577-019-0228-2

Chai L et al (2024) Exercise alleviates fluoride-induced learning and memory impairment in mice: role of miR-206-3p and PREG. Biol Trace Elem Res. https://doi.org/10.1007/s12011-024-04068-w

Chen X et al (2021) Nrf2 epigenetic derepression induced by running exercise protects against osteoporosis. Bone Res 9:15. https://doi.org/10.1038/s41413-020-00128-8

Diener C, Keller A, Meese E (2022) Emerging concepts of miRNA therapeutics: from cells to clinic. Trends Genet 38:613–626. https://doi.org/10.1016/j.tig.2022.02.006

Dixon SJ et al (2012) Ferroptosis: an iron-dependent form of nonapoptotic cell death. Cell 149:1060–1072. https://doi.org/10.1016/j.cell.2012.03.042

Gomez-Cabrera MC, Domenech E, Vina J (2008) Moderate exercise is an antioxidant: upregulation of antioxidant genes by training. Free Radic Biol Med 44:126–131. https://doi.org/10.1016/j.freeradbiomed.2007.02.001

Gomez-Pinilla F, Thapak P (2024) Exercise epigenetics is fueled by cell bioenergetics: supporting role on brain plasticity and cognition. Free Radic Biol Med 220:43–55. https://doi.org/10.1016/j.freeradbiomed.2024.04.237

Gueritat J et al (2014) Exercise training combined with antioxidant supplementation prevents the antiproliferative activity of their single treatment in prostate cancer through inhibition of redox adaptation. Free Radic Biol Med 77:95–105. https://doi.org/10.1016/j.freeradbiomed.2014.09.009

Guo XX et al (2023a) The role of regulated necrosis in inflammation and ocular surface diseases. Exp Eye Res 233:109537. https://doi.org/10.1016/j.exer.2023.109537

Guo B et al (2023b) Cuproptosis induced by ROS responsive nanoparticles with elesclomol and copper combined with alphaPD-L1 for enhanced cancer immunotherapy. Adv Mater 35:e2212267. https://doi.org/10.1002/adma.202212267

Herman AB, Tsitsipatis D, Gorospe M (2022) Integrated lncRNA function upon genomic and epigenomic regulation. Mol Cell 82:2252–2266. https://doi.org/10.1016/j.molcel.2022.05.027

Horsburgh S, Todryk S, Toms C, Moran CN, Ansley L (2015) Exercise-conditioned plasma attenuates nuclear concentrations of DNA methyltransferase 3B in human peripheral blood mononuclear cells. Physiol Rep 3. https://doi.org/10.14814/phy2.12621

Hsu CC et al (2023) Hypermethylation of ACADVL is involved in the high-intensity interval training-associated reduction of cardiac fibrosis in heart failure patients. J Transl Med 21:187. https://doi.org/10.1186/s12967-023-04032-7

Huang M et al (2024) Preconditioning exercise inhibits neuron ferroptosis and ameliorates brain ischemia damage by skeletal muscle-derived exosomes via regulating miR-484/ACSL4 axis. Antioxid Redox Signal. https://doi.org/10.1089/ars.2023.0492

Jiang X et al (2021) The role of m6A modification in the biological functions and diseases. Signal Transduct Target Ther 6:74. https://doi.org/10.1038/s41392-020-00450-x

Joseph AM et al (2014) Mitochondrial adaptations evoked with exercise are associated with a reduction in age-induced testicular atrophy in Fischer-344 rats. Biogerontology 15:517–534. https://doi.org/10.1007/s10522-014-9526-z

Karlic H et al (2014) The role of epigenetics in the regulation of apoptosis in myelodysplastic syndromes and acute myeloid leukemia. Crit Rev Oncol Hematol 90:1–16. https://doi.org/10.1016/j.critrevonc.2013.10.003

Kayagaki N, Webster JD, Newton K (2024) Control of cell death in health and disease. Annu Rev Pathol 19:157–180. https://doi.org/10.1146/annurev-pathmechdis-051022-014433

Kerr JF, Wyllie AH, Currie AR (1972) Apoptosis: a basic biological phenomenon with wide-ranging implications in tissue kinetics. Br J Cancer 26:239–257. https://doi.org/10.1038/bjc.1972.33

Klutstein M, Nejman D, Greenfield R, Cedar H (2016) DNA methylation in cancer and aging. Cancer Res 76:3446–3450. https://doi.org/10.1158/0008-5472.CAN-15-3278

Koppula P, Zhuang L, Gan B (2021) Cystine transporter SLC7A11/xCT in cancer: ferroptosis, nutrient dependency, and cancer therapy. Protein Cell 12:599–620. https://doi.org/10.1007/s13238-020-00789-5

Kordi N et al (2024) Cuproptosis and physical training: a review. Clin Hemorheol Microcirc. https://doi.org/10.3233/CH-242329

Li JY et al (2023) Suberoylanilide hydroxamic acid upregulates reticulophagy receptor expression and promotes cell death in hepatocellular carcinoma cells. World J Gastroenterol 29:5038–5053. https://doi.org/10.3748/wjg.v29.i34.5038

Liang Q et al (2020) Role of muscle-specific histone methyltransferase (Smyd1) in exercise-induced cardioprotection against pathological remodeling after myocardial infarction. Int J Mol Sci 21. https://doi.org/10.3390/ijms21197010

Liang J et al (2023) MicroRNA profiling of different exercise interventions for alleviating skeletal muscle atrophy in naturally aging rats. J Cachexia Sarcopenia Muscle 14:356–368. https://doi.org/10.1002/jcsm.13137

Ling C, Ronn T (2019) Epigenetics in human obesity and type 2 diabetes. Cell Metab 29:1028–1044. https://doi.org/10.1016/j.cmet.2019.03.009

Liu S, Zheng F, Cai Y, Zhang W, Dun Y (2018) Effect of long-term exercise training on lncRNAs expression in the vascular injury of insulin resistance. J Cardiovasc Transl Res 11:459–469. https://doi.org/10.1007/s12265-018-9830-0

Liu K et al (2022) Exercise training ameliorates myocardial phenotypes in heart failure with preserved ejection fraction by changing N6-methyladenosine modification in mice model. Front Cell Dev Biol 10:954769. https://doi.org/10.3389/fcell.2022.954769

Liu X et al (2023) Actin cytoskeleton vulnerability to disulfide stress mediates disulfidptosis. Nat Cell Biol 25:404–414. https://doi.org/10.1038/s41556-023-01091-2

Liu Z et al (2024) Disulfidptosis signature predicts immune microenvironment and prognosis of gastric cancer. Biol Direct 19:65. https://doi.org/10.1186/s13062-024-00518-6

Luo Z et al (2022) Exercise-induced IL-15 acted as a positive prognostic implication and tumor-suppressed role in pan-cancer. Front Pharmacol 13:1053137. https://doi.org/10.3389/fphar.2022.1053137

Luo HY et al (2024) Hepatic Klf10-Fh1 axis promotes exercise-mediated amelioration of NASH in mice. Metabolism 155:155916. https://doi.org/10.1016/j.metabol.2024.155916

McGee SL (2007) Exercise and MEF2-HDAC interactions. Appl Physiol Nutr Metab 32:852–856. https://doi.org/10.1139/H07-082

McGee SL, Hargreaves M (2019) Epigenetics and exercise. Trends Endocrinol Metab 30:636–645. https://doi.org/10.1016/j.tem.2019.06.002

McGee SL, Hargreaves M (2020) Exercise adaptations: molecular mechanisms and potential targets for therapeutic benefit. Nat Rev Endocrinol 16:495–505. https://doi.org/10.1038/s41574-020-0377-1

Newton K, Strasser A, Kayagaki N, Dixit VM (2024) Cell death. Cell 187:235–256. https://doi.org/10.1016/j.cell.2023.11.044

Nishiyama A, Nakanishi M (2021) Navigating the DNA methylation landscape of cancer. Trends Genet 37:1012–1027. https://doi.org/10.1016/j.tig.2021.05.002

Patra S et al (2019) Dysregulation of histone deacetylases in carcinogenesis and tumor progression: a possible link to apoptosis and autophagy. Cell Mol Life Sci 76:3263–3282. https://doi.org/10.1007/s00018-019-03098-1

Saleem A, Hood DA (2013) Acute exercise induces tumour suppressor protein p53 translocation to the mitochondria and promotes a p53-Tfam-mitochondrial DNA complex in skeletal muscle. J Physiol 591:3625–3636. https://doi.org/10.1113/jphysiol.2013.252791

Shang JL, Cheng Q, Duan SJ, Li L, Jia LY (2018) Cognitive improvement following ischemia/reperfusion injury induced by voluntary running-wheel exercise is associated with LncMALAT1-mediated apoptosis inhibition. Int J Mol Med 41:2715–2723. https://doi.org/10.3892/ijmm.2018.3484

Sleiman SF et al (2016) Exercise promotes the expression of brain derived neurotrophic factor (BDNF) through the action of the ketone body beta-hydroxybutyrate. elife 5. https://doi.org/10.7554/eLife.15092

Su H et al (2021) miR-9-5p regulates immunometabolic and epigenetic pathways in beta-glucan-trained immunity via IDH3alpha. JCI Insight 6. https://doi.org/10.1172/jci.insight.144260

Sun L et al (2023a) Lactylation of METTL16 promotes cuproptosis via m(6)A-modification on FDX1 mRNA in gastric cancer. Nat Commun 14:6523. https://doi.org/10.1038/s41467-023-42025-8

Sun Q et al (2023b) Cuproptosis-related LncRNA signatures as a prognostic model for head and neck squamous cell carcinoma. Apoptosis 28:247–262. https://doi.org/10.1007/s10495-022-01790-5

Swiatowy WJ et al (2021) Physical activity and DNA methylation in humans. Int J Mol Sci 22. https://doi.org/10.3390/ijms222312989

Tetti M, Gong S, Veglio F, Reincke M, Williams TA (2022) Primary aldosteronism: pathophysiological mechanisms of cell death and proliferation. Front Endocrinol (Lausanne) 13:934326. https://doi.org/10.3389/fendo.2022.934326

Thirupathi A et al (2021) Effect of different exercise modalities on oxidative stress: a systematic review. Biomed Res Int 2021:1947928. https://doi.org/10.1155/2021/1947928

Tsvetkov P et al (2022) Copper induces cell death by targeting lipoylated TCA cycle proteins. Science 375:1254–1261. https://doi.org/10.1126/science.abf0529

Vasudevan SO, Behl B, Rathinam VA (2023) Pyroptosis-induced inflammation and tissue damage. Semin Immunol 69:101781. https://doi.org/10.1016/j.smim.2023.101781

Vu LP et al (2017) The N(6)-methyladenosine (m(6)A)-forming enzyme METTL3 controls myeloid differentiation of normal hematopoietic and leukemia cells. Nat Med 23:1369–1376. https://doi.org/10.1038/nm.4416

Wang L et al (2022) METTL14 is required for exercise-induced cardiac hypertrophy and protects against myocardial ischemia-reperfusion injury. Nat Commun 13:6762. https://doi.org/10.1038/s41467-022-34434-y

Wang H et al (2023a) m6A methyltransferase WTAP regulates myocardial ischemia reperfusion injury through YTHDF1/FOXO3a signaling. Apoptosis 28:830–839. https://doi.org/10.1007/s10495-023-01818-4

Wang L et al (2023b) Exercise-induced circular RNA circUtrn is required for cardiac physiological hypertrophy and prevents myocardial ischaemia-reperfusion injury. Cardiovasc Res 119:2638–2652. https://doi.org/10.1093/cvr/cvad161

Wang L et al (2024) Endurance exercise preconditioning alleviates ferroptosis induced by doxorubicin-induced cardiotoxicity through mitochondrial superoxide-dependent AMPKalpha2 activation. Redox Biol 70:103079. https://doi.org/10.1016/j.redox.2024.103079

Wu Y et al (2020) The epigenetic regulators and metabolic changes in ferroptosis-associated cancer progression. Mol Cancer 19:39. https://doi.org/10.1186/s12943-020-01157-x

Wu B et al (2022) Aerobic exercise promotes the expression of ATGL and attenuates inflammation to improve hepatic steatosis via lncRNA SRA. Sci Rep 12:5370. https://doi.org/10.1038/s41598-022-09174-0

Xu Z, Qin Y, Lv B, Tian Z, Zhang B (2022) Effects of moderate-intensity continuous training and high-intensity interval training on testicular oxidative stress, apoptosis and m6A methylation in obese male mice. Antioxidants (Basel) 11. https://doi.org/10.3390/antiox11101874

Yang Q et al (2023) Exercise mitigates endothelial pyroptosis and atherosclerosis by downregulating NEAT1 through N6-methyladenosine modifications. Arterioscler Thromb Vasc Biol 43:910–926. https://doi.org/10.1161/ATVBAHA.123.319251

Zhang S et al (2017) m(6)A demethylase ALKBH5 maintains tumorigenicity of glioblastoma stem-like cells by sustaining FOXM1 expression and cell proliferation program. Cancer Cell 31:591–606 e596. https://doi.org/10.1016/j.ccell.2017.02.013

Zhang G, Sun J, Zhang X (2022) A novel Cuproptosis-related LncRNA signature to predict prognosis in hepatocellular carcinoma. Sci Rep 12:11325. https://doi.org/10.1038/s41598-022-15251-1

Zhang RN, Jing ZQ, Zhang L, Sun ZJ (2023) Epigenetic regulation of pyroptosis in cancer: molecular pathogenesis and targeting strategies. Cancer Lett 575:216413. https://doi.org/10.1016/j.canlet.2023.216413

Zhang P et al (2024) Up-regulation of miR-126 via DNA methylation in hypoxia-preconditioned endothelial cells may contribute to hypoxic tolerance of neuronal cells. Mol Biol Rep 51:808. https://doi.org/10.1007/s11033-024-09774-1

Zheng P, Zhou C, Ding Y, Duan S (2023a) Disulfidptosis: a new target for metabolic cancer therapy. J Exp Clin Cancer Res 42:103. https://doi.org/10.1186/s13046-023-02675-4

Zheng T, Liu Q, Xing F, Zeng C, Wang W (2023b) Disulfidptosis: a new form of programmed cell death. J Exp Clin Cancer Res 42:137. https://doi.org/10.1186/s13046-023-02712-2

Zhou S, Liu J, Wan A, Zhang Y, Qi X (2024) Epigenetic regulation of diverse cell death modalities in cancer: a focus on pyroptosis, ferroptosis, cuproptosis, and disulfidptosis. J Hematol Oncol 17:22. https://doi.org/10.1186/s13045-024-01545-6

Chapter 15
Perspectives on Exercise, Epigenetics, and Human Health: Challenges, Opportunities, and the Road Ahead

Yang Hu and Feng Gao

Abstract Accumulating evidence highlights the essential role of exercise in promoting health and preventing chronic diseases. Recent research reveals the intricate mechanisms underlying these benefits, with epigenetics emerging as a key player. Despite these promising insights, our understanding of how exercise influences the epigenome remains incomplete. Fundamental questions persist regarding the specificity, stability, and functional consequences of exercise-induced epigenetic modifications across different organs and life stages. Nevertheless, the field of exercise epigenetics will be at the forefront of biomedical research in the years ahead. By elucidating how physical activity induces these epigenetic modifications, we can gain deeper insights into its role in shaping our health trajectories. Furthermore, the translational potential of exercise epigenetics, particularly in identifying epigenetic biomarkers, could enhance our understanding of individual variability in response to exercise. These insights may inform precision medicine and guide drug development, further underscoring the importance of integrating exercise into holistic healthcare strategies.

Keywords Epigenetics · Physical exercise · Human health

Abbreviations

AMPK	AMP-activated protein kinase
DNA	Deoxyribonucleic acid
DNMTs	DNA methyltransferases
HDACs	Histone deacetylases
ncRNAs	Non-coding RNAs

Y. Hu · F. Gao (✉)
Key Laboratory of Aerospace Medicine of the Ministry of Education, School of Aerospace Medicine, Fourth Military Medical University, Xi'an, China
e-mail: fgao@fmmu.edu.cn

F. Gao, X. Zhang (eds.), *Exercise, Epigenetics and Human Health*, Epigenetics and Human Health 15, https://doi.org/10.1007/978-3-032-08045-5_15

PGC-1α	Peroxisome proliferator-activated receptor gamma coactivator 1-alpha
RNA	Ribonucleic acid
ROS	Reactive oxygen species

15.1 The Underestimated Impact of Epigenetic Modulations in Exercise Research

Despite growing evidence supporting the health benefits of regular physical activity, the profound impact of exercise on epigenetic modifications remains largely underexplored. Only a limited number of studies have investigated the epigenetic mechanisms underlying metabolic and immunological adaptations to exercise (Zhang et al. 2024b, Wu et al. 2021). The number of publications on this topic increased from just 13 in 2003 to 239 by 2022, indicating a growing interest but also highlighting that the field is still in its early stages. While the importance of epigenetics in health and disease is increasingly recognized, much of the discourse has yet to fully address how exercise-induced epigenetic modifications may influence gene expression, metabolic pathways, and long-term health outcomes (Wu et al. 2024).

Epigenetics involves heritable yet reversible changes in gene expression that do not alter DNA sequence. The primary mechanisms of epigenetic modification include DNA methylation, histone modifications, and non-coding RNA regulation. These changes can be triggered by environmental factors, lifestyle choices, and pathological processes, leading to variations in gene activity that influence cellular functions and overall health. Numerous studies suggest that exercise can modulate these epigenetic markers, influencing genes involved in metabolic processes, inflammation, and muscle adaptation. For example, aerobic exercise has been shown to cause DNA methylation changes in the promoter regions of genes related to oxidative stress and inflammation, providing a biological basis for exercise's potential to mitigate chronic disease risk (Zhang et al. 2024a). Furthermore, resistance training has also been shown to alter epigenetic markers, with studies highlighting that higher levels of physical activity are associated with improved metabolic health such as enhancing insulin sensitivity (Abraham et al. 2023).

Despite these promising findings, epigenetic modulation is frequently overlooked in discussions of exercise physiology. This omission represents a missed opportunity to integrate a more comprehensive, mechanism-based framework into exercise prescription and public health initiatives.

Most existing studies tend to focus on correlation analysis, which limits our understanding of whether epigenetic modifications are a direct consequence of exercise or a secondary effect of enhanced metabolic processes, such as improved insulin sensitivity. For instance, the hypomethylation observed in the PGC-1α promoter region within skeletal muscle following exercise may be mediated by AMPK activation; however, the mechanisms by which upstream signals, including calcium ion oscillations and ROS fluctuations, lead to the inhibition of methyltransferase

(DNMTs) activity remain poorly understood. Recognizing the lasting influence of exercise on gene expression could enhance public motivation to engage in regular physical activity. To harness full potential of exercise in promoting health and preventing disease, it is urgent to increase awareness among healthcare professionals and to prioritize research that elucidates how different exercise modalities and intensities specifically influence epigenetic changes.

Taken together, the epigenetic effects of exercise represent a critical yet underestimated dimension of its impact on health. As research advances, it is essential for the scientific community, health practitioners, and the public to recognize the profound impact of exercise beyond the genetic level. Integrating this knowledge into practice will lead to more effective public health strategies, emphasizing exercise as a fundamental component of wellness and disease prevention. Ultimately, appreciating the interplay between exercise and epigenetics will enhance our insights into health promotion and motivate individuals to adopt lifelong exercise habits for healthy aging.

15.2 Long-Term Effects of Exercise: The Critical Role of Epigenetic Modulation

As researchers delve deeper into the relationship between exercise and health, understanding the long-term effects of physical activity, particularly how early-life exercise impacts health later in life and aging (Feng et al. 2025), becomes increasingly crucial. While the immediate benefits of exercise are widely recognized, the long-term implications extend far beyond these transient outcomes. Notably, exercise induces significant epigenetic modifications that can have lasting effects on gene expression, cellular function, and overall health throughout the lifespan. Emphasizing the enduring impact of exercise through epigenetic modulation is essential to foster a paradigm shift in how we perceive the role of regular exercise, especially starting from early-life period, in long-term health promotion.

Epigenetic modifications provide a biological mechanism through which environmental factors, including lifestyle choices such as physical activity, can influence gene expression and, consequently, disease risks. The effects of exercise on epigenetics are particularly profound when considering chronic diseases, metabolic disorders, and the aging process. The ability of exercise to induce epigenetic changes presents a unique opportunity not only for disease prevention but also for promoting healthier aging, thereby enhancing the quality of life across different life stages. Through multidimensional epigenetic modifications, regular exercise can lower the risk of chronic diseases such as obesity, cardiovascular disease, and cancers (Seaborne et al. 2018). Furthermore, exercise has been shown to influence the aging process at the epigenetic level. Specific epigenetic alterations that promote genomic stability, enhance DNA repair mechanisms, and reduce inflammation. By fostering a more “youthful” epigenetic profile, exercise may delay the onset of age-related decline, promote healthspan, the period of life spent in good health, and contribute to increased longevity and improved quality of life in older adults. Moreover, sustained physical activity has been associated with enhanced long-term mental health outcomes, potentially mediated through epigenetic mechanisms.

As the understanding of the intricate relationship between exercise and epigenetics continues to evolve, it is increasingly important to shift prevailing perspectives, viewing exercise not merely as a tool for short-term fitness but as a fundamental component of lifelong health. Recognizing and integrating the often-overlooked molecular outcomes of consistent physical activity will be critical to advancing a culture that embraces lifelong exercise as a central pillar of holistic well-being.

15.3 Translational Advances: Bridging Epigenetic Research and Practical Applications

The exploration of the relationship between exercise and epigenetics has opened new avenues for translational research, offering opportunities to leverage epigenetic insights into practical health solutions, such as biomarkers for health assessment and innovative approaches to drug development. Translational studies aim to bridge the gap between basic research and real-world applications, ensuring that discoveries in epigenetics are effectively translated into clinically relevant strategies for disease prevention, health promotion, and therapeutic interventions (Chaffer et al. 2022). The integration of epigenetic insights with exercise science not only advances biomarker development but also influence drug discovery, potentially leading to improved health outcomes.

In the context of exercise and epigenetics, biomarkers can assist in assessing health status, predicting disease risk, and monitoring responses to physical activity. Identifying epigenetic biomarkers related to exercise offers substantial potential for personalized medicine, facilitating tailored interventions based on an individual's unique epigenetic profile. Longitudinal studies have demonstrated that regular exercise can lead to specific changes in DNA methylation and histone modifications associated with metabolic health. These biomarkers could eventually be utilized in practical settings to monitor individuals' adherence to exercise regimens and to optimize personal training programs based on their biological responses. By analyzing alterations in methylation patterns linked to exercise, researchers have begun to investigate how exercise-induced epigenetic modifications may identify individuals at higher risk for conditions such as diabetes or cardiovascular disease. For example, epigenetic changes in genes associated with inflammation and insulin regulation could provide insights into an individual's susceptibility to diseases, paving the way for preemptive health interventions (Grazioli et al. 2017).

Beyond biomarker identification, the insights gained from studying the epigenetic effects of exercise can inform drug development, particularly in designing therapeutics that mimic the beneficial effects of physical activity. Understanding the underlying mechanisms through which exercise induces epigenetic changes opens new avenues for pharmacological interventions, potentially leading to innovative treatments for metabolic disorders, and other chronic diseases. Epigenetic drugs, such as DNA methyltransferase inhibitors and histone deacetylase inhibitors, have shown promise in preclinical and clinical studies. Translating exercise-induced epigenetic insights could guide the development of new compounds designed to activate similar pathways, such as those involved in energy metabolism, muscle growth and anti-inflammatory responses, offering therapeutic alternatives for individuals unable to engage in regular physical activity due to health limitations. Additionally, combining lifestyle interventions with pharmacological treatments may amplify therapeutic efficacy for individuals with chronic diseases.

However, several challenges must be addressed to fully realize the potential of these translational applications. The complexity of epigenetic modifications and their variability among individuals pose significant challenges in establishing standardized biomarkers and therapeutic targets. Variability arises from environmental influences, as well as intrinsic factors such as age, sex, and genetic background, all of which can modulate epigenetic responses to exercise. Standardization issues, such as differences in measurement platforms (e.g., Illumina EPIC array vs. Nanopore sequencing), further complicate translation into clinical practice. In addition, much of the current evidence stems from preclinical models or small human cohorts, limiting generalizability. Large-scale, longitudinal studies across diverse populations are essential to validate the utility of epigenetic biomarkers and exercise-related therapeutic interventions.

Translational research in epigenetics and exercise holds promise for advancing personalized medicine, disease prevention and health promotion. Recognizing the potential to harness exercise-induced epigenetic changes as both biomarker and therapeutical targets will be important in moving toward a future where physical activity is not only a lifestyle choice but also a foundational pillar of personalized healthcare.

15.4 Closing Remarks and Future Directions

The burgeoning field of exercise epigenetics presents a transformative opportunity to deepen our understanding of how physical activity influences gene expression and overall health. As highlighted throughout this chapter, exercise induces significant epigenetic modifications that contribute not only to immediate physiological benefits but also to profound long-term effects on metabolic health, disease prevention, and healthy aging. Recognizing the intricate interplay between exercise and epigenetics is essential for harnessing the full potential of physical activity as a vital component of health promotion and disease management.

The implications for public health are substantial. Incorporating epigenetic insights into personalized fitness recommendations and preventive interventions will enhance our ability to monitor health, and predict disease risk, paving the way for proactive health measures that prioritize prevention over treatment. Furthermore, translational research linking exercise-induced epigenetic changes to drug development holds promise for creating innovative therapies that mimic these beneficial effects, providing new avenues for treating chronic diseases and improving health outcomes for individuals unable to engage in regular physical activity.

Looking ahead, fostering interdisciplinary collaboration among researchers, clinicians, and public health practitioners will be essential to integrate epigenetic insights into practical applications. This approach supports a cultural and scientific shift, redefining exercise not as just a reactive lifestyle adjustment, but as a proactive, biologically encoded intervention for lifelong wellness. The future of exercise science will transcend macroscopic descriptions of "dose-effect" relationships, evolving toward the decoding of each individual's unique epigenetic language. By integrating dynamic monitoring, artificial intelligence, and epigenetic editing technologies, exercise prescriptions can be dynamically optimized in response to an individual's real-time epigenetic feedback. Additionally, universal exercise programs tailored to diverse genetic and epigenetic backgrounds can be developed to help address and narrow the health disparity gap and promote equity in preventative care. By embracing this epigenetic perspective, exercise becomes a means of rewriting the biological script toward a healthier and more robust life.

Compliance with Ethical Standards

Conflict of Interest All authors declare no conflict of interest.

Ethical Approval This chapter is a review of previously published accounts; as such, no animal or human studies were performed.

References

Abraham MJ, El Sherbini A, El-Diasty M, Askari S, Szewczuk MR (2023) Restoring epigenetic reprogramming with diet and exercise to improve health-related metabolic diseases. Biomolecules 13(2). https://doi.org/10.3390/biom13020318. Available at: https://www.ncbi.nlm.nih.gov/pubmed/36830687

Chaffer TJ, Broering FE, Sedraoui S (2022) Histone-mediated skeletal muscle adaptations to exercise: insights into a new era of epigenome-based personalized medicine. J Physiol 600(17):3903–3904. https://doi.org/10.1113/JP283434. Available at: https://www.ncbi.nlm.nih.gov/pubmed/35908267

Feng M, Li M, Lou J, Wu G, Gao T, Wu F, Tan Y, Zhang N, Zhao Y, Zhao L, Li J, Shi C, Zhang X, Liu J, Gao F (2025) Early-life exercise extends healthspan but not lifespan in mice. Nat Commun 16(1):6328. https://doi.org/10.1038/s41467-025-61443-4. Available at: https://pubmed.ncbi.nlm.nih.gov/40634291

Grazioli E, Dimauro I, Mercatelli N, Wang G, Pitsiladis Y, Di Luigi L, Caporossi D (2017) Physical activity in the prevention of human diseases: role of epigenetic modifications. BMC Genomics 18(Suppl 8):802. https://doi.org/10.1186/s12864-017-4193-5. Available at: https://www.ncbi.nlm.nih.gov/pubmed/29143608

Seaborne RA, Strauss J, Cocks M, Shepherd S, O'Brien TD, van Someren KA, Bell PG, Murgatroyd C, Morton JP, Stewart CE, Sharples AP (2018) Human skeletal muscle possesses an epigenetic memory of hypertrophy. Sci Rep 8(1):1898. https://doi.org/10.1038/s41598-018-20287-3

Wu G, Zhang X, Gao F (2021) The epigenetic landscape of exercise in cardiac health and disease. J Sport Health Sci 10(6):648–659. https://doi.org/10.1016/j.jshs.2020.12.003

Wu H, Hu Y, Jiang C, Chen C (2024) Global scientific trends in research of epigenetic response to exercise: a bibliometric analysis. Heliyon 10(4):e25644. https://doi.org/10.1016/j.heliyon.2024.e25644. Available at: https://www.ncbi.nlm.nih.gov/pubmed/38370173

Zhang J, Tian Z, Qin C, Momeni MR (2024a) The effects of exercise on epigenetic modifications: focus on DNA methylation, histone modifications and non-coding RNAs. Hum Cell 37(4):887–903. https://doi.org/10.1007/s13577-024-01057-y. Available at: https://www.ncbi.nlm.nih.gov/pubmed/38587596

Zhang N, Wang X, Feng M, Li M, Wang J, Yang H, He S, Xia Z, Shang L, Jiang X, Sun M, Wu Y, Ren C, Zhang X, Li J, Gao F (2024b) Early-life exercise induces immunometabolic epigenetic modification enhancing anti-inflammatory immunity in middle-aged male mice. Nat Commun 15(1):3103. https://doi.org/10.1038/s41467-024-47458-3

MIX
Papier aus verantwortungsvollen Quellen
Paper from responsible sources
FSC® C105338

If you have any concerns about our products,
you can contact us on
ProductSafety@springernature.com

In case Publisher is established outside the EU,
the EU authorized representative is:
Springer Nature Customer Service Center GmbH
Europaplatz 3, 69115 Heidelberg, Germany

Printed by Libri Plureos GmbH
in Hamburg, Germany